*About this Series*

The aim of this series is to publish a Reference Library, including novel advances and developments in all aspects of Intelligent Systems in an easily accessible and well structured form. The series includes reference works, handbooks, compendia, textbooks, well-structured monographs, dictionaries, and encyclopedias. It contains well integrated knowledge and current information in the field of Intelligent Systems. The series covers the theory, applications, and design methods of Intelligent Systems. Virtually all disciplines such as engineering, computer science, avionics, business, e-commerce, environment, healthcare, physics and life science are included.

More information about this series at http://www.springer.com/series/8578

# Intelligent Systems Reference Library

Volume 115

**Series editors**

Aboul Ella Hassanien · Mohamed Mostafa Fouad
Azizah Abdul Manaf · Mazdak Zamani
Rabiah Ahmad · Janusz Kacprzyk
Editors

# Multimedia Forensics and Security

## Foundations, Innovations, and Applications

*Editors*
Aboul Ella Hassanien
Scientific Research Group in Egypt (SRGE), Faculty of Computers and Information, Department of Information Technology
Cairo University
Giza
Egypt

Mohamed Mostafa Fouad
Scientific Research Group in Egypt (SRGE)
Arab Academy for Science, Technology, and Maritime Transport
Giza
Egypt

Azizah Abdul Manaf
Advanced Informatics School
Universiti Teknologi Malaysia
Kuala Lumpur
Malaysia

Mazdak Zamani
Advanced Informatics School
Universiti Teknologi Malaysia
Kuala Lumpur
Malaysia

Rabiah Ahmad
Universiti Teknikal Malaysia Melaka (UTem)
Malacca City
Malaysia

Janusz Kacprzyk
Systems Research Institute
Polish Academy of Sciences
Warsaw
Poland

ISSN 1868-4394 ISSN 1868-4408 (electronic)
Intelligent Systems Reference Library
ISBN 978-3-319-83026-1 ISBN 978-3-319-44270-9 (eBook)
DOI 10.1007/978-3-319-44270-9



Printed on acid-free paper

This Springer imprint is published by Springer Nature
The registered company is Springer International Publishing AG Switzerland

# Preface

Digital forensics is the process of uncovering and interpreting electronic data. The goal of the process is to preserve any evidence in its most original form while performing a structured investigation by collecting, identifying and validating the digital information for the purpose of reconstructing past events. However, the emergence of the cloud computing structures and services, where the information is stored on anonymous data centers scattered around the world, makes the digital forensics pose more challenges for law enforcement agencies. Other problems are the variety formats and the exponential growth of data that need to be analyzed in reasonable time to conduct a forensics decision.

Although the trust is a fiduciary relationship between the law enforcement agencies and the cloud service providers, still there is fear that the information on cloud servers can be altered or hidden without a trace. Agencies are collecting unencrypted as well as encrypted content. This encrypted content presents another limitation for forensic investigators.

The objective of this book is to provide the researchers of computer science and information technology the challenges in the fields of digital forensics, which are required to achieve necessary knowledge about this emerging field. The book goes through defining the cloud computing paradigm and its impacts over the digital forensic science, to the proposal of some authentication and validation approaches.

The book is organized into three parts: Part I introduces the challenges facing the digital forensics in the new computing paradigm; the cloud computing. This section provides the characteristics and the limitations attached to the forensic analysis in such paradigm. Part II focuses on the forensics in multimedia and provides the application of watermarking as an authentication and validation technique. Finally,

Part III gives a number of recent innovations in the digital forensics field. These innovations include the data processing, the biometrics evaluations, the cryptography in Internet of Things, and the smart phone forensics.

| | |
|---|---|
| Giza, Egypt | Aboul Ella Hassanien |
| Giza, Egypt | Mohamed Mostafa Fouad |
| Kuala Lumpur, Malaysia | Azizah Abdul Manaf |
| Kuala Lumpur, Malaysia | Mazdak Zamani |
| Malacca City, Malaysia | Rabiah Ahmad |
| Warsaw, Poland | Janusz Kacprzyk |

# Contents

# Contributors

**Ahmed E. Abdel Raouf** Faculty of Computer and Information Sciences, Ain Shams University, Cairo, Egypt

**Shahidan M. Abdullah** Advanced Informatics School (AIS), Universiti Teknologi Malayisa, Kuala Lumpur, Malaysia

**Alshaimaa Abo-alian** Faculty of Computer and Information Sciences, Ain Shams University, Cairo, Egypt

**Bechir Alaya** Department of Management Information Systems, CBE Qassim University, Buraidah, Saudi Arabia; Higher Institute of Technological Studies, University of Gabes, Gabes, Tunisia

**Farnaz Arab** Kean University, Union, NJ, USA

**Amira Sayed A. Aziz** Université Française d'Egypte, Cairo, Egypt

**Nagwa L. Badr** Faculty of Computer and Information Sciences, Ain Shams University, Cairo, Egypt

**G.M. Bhat** Department of Electronics and Instrumentation Technology, University of Kashmir, Srinagar, India

**Imed Bouchrika** Faculty of Science and Technology, University of Souk Ahras, Souk Ahras, Algeria

**Jasmin Cosic** Ministry of Interior, University of Bihac, Bihac, Bosnia and Herzegovina

**Ashraf Darwish** Faculty of Science, Computer Science Department, Helwan University, Cairo, Egypt

**Maged M. El-Gendy** Faculty of Science, Computer Science Department, Helwan University, Cairo, Egypt

**Mohamed Mostafa Fouad** Arab Academy for Science, Technology, and Maritime Transport, Cairo, Egypt

**Safaa I. Hajeer** Ain Shams University, Cairo, Egypt

**Chaffa Hamrouni** LITIS-Lab, University of Le Havre, UFR Sciences et Techniques, Le Havre Cedex, France; MAC'S-Lab, National Engineering School of Gabes, Zerig, Gabes, Tunisia

**Aboul Ella Hassanien** Faculty of Computers and Information, Cairo University, Cairo, Egypt

**Rasha M. Ismail** Ain Shams University, Cairo, Egypt

**Yashar Javadianasl** AIS, UTM, Kuala Lumpur, Malaysia

**Sasan Karamizadeh** Advanced Informatics School (AIS), Universiti Teknologi Malayisa, Kuala Lumpur, Malaysia

**Lamri Laouamer** Department of Management Information Systems, CBE Qassim University, Buraidah, Saudi Arabia; Lab-STICC (UMR CNRS 6285), University of Western Brittany, Brest Cedex, France

**Nazir A. Loan** Department of Electronics and Instrumentation Technology, University of Kashmir, Srinagar, India

**Azizah Abd Manaf** AIS, UTM, Kuala Lumpur, Malaysia

**Sergii Mashtalir** Kharkiv National University of Radio Electronics, Kharkiv, Ukraine

**M.E. Megahed** Faculty of Computer and Information Sciences, Ain Shams University, Cairo, Egypt

**Qurban A. Memon** UAE University, Al-Ain, United Arab Emirates

**Olena Mikhnova** Kharkiv Petro Vasylenko National Technical University of Agriculture, Kharkiv, Ukraine

**Fatma Mohamed** Faculty of Computer and Information Sciences, Ain Shams University, Cairo, Egypt

**Nihel Msilini** MAC'S, National Engineering School of Gabes, University of Gabes, Zerig, Gabes, Tunisia

**Parham Nooralishahi** Department of Computer Science and Information Technology, University of Malaya, Kuala Lumpur, Malaysia

**Shabir A. Parah** Department of Electronics and Instrumentation Technology, University of Kashmir, Srinagar, India

**Amir Sadeghian** Advanced Informatics School, Universiti Teknologi Malaysia, Kuala Lumpur, Malaysia

**Jafar Shayan** Advanced Informatics School (AIS), Universiti Teknologi Malayisa, Kuala Lumpur, Malaysia

**Javaid A. Sheikh** Department of Electronics and Instrumentation Technology, University of Kashmir, Srinagar, India

**Mohamed Fahmy Tolba** Faculty of Computer and Information Sciences, Ain Shams University, Cairo, Egypt

**Mazdak Zamani** Department of Computer Science, Kean University, Union, NJ, USA

# Part I
# Forensic Analysis in Cloud Computing

# Cloud Computing Forensic Analysis: Trends and Challenges

**Amira Sayed A. Aziz, Mohamed Mostafa Fouad and Aboul Ella Hassanien**

**Abstract** Computer forensics is a very important field of computer science in relation to computer, mobile and Internet related crimes. The main role of Computer forensic is to perform crime investigation through analyzing any evidence found in digital formats. The massive number of cybercrimes reported recently, raises the importance of developing specialized forensic tools for collecting and studying digital evidences in the digital world, in some situation even before they are lost or deleted. The emergence of the new Cloud Computing paradigm with its unique structures and various service models, had added more challenge to digital forensic investigators to gain the full access and control to the spread cloud resources. While, the current chapter starts to lay the importance of digital forensics as whole, it specially focuses on their role in cybercrimes investigations in the digital cloud. Therefore, the chapter goes through the definition of the basic concepts, structures, and service models of the cloud computing paradigm. Then, it describes the main advantages, disadvantages, challenges that face the digital forensic processes, and techniques that support the isolation and preservation of any digital evidences. Finally, the chapter stresses on a number of challenges in the cloud forensic analysis still open for future research.

Scientific Research Group in Egypt (SRGE).

A.S.A. Aziz (✉)
Université Française d'Egypte, Cairo, Egypt
e-mail: amiraabdelaziz@gmail.com
URL: http://www.egyptscience.net

M.M. Fouad
Arab Academy for Science, Technology, and Maritime Transport, Cairo, Egypt

A.E. Hassanien
Faculty of Computers and Information, Cairo University, Cairo, Egypt


A.E. Hassanien et al. (eds.), *Multimedia Forensics and Security*,
Intelligent Systems Reference Library 115, DOI 10.1007/978-3-319-44270-9_1

## 1 Introduction

The Cloud Computing, is one of the fastest growing technologies that attracts researchers to add and improve its services [1, 2]. Organizations benefit from this technology by replacing traditional IT hardware and data centers with remote, on-demand paid hardware and software services that are configured for their particular needs, managed and hosted by the organization users or even a third party. This increases the organization's flexibility and efficiency, without the need to have a dedicated IT staff or owning special hardware equipment or software licenses.

However, cloud computing security is still an open research issue, and malicious users take advantage of this lack of advanced security mechanisms. According to a Forbes magazine report in 2015 [3], "The cybersecurity space is arguably the hottest and fastest growing tech sector." The worldwide cybersecurity market estimates a range from $77 billion in 2015 to $170 billion by 2020. In the Guardian [4], they stated that "The sharp rise in the headline figures is due to the inclusion of an estimated 5.1 million online fraud incidents and 2.5 million cybercrime offences for the first time." The statistics of 2015 stated in Hackmageddon [5]—an Information Security Timelines and Statistics website—shows that 2015 has reported a more sustained activity in cybercrime. Figure 1 [5] shows that cybercrime is the major motivations behind attacks and intrusions, it even increased compared to year 2014.

Figure 2 shows the targeted sectors by cybercrime and attacks, while Fig. 3 shows different attack techniques followed by criminals in 2015 versus 2014 [5].

Based on previous statistics, it becomes a necessity to conduct a digital forensic investigations once an attack takes place. Computer forensics has emerged to assist

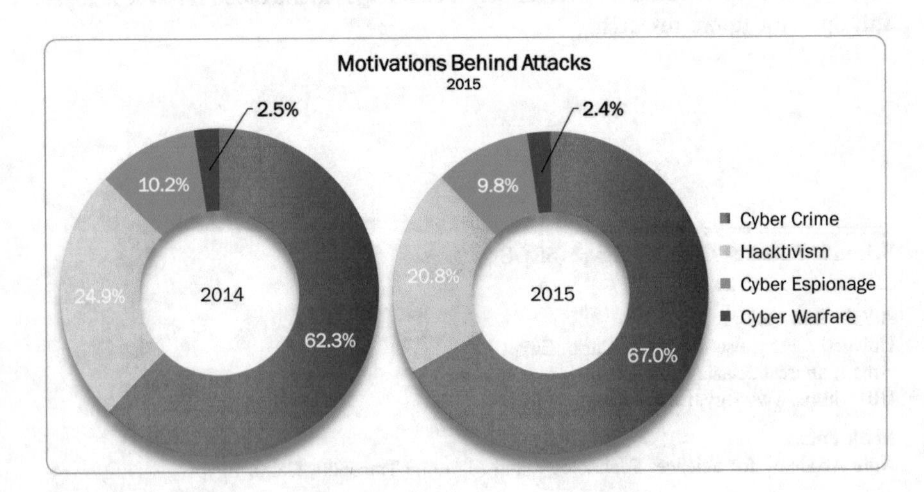


**Fig. 1** Motivation behind online attacks [5]

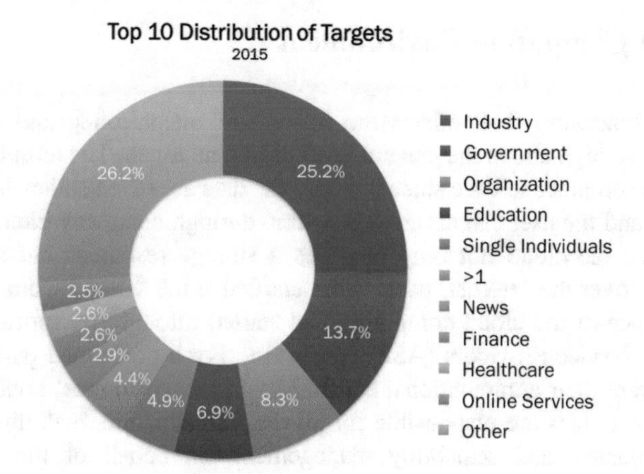


**Fig. 2** Targeted sectors by cybercrime [5]

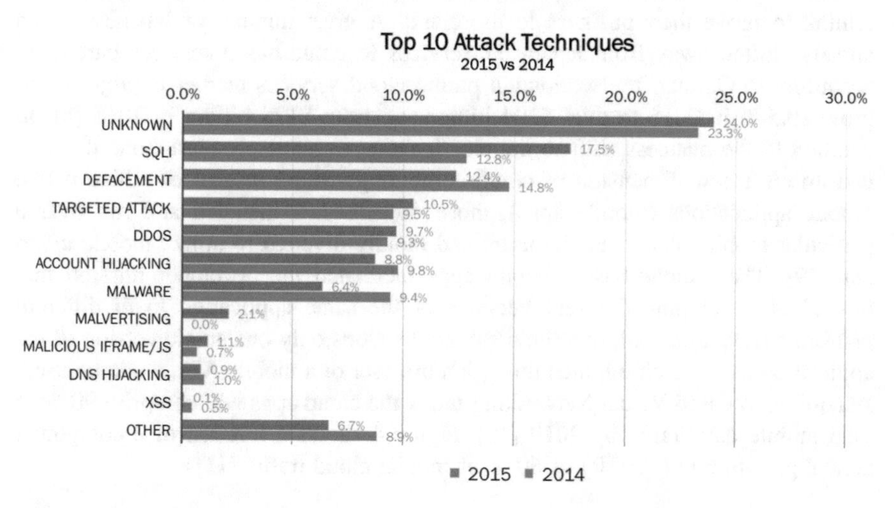


**Fig. 3** Attack techniques [5]

law enforcement and provide them with means to investigate cybercrimes and online attacks through the digital world. Live digital forensic is required to be able to collect and analyze evidences before they are lost or deleted. Investigators need more tools to help them conduct digital forensics in the cloud [6].

This chapter provides a quick review of basic cloud computing concepts and structures. Then, it describes the general model of digital forensic process. Finally, the chapter goes through the analysis process in the cloud environment, along with current challenges and open research topics.

## 2 Cloud Computing Environment

Cloud is a buzzword that reflects the floating of uncontrolled and unstructured density of mist high above the general level of human touch. The term has the same metaphor in computer science since it means the data are saved somewhere, through the internet and the user can access it any time through using any internet enabling device. Since the cloud not only provides a storage resources but also provide computation over the internet, users often entitled it the "Cloud Computing". The real emergence of the cloud computing had started through the appearance of the Application Service Providers (ASP) companies. For a predefined paid fees, these companies rent their computational capabilities to run customers' applications [7]. The ASP companies are responsible for all the infrastructure, including hardware, software, updates, and scalability management, on behalf of their customers. Therefore, the cloud computing had removed the fear of hardware scalabilities, hiring or training new employees, and purchasing software licenses.

The advances in cloud computing technologies have made several organizations rethink to move their business to the cloud. A great number of businesses had already shifted away from legacy IT services to cloud-based services paradigm; according to Gartner, the worldwide public cloud services market is projected to grow 16.5 % in 2016 to total $204 billion, up from $175 billion in 2015 [8]. In addition to the business shift to the cloud, the emergence of smartphone devices, had arisen a new dimension of cloud computing; where instead of utilize native mobile applications (mobile apps), those downloaded, installed and run over a particular mobile platforms, the users had rapidly diverted to utilize mobile cloud apps [9]. These cloud-based mobile apps facilitated the developer mission that instead of developing different versions of the same application to fit different platforms (IOS, Android, or Windows), he develops only one single version of the application over the cloud, then through a browser or a mobile API, it can be used. According to Cisco Visual Networking Index the cloud apps will comprise 90 % of total mobile data traffic by 2019 [10]. Figure 4 shows a forecast of a compound annual growth rate (CAGR) of 60 % of mobile cloud traffic [11].

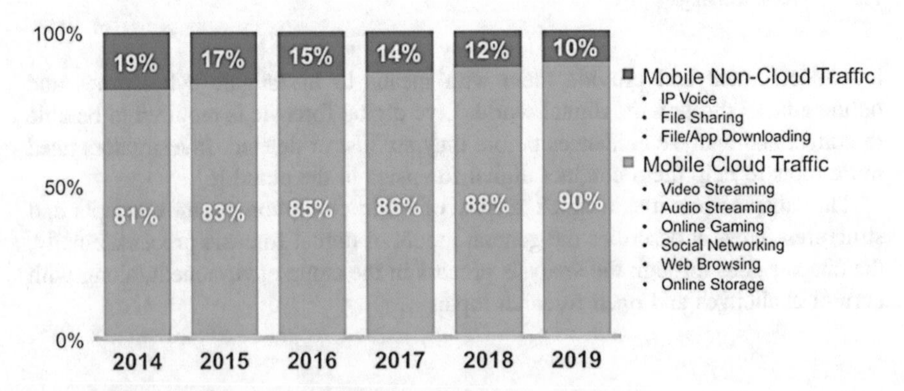


**Fig. 4** Mobile cloud traffic forecasting from 2014 till 2019 [11]

## *2.1 The Cloud Models*

The cloud models can be differentiated based on general features such as the level of security, control, and cost effectiveness [12]. The cloud model is either private, public, or hybrid.

***Public Cloud Model***: is where the infrastructure is made available for multiple customers. The owner of the cloud is either a single or multiple organizations. The main feature of that cloud model is its cost effectiveness to customers. Customers are only paying for the resources they acquired. The location independence and flexibility to access the cloud from any internet enabled device are other added values to the spreading of the public cloud. Microsoft, Google and Amazon are of the big companies offering their infrastructure to public customers.

***Private Cloud Model***: provides a shared pool of resources and services under the control of a single organization. The customer of the private cloud could be the same organization or a third party. Private clouds offer a great security and privacy at greater cost such as the NIRIX's oneServer [13] which provides a dedicated servers to host e-commerce applications, websites, or web-based business applications for either internal or external access. In brief, while private clouds offer a best suitable solution to securing critical data, investment in configuring and maintaining private clouds is more expensive than public clouds. In addition, the scalability of the private clouds is constrained to the acquired resources.

Hybrid cloud model gets the best of both public and private models (Fig. 5). It is considered a good business strategy to cut down the cost through utilizing the public clouds for some applications while maintaining the private data over their private cloud. The integration of both clouds is a major concern since it requires a strict security requirements to control which information should flow to the public cloud.

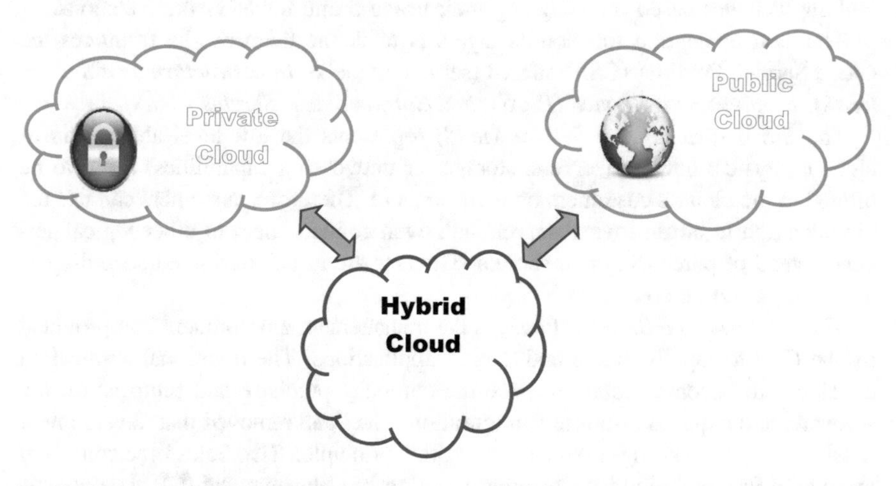


**Fig. 5** Basic cloud computing models

## 2.2 The Cloud Structure and Services

Cloud structure is composed of building blocks (layers). The perception of the cloud structure is based on either the cloud functionality, or on the resources it offered.

Based on the cloud functionality, the cloud structure composed of four layers, each layer providing a distinct level of functionality. According to [14] the layers are: *hardware components (datacenters), infrastructure, platform, and application (API) layer*. Each layer is served by the layer below and it serves the above layer.

The *hardware components layer* is referred as datacenters layer since it is the base layer that including all hardware components that usually exist in the datacenters including: servers, storage mediums, communication devices, and power resources.

The *Infrastructure layer* is the dynamic assignment layer that uses virtualization management principles to partitioning the hardware resources among the customers. The **Linux-VServer** [15], and VMware [16] are widely applied virtualization software to partitioning a server into multiple logical servers that can run independently for different operating systems and applications.

The *platform layer* is the attached layer to the infrastructure layer which provides flexibility to developers to use the existing API to implement and deploy their applications over the cloud. Hence, there is no need for the developer to deal with the operating system that installed within the logical server; e.g. the App Engine of Google Inc. [17] allows developers to easily build web and mobile backends applications over the Google's cloud with the advantage of automatic scalability of hosted applications based on the usage traffic.

The *application layer* is the most visible layer for the end user at which he uses the service or the application provided by the cloud. Usually, the provided services are not free of charge; such as Alexa [18] that provides analytical digital tools for ranking websites based on analyzing their usage traffic for other organizations.

The next perception for cloud's layers is to define them by the resources the Cloud Service Provider (CSP) offered (service models): *Infrastructure as a Service (IaaS), Platform as a Service (PaaS), and Software as a Service (SaaS).*

The *Infrastructure as a Service* (*IaaS*) represents the low-level abstraction of cloud's physical devices (servers, storage, or networking capabilities) those to be offered to the cloud's customers on their demand. Therefore, customers can use the virtualization to create logical servers and even connect them to other logical servers instead of purchasing real servers. The fees are to be considered according to the resources to be consumed.

The *Platform as a Service* (*PaaS*) is the management environment that provided by the CSP to rapidly create and deploy applications. The traditional application development becomes complex since the company purchase and setup hardware, software, and required in-house configuration. The PaaS removed that development hassle through "just-log-in-and-get-to-work" principle. The Salesforce.com [19] provides a PaaS solution for Customer Relationship Management (CRM) products.

The solution allows non-technical customers to effectively customize functions that exist across their businesses.

Finally, the *Software as a Service* (*SaaS*) is the most accessible layer of the cloud that represents the provision of actual applications or services to the end users. The providers of SaaS should keep running and adapting the increasing number of customers and applications over their cloud. As to imagine the problem that faces those providers, the Statista portal [20] had reported that the Google Play had passed over 1.8 million in last quarter of 2015 apps while Apple iTune had exceeded from 800 apps in July 2008, the month of its launch, to 1.5 million in June 2015.

## 3 Digital Forensics

Digital Forensics is defined as "the use of scientifically derived and proven methods toward the preservation, collection, validation, identification, analysis, interpretation, documentation and presentation of digital evidence derived from digital sources for the purpose of facilitation or furthering the reconstruction of events found to be criminal, or helping to anticipate unauthorized actions shown to be disruptive to planned operations" [21]. Digital forensics is a discipline where law and computer science are combined to collect and analyze data from computers, networks, and storage devices where there is an evidence that would be admissible in a court of law [1, 22].

In the past, digital forensic services were used only in late stages of an investigation, where digital evidence might have been damaged or spoiled. Now, it has become and essential right at the beginnings of all investigation types. Digital forensics emerged out of practitioners' community, where investigators and developers unify their efforts to find solutions to real-world problems. The first Digital Forensic Research Workshop (DFRWS) was held in 2001 to construct a community that applies scientific method to find solutions driven by practitioners' requirements and needs. Different organizations have contributed to establish an academic and scientific basis for the Digital Forensic research. The Scientific Working Group on Digital Evidence (SWGDE) released many documents concerning standards, best practices, validation processes. The US National Institute of Standards and Technology (NIST) started the Computer Forensic Tool Testing (CFTT) Project in 2001 that has established and executed protocols for digital forensic tools [23, 24].

Physical media, operating system, file system, and user-level applications produce artifacts where digital evidence is created and left behind. They are used by investigators to extract information that would help them understand the past behavior in the digital environment. Relatively, most digital forensics research has focused on where artifacts exist and why, and how to recover them. Two research areas have gone through notable growth: *Data Carving* and *Memory Analysis*, which contribute more to the forensic analysis phase. At the same time, the response and data collection phases were still away from attention and have unanswered questions, as more priority is given to the analysis [23, 25].

Two categories of digital forensics exist: *Static Forensics* and *Live Forensics*. Static Forensics is offline forensics where analysis is performed on data acquired from storage devices and hard drives obtained using traditional formalized procedures. Live Forensics is where analysis of any relevant data is done while the system being analyzed is running [22].

Certain issues may face the digital forensic process, according to the service model provided by the cloud [26]:

- ***SaaS Model Environment***—As it was previously described in Sect. 2, the customer does not have any control over the platform, the infrastructure, or the operating system in such environment. Only control over some application settings may be granted, therefore, they do not have any chance of analyzing possible incidents. The CSP merely is the main source of data investigation, even with the availability of log information. The situation might be better in the SaaS over the private cloud, where the customer and the CSP belong to the same entity.
- ***PaaS Model Environment***—there is a main advantage in this model, where the customer has full control over running applications, which means that logging mechanisms can be deployed to gather information and transfer it to a third party or locally analyzed. However still, CSP has to do necessary configuration to control the underlying runtime environment—hence, give the ability to developers collect some diagnostic data.
- ***IaaS Model Environment***—in this model the customer has the full control to install and set up images for forensic purposes. Snapshots of virtual machines can be employed, to provide data for the investigation process, a snapshot clones the virtual machine by one click, including the system's memory. Also, the system itself can be prepared for forensic investigation purposes logging lively continuous information of users, open ports, running processes, registry information, and other forensic analysis processes.

### *3.1 Digital Forensics Common Phases*

Four principles of digital forensic practice were proposed by the Association of Chief Police Officers, and these are [6]:

1. Data extracted or held on a computer or any storage media should not be changed or modified by any action taken by the investigators, so that the evidence would remain reliable in court.
2. An expert and dependent person should be in charge of accessing original data on computer or storage media if needed, so that person would be qualified to do so and be able to explain their relevance and implications in the investigated case.

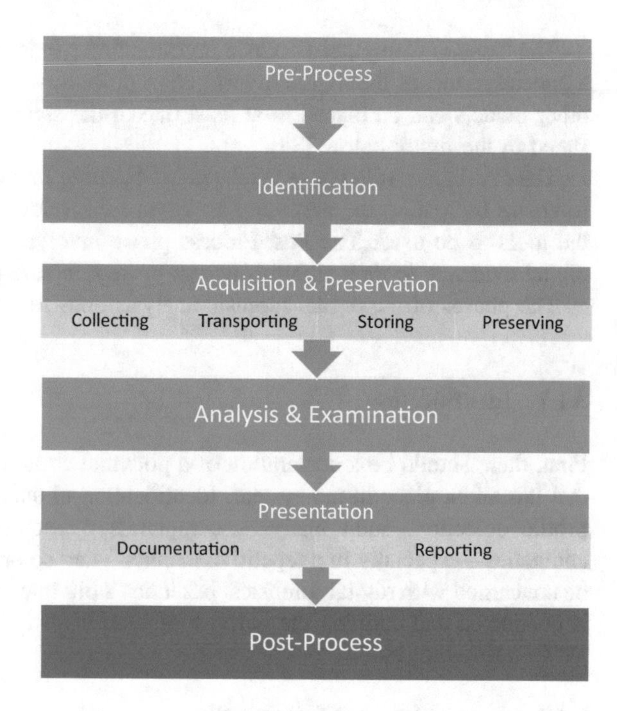


**Fig. 6** Digital forensic process model

3. It should be available to create and preserve audit trails and records of all procedures and processes applied to digital evidence so that if an independent party examined these processes, they should achieve the same results.
4. A case officer is assigned to the investigation to hold the overall responsibility for ensuring that law and legal principles are applied as followed.

The primary model of digital forensics stated in 1984 that it consists of four phases that were presented by Pollitt in [27, 28]: *Acquisition, Identification, Evaluation*, and *Admission*, as shown in Fig. 6. Through the beginning of 2000s some phases were added to the model or changed, where additional steps were needed to be added to the main process through the development of the digital forensics. The additional phases included: *Preservation, Collection, Examination, Analysis*, and *Presentation*. Considerable research adapted specific conception of the model, where some added a *Traceback* phase where investigators are able to trace back all the way to the actual devices used by the criminals [29]. Others added a *Planning* phase to ensure the success of the investigation; this was in Extended Model of Cybercrime Investigation (EMCI) [30], Computer Forensic Field Triage Process Model (CFFTPM) [31], and Digital Forensic Model based on Malaysian Investigation Process DFMMIP [32], to improve the investigation process by prior planning of all the phases to follow. Furthermore, others added a *Proof and Defense* phase (in EMCI [30] and DFMMIP [31]). In previous phase, the investigators are required to present proof for the used evidence in the investigation, to support the presented case.

The Generic Computer Forensic Investigation Model (GCFIM) was proposed as a general model of the digital investigation process, where recommended phases in other models can be placed in at least one of the stated phases in that model—as stated in the figure below [33].

The *Pre-Processing* phase is related to obtaining forensic data and requesting for forensics by getting the necessary approval from relevant authority and setting up the tools to be used. The Post-Process phase involves the return of physical and digital evidence to their rightful owners or kept in safe place if necessary.

The phases of the digital forensic analysis goes as follows [1, 2, 6].

### 3.1.1 Identification

First, there should be a declaration of a potential committed crime or improper act that has taken place in the system. Identification of such crime may be a result of profile detection, audit analysis, complaints by some individuals, or detected anomalies—especially in a repetitive manner—and so on. This phase may not only be concerned with digital forensics, but it has a big impact on how the investigation is conducted and defining the purpose of such investigation.

### 3.1.2 Acquisition and Preservation

Relevant data to the identified crime or illegal act should be collected and preserved so that it will not be lost, manipulated, or modified in anyway. Specialized tools should be used and approved methods should be followed in this phase such as the Forensic Explorer (FEX) software [34]. A challenging topic for investigators is the massive amount of data that they might have to collect and deal with. Investigators should also keep a roadmap or a registry of evidence was collected, analyzed, and reserved for the presentation in court—which is called Chain of Custody. This provides a proper documentation of how evidence was gathered and handled, by whom and when. The preservation is concerned with keeping a timeline of the collected evidences to be able later to create the sequence of events involved in the attack. This can be done by collecting timing information from timestamps in meta-data or different log files of applications and networks.

### 3.1.3 Analysis

Once data has been collected and preserved, they should go through examination and analysis to extract important patterns required by investigation process out of the collected data. Many software tools, such as Digital Forensics Framework [35], Open Computer Forensics Architecture [36] and EnCase [37] are used by an investigator for pattern matching, filtering, searching, discovering attempts to delete data, or recover lost data. During the analysis phase, a scenario is developed based on the evidence and their timeline to explain how a crime was committed. If possible, certain users or user account might be associated with certain evidence or event. Evidences also should be subjected to validation to assure they have not been altered or manipulated before the examination phase.

### 3.1.4 Presentation

Finally, reports should be prepared, in order to summarize the conclusions and to provide explanations for these conclusions through evidence collection and examination. Then, these reports are submitted to court of law, and the investigator would be subject to expert testimony and cross-validation.

## *3.2 Digital Evidence*

The National Institute of Justice defined a digital evidence as any data or information that is of value to an ongoing investigation, and that is stored on, received, or transmitted by an electronic device. A digital evidence can be found on the Internet, on stand-alone computers, or on mobile phones. The Internet offers global access to information and computers, which gives the criminals the ability to access different information sources, banks, governmental networks, and other systems they could sabotage. The communication activities with time stamps and traces left on accessed computers can be used as digital evidence for investigations of digital crimes. Also, any stored data on computers can be used as digital evidence, where they are located on the physical hard drive and removable media. For mobile phones, their tracking capability would turn the mobile phones into key evidence in many cases [6, 38].

The data to be collected and acquired may be available in three different statuses [26]. Figure 7 illustrates these statuses. Data at *rest* means they stored in a database or a specific file format, allocated in disk space. Data in *motion* refers to data that transferred between entities. Data in *execution* is data loaded into memory and executed as a process. For each state, different techniques are applied to acquire the data. Data at rest can be extracted by investigators from hard disks, even if they are deleted—as long as they are not de-allocated, they can be retrieved by some software applications. Data in motion usually leave traces on systems and network devices through protocols applied for data transfer on networks. Hence, these traces can be collected and used by the investigators. For data in execution, snapshot technology can be used, where process information, machine instruction, and memory data can be analyzed.

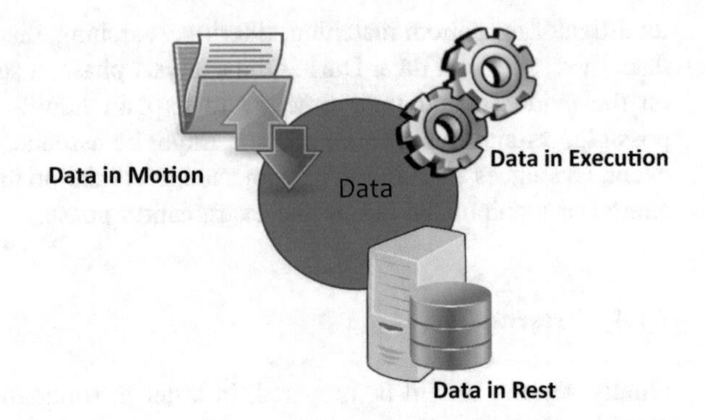


**Fig. 7** Data acquired statuses

Hence, there are three possible sources of data to be collected as artifacts and used later as evidence: Virtual Cloud Instance, Network Layer, and Client System [26]. A **Virtual Cloud Instance** is where an incident took place, hence it can be a potential starting point or an investigation. Based on the type of service the client is using (IaaS, SaaS, or PaaS), instance can be accessed by the CSP only or also through the customer. Snapshots are powerful tool a customer can use to save specific states of the virtual machine. **Network Layer** (and other ISO/OSI layers) can provide different information on communication between instances inside and outside the cloud. Unfortunately there is a problem of the log data that can be provided by the CSP for investigations in the case of an incident, which is explained later in challenges. The **Client System**, it completely depends on the service used to whether any data can be provided or not. In most cases, the browser provides a source of data since it is the application used to connect clients to the cloud services.

Pre- and post-crime information might be used as crime evidence, especially if the crime was completely committed through digital means. In digital world, there is always an electronic trail of activities and information left behind to be seized and exploited. The most important thing for investigators is to follow proper procedures so that evidences would not be lost, damaged, or manipulated that it will not be admissible in courts of law.

A digital evidence should fulfill some characteristics to be legally considered in an investigation [39]:

- Authentic—original and related to the investigated crime.
- Reliable—collected using reliable procedures that if it run by an independent party would give the same results.
- Complete—neither corrupted nor manipulated.
- Believable—convincing and making sense to an ordinary juries.
- Admissible—collected using common law procedures, following agreeable policies.

### *3.3 Digital Forensic Analysis*

Digital forensic analysis process is held by investigators using different analytical tools to turn extracted data from artifacts into usable information that would hopefully lead to the criminal who committed the crime. The majority of the computational approaches existing rely on literal pattern searching and matching with necessary indexing techniques to speed up the search and match [23]. These results in two major problems: underutilizing available computational power and the overhead resulting from the information retrieval. Existing platforms and user-class workstations handle more complicated and advanced operations that simple matching. Techniques that can be employed and implemented to reduce the human participations and burden in this is of high importance, even if it requires more computational time. Artificial intelligence, information science, data mining, and information retrieval are existing algorithms that supporting the gain of better results [40–42]. These algorithms will help the investigators to obtain hidden knowledge and reveal data trends and undetectable information those are hardly to be noticeably by common human observation.

Investigators trained specifically to deal with digital evidences should be the only ones examining these evidences. The thing is, there is no single path to gain this expertise, qualifications and certifications are not standardized nationally or internationally. What happens is that investigators take an interest in the area and they learn what they can—and even in this case, an investigator might be a specialist in cell phones for example, but not in social media or bank fraud. He could be of knowledgeable enough to investigate stolen identities on the Internet, but has no experience in tracing left traces on digital devices and extracting histories of exchanged information through communication devices [38].

Certified Digital Media Examiners are investigators that have proper knowledge, training, and experience to have the ability to exploit sensitive data. Many certifications are offered through the Digital Forensics Certification Board (DFCB), which is an independent certifying organization for digital evidence examiners, and the National Computer Forensics Academy at some colleges in the US. In some states, investigation laboratories include a section for digital crimes—such as Internet Crimes Against Children (ICAC), Joint Terrorism Task Force (JTTF), and Narcotics and Property crimes [1].

Collected evidence should be examined inside laboratories by qualified analysts, who should follow these steps to retrieve and analyze data. The analyst should make sure the collected evidence is not to be contaminated by creating a copy of the original storage device, using clean storage media to prevent any possible contamination. Wireless devices should be isolated to prevent connection to any networks or prohibit any possible exploitation of the device. If necessary, write-blocking software should be installed to make sure the date will not be lost, altered, or over-written.

# 4 Forensic Analysis in the Cloud Environment

Cloud Forensics combines cloud computing and digital forensics. It is concerned with computer forensics with some consideration to network/intrusion forensics. Computer forensic focuses on using procedures to create audit trails based on the residing data. Network forensic focuses on analyzing network traffic and gathering information by monitoring that traffic to extract or collect information that might be considered a possible evidence. Intrusion forensic is concerned with investigating possible intrusions to computers or networks [39, 43].

A cloud crime is any crime where a cloud might be the object of, subject of, or tool used in committing that crime [43]. The cloud is: the **object** when the CSP is the target of the crime act, the **subject** when it is where the crime was committed, and the **tool** if it is used to conduct or plan the crime (where a cloud that is used to attack another cloud is called a Dark Cloud). Cloud forensics is not necessarily carried on when there's only a crime. There are several usages of the cloud forensics [43, 44], which include:

- *Investigations*—when a cloud crime or a security violation takes place, where there is collaboration with the law force to investigate suspected online transactions and operations to provide admissible evidence to the court.
- *Troubleshooting*—to target different problems and trace events and hosts to find the root cause of certain events, preventing repeated incidents, assessing the performance of the cloud, or resolving functional and operational issues in the loud services and applications.
- *Log Monitoring*—to collect and analyze different activities log entries in the cloud for auditing, regulating, monitoring the cloud services and other efforts.
- *Data and System Recovery*—Recovering lost data (deleted accidentally or intentionally), recovering systems after an attack or damage, acquiring cloud data that need to be retired or sanitized.
- *Regulatory Compliance*—for organizations to make sure they comply with different requirements of security and safety of data, and maintain audit records if a certain policy is violated and notifying proper parties when incidents take place.

Because of the concerns about data privacy and security, a corporate security team would want to investigate security incidents through their own team, instead of relying on a third party. The cloud entity should provide some roles in their staff (internal and external) to establish a forensic capability [43]:

- *Investigators*—who will be responsible for examining different allegations and deal with external law enforcement and interact with them in forensic investigations.
- *IT professionals*—including security administrators, ethical hackers, cloud security architects, and technical staff, to support the investigations accessing crime scenes and collect data on their behalf.

- *Incident handlers*—who should have appropriate expertise to handle different security incidents of different levels, such as data leakage or loss, unauthorized access, and breach of confidentiality, etc.
- *Legal advisors*—who are familiar with laws of multi-jurisdiction and multi-tenant, and ensure that no laws and regulations are violated throughout the forensics process. They also are responsible for clarifying the procedures to be followed in the investigation.
- *External assistants*—to help along with the internal staff with different relative actions and forensic tasks that should be performed for the investigations under relevant policies and agreements.

Both Static and Live forensics face challenges in the cloud environment. However, Live investigations on running virtual machines provide investigation data that might not be available after the shutdown. The need to shut down the system for data acquisition and guaranteeing that the data centers are within physical reach to collect data is one challenge. Another challenge is the difficulty to catch up with changing paradigms of the dynamic structure of the cloud environment, which makes it more difficult to locate information for data acquisition. The data being stored itself could be encrypted by the cloud users, so the investigators would not be able to deal with the acquired data. Even more, the investigators might not be able to associate the date stored in the cloud with specified users' identities—either because they cannot find a connection between the data and the users or because users are using aliases on the cloud that cannot be connected to their real identities [1, 26].

## 4.1 Isolation of Crime Scene in Cloud

In real world crimes forensic investigation, a crime scene is isolated so that evidence would not be lost or contaminated, and to be accessed by only authorized personnel. Same happens in digital crime investigation process, where storage devices and processing units are isolated and kept away from reach to provide safety and integrity to collected evidences. In the cloud environment, there should be methods to be applied in place to protect confidentiality and privacy of users and their data. Hence, there are existing methods that either move the suspicious instance to another node or move away irrelevant instances to other nodes [44]. The forensic analysis techniques then differ according to the type of analysis done: live or dead analysis. In live forensics, the instance should be stopped so that the evidence will not be tampered with, while in dead forensics other nodes should be protected from the power outage.

Different techniques exist for the isolation, they are described in the following list [44, 45]:

- *Instance Relocation*—the instance is moved inside the cloud, manually by the administrators or automatically by the operating system. This can be done in three ways: ending the existing instance and starting a new one, creating a new instance then terminating the old instances, or the instance is logically moved where data is moved to another node with destroying the instance itself.
- *Server Farming*—where a server farm is a multi-node system, if a single node fails the remaining instances will continue to function. This would facilitate the isolation of the suspicious instance without having to stop the functionality. This should be implemented by the cloud operating system creators in order to be executable.
- *Failover*—this uses the fact of an existing replica server, which acts as a backup of the original server if the primary server fails. Simply, replicating the same units existing in the replica server by the digital forensic team. This also requires the collaboration of the cloud operating system manufacturers.
- *Address Relocation*—when network traffic is redirected to other computer, for the uninvolved nodes so that the suspicious instance would be isolated from them.
- *Sandboxing*—this is a technique usually used for applications, to control the running environment, so that running programs cannot affect other programs outside the sandbox. In the cloud, a virtual sandbox is created either by the cloud operating system or a launched application by the investigator, to monitor the network traffic to the application and block if needed.
- *Man in the Middle*—this is similar to the interception network threats, where an entity places itself between two connected ends. Interruption, modification, or fabrication can then take place to manipulate the transferred data. In the same manner, an entity can be places between the cloud instance and the cloud hardware, which analyzes the data going between the instance and the hardware, without the instance being aware of the entity. This entity must be added by the creators of the cloud software.
- *Let is Hope for the Best*—in this method, the node is turned off and moved to the investigation environment, where images are taken of hard drives and analyzed. The main problem is that a node may contain many instances that can be lost, and the running information might be lost too.

## 4.2 Pros and Cons of Forensic Analysis in Cloud

In spite of the challenges that face digital forensics in the cloud environment, investigations could benefit from services and resources provided by cloud system. These advantages include [39]:

- Centralized data is a main benefit where data is stored in one place, which results in quicker and coordinated response to incidents.

- Powerful services offered by the cloud can be of good use to the investigator, used in compute intense jobs—such as cracking passwords or decrypted text, or examining captured images.
- Investigators could authenticate stored data or disk images using hashed authentication techniques inbuilt in the cloud system.
- Logging can be performed on different levels using the large scale of cloud storage.
- Under virtualization, there could be an advantage of using snapshots captured of users' virtual machine instead of images of the environment, where they are easier to collect and requires less storage.

Some drawbacks in the perspective of digital forensics of the cloud environment can be briefed as following:

- Data acquisition is the hardest phase that can be held in the cloud system. To locate data exactly and actually acquiring the data are difficult tasks in such a distributed environment. It is also difficult to maintain a chain of custody related to such case.
- Acquiring and handling evidence from a cloud that reside in remote data centers makes it almost impossible to satisfy principles of collecting and acquiring data, which may affect the admissibility of an evidence in court of law.
- Data stored elsewhere in such inaccessible way prevents the reconstruction of the crime scene and piecing together a sequence of events and creating a timeline.
- The aspect of computer forensics for both legal and people perspectives makes a big obstacle. Digital evidences have to be introduced in a court in front of a jury that should judge based on presented evidences. Investigators have to present and explain the acquiring process of the evidence, along with its meaning to the investigated crime. This can prove challenging, especially that an average jury will have a basic knowledge of dealing with home PC—let alone grasping the concept of cloud computing and remote services.

### 4.3 Challenges of Analysis and Examination

The analysis of digital evidences faces a lot of challenges in the cloud environment. First of all, there is the volume of resources and objects to be examined, where they are huge faced with the limitation of processing and examining tools. There is no standard program to be used for data extraction; since the data extraction format may differ according to the service model used and what can be accessed by the customer. Extracted data out of evidences are in an unstructured fashion, which makes it difficult to reconstruct a stream of events for the investigation. The reconstructed scenarios are very valuable and crucial for the investigation to be able to recreate the crime and follow suspects [43, 46].

The challenges the investigators may face while they examine and analyze the evidences include:

- **Dependency on CSPs**: to acquire different activity logs, investigators rely extensively on the CSPs, where the availability of the logs depends on the service model.
- **Virtualization**: This is a basic technique used in the cloud environment to fully use the resources of the service provider by distributing the resources between users as if each user is using these resources separately. The shared resources could: servers, storage, software, platform, or infrastructure. This is done by running several instances of their servers as virtual machines. Mirrored data over multiple jurisdictions and the lack of information about data physical locations introduce more challenges to the investigators. Sometimes a CSP do not even have precise information about the actual locations of the data centers, or they might be located in a different country or geographical region. This requires strong international cooperation and the initiation of policies and procedures to be followed by forensic investigators without violation of laws and regulations.
- **Lack of examination tools**: It is important to accurately trace each event of the crime to create the logical order of the crime. It becomes more difficult if each event took place in a different country. Tools then are needed for computer and network forensics for acquiring and analyzing the forensic data in a timely fashion. Also, current tools are somehow limited, some aspects should be taken into consideration for future development of forensic tools. Ease of use is a major concern, where the tools should not be very technical and at the same time they should be customizable for use by experienced practitioners. Data visualization, automated analysis, and temporal presentation of the data instead of structural presentation is to be considered also, to provide information and knowledge instead of just displaying data.
- **Evidence distribution over multiple tenants**: The spread of activities and evidences over multiple digital resources introduces a problem or investigators, technically and legally. Because of the shared infrastructure between users, they do not have access to physical storage disks -instead, they access virtualized disks. Hence, it is a challenge to separate and extract the resources needed for an investigation without breaching the confidentiality of other users who might not be involved and sharing the same infrastructure.
- **Reconstruction of crime scene**: This is crucial, especially under virtualization that is a basic technique used in the cloud environment.
- **Anonymity of customers**: The cloud services facilitates anonymity due to its weak registration system that supports easy-to-go features of the cloud. This makes is difficult for the investigators to identify the true suspects who have their identities concealed.
- **Cryptography of data**: Customers sometimes use end-to-end customized encryption techniques to secure their stored data, which creates another challenge. These needs some sort of standardized agreements between the CSPs,

consumers, and enforcement agencies to help investigations by providing the investigators with cryptographic keys to decrypt the data.

## 4.4 Conclusions and Open Research Issues

Currently, the cloud computing has become a more mature paradigm that provides scalable storage, and unlimited computational capabilities at low cost. These features are acquired form a large pool of shared computing resources (e.g., servers, storage mediums, applications, and services). There are three well-known service models for the cloud: Infrastructure as a Service (IaaS), Platform as a Service (PaaS), and Software as a Service (SaaS). These models define the application architecture and the customer authorization over the cloud.

However, still there are a number of unsolved limitations and challenges attached directly to this new computing paradigm. Along with the security, and the privacy challenges, comes another important challenge is the conduction of digital forensics analysis in such paradigm.

Therefore, this chapter provides introductions about several preliminaries (concepts) related to the cloud computing. Through these concepts the chapter reveals the appropriate procedures to performing digital forensics and investigations to collect possible evidences of a conducted cybercrime. With lists of advantages and disadvantages of digital forensics in cloud environment, the chapter is concluding by the following open research issues (not only technical), to be available for more research [46, 47]:

- More investigation for the dependence on the cloud service providers.
- Power forensic tool development.
- Timeline analysis across multiple sources and evidence correlation, timeline analysis for logs.
- Guideline of global unity to overcome cross border issue.
- Provide some sort of standardization for the data extractions for forensic analysis.
- How to prepare jury's technical comprehension.
- Preserving Chain of Custody.
- Security, decryption and visualization of logs.

## References

1. Sibiya, G., Venter, H.S., Fogwill, T.: Digital forensic framework for a cloud environment. In: The Proceedings of the 2012 Africa Conference, pp. 1–8 (2012)
2. Patrascu, A., Patriciu, V.V.: Beyond digital forensics. A cloud computing perspective over incident response and reporting. In: The IEEE 8th International Symposium on Applied Computational Intelligence and Informatics (SACI), pp. 455–460. IEEE (2013)

3. Morgan, S.: The business of cybersecurity: 2015 market size, cyber crime, employment, and industry statistics. http://www.forbes.com/sites/stevemorgan/2015/10/16/the-business-of-cybersecurity-2015-market-size-cyber-crime-employment-and-industry-statistics/. Accessed April 2016
4. Travis, A.: Crime rate in England and Wales soars as cybercrime is included for first time. http://www.theguardian.com/uk-news/2015/oct/15/rate-in-england-and-wales-soars-as-cybercrime-included-for-first-time. Accessed April 2016
5. Passeri, P.: 2015 cyber attacks statistics. http://www.hackmageddon.com/2016/01/11/2015-cyber-attacks-statistics/. Accessed April 2016
6. Grispos, G., Storer, T., Glisson, W.B.: Calm before the storm: the challenges of cloud. In: Emerging Digital Forensics Applications for Crime Detection, Prevention, and Security 4, pp. 28–48 (2013)
7. Tao, L.: Shifting paradigms with the application service provider model. Computer **34**(10), 32–39 (2001)
8. Gartner Inc.: Gartner says worldwide public cloud services market is forecast to reach $204 billion in 2016. http://www.gartner.com/newsroom/id/3188817. Accessed April 2016
9. Mohamed, A., Fouad, M.M.M., Elhariri, E., El-Bendary, N., Zawbaa, H.M., Tahoun, M., Hassanien, A.E.: RoadMonitor: an intelligent road surface condition monitoring system. In: Advances in Intelligent Systems and Computing, vol. 323, pp. 377–387. Springer (2014)
10. Cisco Visual Networking Index: Global mobile data traffic forecast update, 2015–2020 white paper. Document ID1454457600809267, issued February 3, 2016. http://www.cisco.com/c/en/us/solutions/collateral/service-provider/visual-networking-index-vni/mobile-white-paper-c11-520862.pdf. Accessed April 2016
11. Columbus, L.: Roundup of cloud computing forecasts and market estimates Q3 update, 2015. http://www.forbes.com/sites/louiscolumbus/2015/09/27/roundup-of-cloud-computing-forecasts-and-market-estimates-q3-update-2015/#7950ac916c7a. Accessed April 2016
12. Puthal, D., Sahoo, B.P.S., Mishra, S., Swain, S.: Cloud computing features, issues, and challenges: a big picture. In: International Conference on Computational Intelligence and Networks (CINE), pp. 116–123. IEEE (2015)
13. Nirix Inc. www.nirix.com/oneserver/. Accessed April 2016
14. Zhang, Q., Cheng, L., Boutaba, R.: Cloud computing: state-of-the-art and research challenges. J. Internet Serv. Appl. **1**, 7–18 (2010). doi:10.1007/s13174-010-0007-6
15. Linux-VServer. http://linux-vserver.org/Welcome_to_Linux-VServer.org. Accessed April 2016
16. VMware ESXi. https://www.vmware.com/products/esxi-and-esx/overview. Accessed April 2016
17. Google App Engine. https://appengine.google.com/. Accessed April 2016
18. Alexa—actionable analytics for the web. www.alexa.com. Accessed April 2016
19. Salesforce company. https://www.salesforce.com/. Accessed April 2016
20. Statista-the Statistics Portal. www.statista.com/statistics/276623/number-of-apps-available-in-leading-app-stores/. Accessed May 2016
21. Palmer, G.: A road map for digital forensic research. In: First Digital Forensic Research Workshop, Utica, New York, pp. 27–30 (2001)
22. Li, X., Seberry, J.: Forensic Computing, Lecture Notes in Computer Science, vol. 2904. Springer (2003). ISBN: 978-3-540-20609-5
23. Beebe, N.: Digital forensic research: the good, the bad and the unaddressed. In: Advances in Digital Forensics V, vol. 306, pp. 17–36. Springer, Heidelberg (2009)
24. Palmer, G.L.: Forensic analysis in the digital world. Int. J. Digital Evid. **1**(1), 1–6 (2002)
25. Carrier, B.: File System Forensic Analysis. Addison-Wesley Professional Publisher (2005). ISBN: 0-321-26817-2
26. Birk, D.: Technical challenges of forensic investigations in cloud computing environments. In: Workshop on Cryptography and Security in Clouds, pp. 1–6 (2011)
27. Pollitt, M.: Computer forensics: an approach to evidence in cyberspace. Proc. Natl. Inf. Syst. Secur. Conf. **2**, 487–491 (1995)

28. Pollitt, M.M.: An ad hoc review of digital forensic models. In: Second International Workshop on Systematic Approaches to Digital Forensic Engineering (SADFE), pp. 43–54. IEEE (2007)
29. Baryamureeba, V., Tushabe, F.: The enhanced digital investigation process model. In: Proceedings of the Fourth Digital Forensic Research Workshop, pp. 1–9 (2004)
30. Ciardhuáin, S.Ó.: An extended model of cybercrime investigation. Int. J. Digital Evid. **3**(1), 1–22 (2004)
31. Rogers, M.K., Goldman, J., Mislan, R., Wedge, T., Debrota, S.: Computer forensics field triage process model. In: Proceedings of the conference on Digital Forensics, Security and Law, p. 27. Association of Digital Forensics, Security and Law (2006)
32. Perumal, S.: Digital forensic model based on Malaysian investigation process. Int. J. Comput. Sci. Netw. Secur. **9**(8), 38–44 (2009)
33. Yusoff, Y., Ismail, R., Hassan, Z.: Common phases of computer forensics investigation models. Int. J. Comput. Sci. Inf. Technol. (IJCSIT) **3**(3), 17–31 (2011)
34. Forensic Explorer software. http://www.forensicexplorer.com/
35. Digital Forensics Framework. http://www.digital-forensic.org/
36. Open Computer Forensics Architecture. https://sourceforge.net/projects/ocfa/
37. EnCase Tool. https://www.guidancesoftware.com/encase-forensic
38. A simplified guide to digital evidence. http://www.crime-scene-investigator.net/simplified-guide-to-digital-evidence.html
39. Reilly, D., Wren, C., Berry, T.: Cloud computing: pros and cons for computer forensic investigations. Int. J. Multimedia Image Proces. (IJMIP) **1**(1), 26–34 (2011)
40. Aziz, A.S.A., Azar, A.T., Salama, M.A., Hassanien, A.E., Hanafy, S.E.O.: Genetic algorithm with different feature selection techniques for anomaly detectors generation. In: Federated Conference on Computer Science and Information Systems (FedCSIS), pp. 769–774. IEEE (2013)
41. Fouad, M.M., Gaber, T., Ahmed, M., Oweis, N.E., Snasel, V.: Big data pre-processing techniques within the wireless sensors networks. In: Proceedings of the Second International Afro-European Conference for Industrial Advancement (AECIA), pp. 667–677. Springer (2015)
42. Fouad, M.M., Oweis, N.E., Gaber, T., Ahmed, M., Snasel, V.: Data mining and fusion techniques for WSNs as a source of the big data. In: Procedia Computer Science, vol. 65, pp. 778–786 (2015)
43. Ruan, K., Carthy, J., Kechadi, T., Crosbie, M.: Cloud forensics. In: Advances in Digital Forensics VII, vol. 361, pp. 35–46. Springer, Heidelberg (2011)
44. European Network and Information Security Agency: Cloud computing: benefits, risks and recommendations for information security. ENISA (2009)
45. Delport, W., Köhn, M., Olivier, M.S.: Isolating a cloud instance for a digital forensic investigation. In: ISSA (2011)
46. Alqahtany, S., Clarke, N., Furnell, S., Reich, S.: Cloud forensics: a review of challenges, solutions and open problems. In: 2015 International Conference on Cloud Computing (ICCC), pp. 1–9. IEEE (2015)
47. Zawoad, S., Hasan, R.: Cloud forensics: a meta-study of challenges, approaches, and open problems, pp. 1–15 (2013). arXiv:1302.6312

# Data Storage Security Service in Cloud Computing: Challenges and Solutions

**Alshaimaa Abo-alian, Nagwa L. Badr and Mohamed Fahmy Tolba**

**Abstract** Cloud computing is an emerging computing paradigm that is rapidly gaining attention as an alternative to other traditional hosted application models. The cloud environment provides on-demand, elastic and scalable services, moreover, it can provide these services at lower costs. However, this new paradigm poses new security issues and threats because cloud service providers are not in the same trust domain of cloud customers. Furthermore, data owners cannot control the underlying cloud environment. Therefore, new security practices are required to guarantee the availability, integrity, privacy and confidentiality of the outsourced data. This paper highlights the main security challenges of the cloud storage service and introduces some solutions to address those challenges. The proposed solutions present a way to protect the data integrity, privacy and confidentiality by integrating data auditing and access control methods.

## 1 Introduction

Cloud computing can be defined as a type of computing in which dynamically scalable resources (i.e., storage, network, and computing) are provided on demand as a service over the Internet. The service delivery model of cloud computing is the set of services provided by cloud computing that is often referred to as an SPI model, i.e., Software as a Service (SaaS), Platform as a Service (PaaS) and Infrastructure as a Service (IaaS). In a SaaS model, the cloud service providers (CSPs) install and operate application software in the cloud and the cloud users can then access the software from cloud clients. The users do not purchase software, but rather rent it for use

A. Abo-alian (✉) · N.L. Badr · M.F. Tolba
Faculty of Computer and Information Sciences, Ain Shams University, Cairo, Egypt
e-mail: a_alian@cis.asu.edu.eg

N.L. Badr
e-mail: dr.nagwabadr@gmail.com

M.F. Tolba
e-mail: fahmytolba@gmail.com



A.E. Hassanien et al. (eds.), *Multimedia Forensics and Security*,
Intelligent Systems Reference Library 115, DOI 10.1007/978-3-319-44270-9_2

on a subscription or pay-per-use model, e.g. Google Docs [1]. The SaaS clients do not manage the cloud infrastructure and platform on which the application is running. In a PaaS model, the CSPs deliver a computing platform which includes the operating system, programming language execution environment, web server and database. Application developers can subsequently develop and run their software solutions on a cloud platform. With PaaS, developers can often build web applications without installing any tools on their computer, and can hereafter deploy those applications without any specialized system administration skills [2]. Examples of PaaS providers are Windows Azure [3] and Google App Engine [4]. The IaaS model provides the infrastructure (i.e., computing power, network and storage resources) to run the applications. Furthermore, it offers a pay-per-use pricing model and the ability to scale the service depending on demand. Examples of IaaS providers are Amazon EC2 [5] and Terremark [6].

Cloud services can be deployed in four ways depending upon the clients' requirements. The cloud deployment models are: public cloud, private cloud, community cloud and hybrid cloud. In the public cloud (or external cloud), a cloud infrastructure is hosted, operated, and managed by a third-party vendor from one or more data centers [2]. The network, computing and storage infrastructures are shared with other organizations. Multiple enterprises can work simultaneously on the infrastructure provided. Users can dynamically provide resources through the internet from an off-site service provider [7]. In the private cloud, cloud infrastructure is dedicated to a specific organization and is managed either by the organization itself or third party service provider. This emulates the concept of virtualization and cloud computing on private networks. Infrastructure, in the community cloud, is shared by several organizations for a shared reason and may be managed by themselves or a third-party service provider. Infrastructure is located at the premises of a third party. Hybrid Cloud consists of two or more different cloud deployment models, bound together by standardized technology, which enables data portability between them. With a hybrid cloud, organizations might run non-core applications in a public cloud, while maintaining core applications and sensitive data in-house in a private cloud [2].

A cloud storage system (CSS) can be considered a network of distributed data centers which typically uses cloud computing technologies like virtualization, and offers some kind of interface for storing data [8]. Data may be redundantly stored at different locations in order to increase its availability. Examples of such basic cloud storage services are Amazon S3 [9] and Rackspace [10]. One fundamental advantage of using a CSS is the cost effectiveness, where data owners avoid the initial investment of expensive large equipment purchasing, infrastructure setup, configuration, deployment and frequent maintenance costs. Instead, data owners pay for only the resources they actually use and for only the time they require them. Elasticity is also a key advantage of using a CSS, as Storage resources could be allocated dynamically as needed, without human interaction. Scalability is another gain of adopting a CSS because Cloud storage architecture can scale horizontally or vertically, according to demand, i.e. new nodes can be added or dropped as needed. Moreover, a CSS offers more reliability and availability, as data owners can access their data from anywhere and at any time. Furthermore, Cloud service providers use several replicated sites for business continuity and disaster recovery reasons.

Despite the appealing advantages of cloud storage services, they also bring new and challenging security threats towards users outsourced data. Since cloud service providers (CSPs) are separate administrative entities, the correctness and the confidentiality of the data in the cloud is at risk due to the following reasons: First, Since cloud infrastructure is shared between organizations, it is still facing the broad range of both internal and external threats for data integrity, for example, outages of cloud services such as the breakdown of Amazon EC2 in 2010 [11]. Second, users no longer physically possess the storage of their data, i.e., data is stored and processed remotely. So, they may worry that their data could be misused or accessed by unauthorized users [12]. For example, a dishonest CSP may sell the confidential information about an enterprise to the enterprise's closest business competitors for profit. Third, there are various motivations for the CSP to behave disloyally towards the cloud users regarding their outsourced data status. For example, the CSP might reclaim storage for monetary reasons by discarding data that has not been or is rarely accessed or even hide data loss incidents to maintain a reputation [13]. In a nutshell, although outsourcing data to the cloud is economically attractive for long-term large-scale storage, the data security in cloud storage systems is a prominent problem. Cloud storage systems do not immediately offer any guarantee on data integrity, confidentiality and availability. As a result, the CSP should adopt data security practices to ensure that their (clients) data is available, correct and safe from unauthorized access and disclosure.

Downloading all the data and checking on retrieval is the traditional method for verifying the data integrity but it causes high transmission costs and heavy I/O overheads. Furthermore, checking data integrity when accessing data is not sufficient to guarantee the integrity and availability of all the stored data in the cloud because there is no assurance for the data that is rarely accessed. Thus, it is essential to have storage auditing services that verify the integrity of outsourced data and to provide proof to data owners that their data is correctly stored in the cloud.

Traditional server-based access control methods such as Access Control List (ACL) [14] cannot be directly applied to cloud storage systems because data owners do not fully trust the cloud service providers. Additionally, traditional cryptographic solutions cannot be applied directly while sharing data on cloud servers because these solutions require complicated key management and high storage overheads on the server. Moreover, the data owners have to stay online all the time to deliver the keys to new users. Therefore, it is crucial to have an access control method that restricts and manages access to data and ensure that the outsourced data is safe from unauthorized access and disclosure [15].

In this paper, an extensive survey of cloud storage auditing and access control methods is presented. Moreover, an evaluation of these methods against different performance criteria is conducted. The rest of the paper is organized as follows. Section 2 overviews various cloud storage auditing methods. Section 3 presents some literature for access control methods in cloud computing. A comparative analysis of some existing data security methods in cloud computing is provided in Sect. 4.

Section 5 discusses the limitations of different data security methods in cloud computing and provides some concluding remarks that can be used in designing new data security practices. Finally, we conclude in Sect. 6.

## 2 Data Storage Auditing Methods

This section first defines the system model and the security model of data storage auditing schemes within cloud computing. Then, the existing data storage auditing methods, that are classified into different categories, are presented.

Data storage auditing can be defined as a method that enables data owners to check the integrity of remote data without downloading the data or explicit knowledge of the entire data [16]. Any system model of auditing scheme consists of three entities as mentioned in [17]:

1. Data Owner: An entity which has large data files to be stored in the cloud and can be either individual consumers or organizations.
2. Cloud Storage Server (CSS): An entity which is managed by a Cloud Service Provider (CSP) and has significant storage space and computation resources to maintain clients data.
3. Third Party Auditor or Verifier (TPA): An entity which has the expertise and capabilities to check the integrity of data stored on CSS.

In the security model of most data auditing schemes, the auditor is assumed to be honest-but-curious. It performs honestly during the entire auditing protocol but it is curious about the received data. So, it is essential to keep the data confidential and invisible to the auditor during the auditing protocol, but the cloud storage server could be dishonest and may launch the following attacks [18]:

1. Replace Attack: The server may choose another valid and uncorrupted pair of data block and data tag $(m_k, t_k)$ to replace the challenged pair of data block and data tag $(m_i, t_i)$, when it has already discarded $m_i$ or $t_i$.
2. Forge Attack: The server may forge the data tag of data block and deceive the auditor if the owners secret tag keys are reused for the different versions of data.
3. Replay Attack: The server may generate the proof from the previous proof or other information, without retrieving the actual owners data.

As illustrated in Fig. 1, a data auditing scheme should basically consist of five algorithms:

1. Key Generation: It is run by the data owner. It takes as input security parameter $1^\lambda$ and outputs a pair of private and public keys $(sk, pk)$.
2. Tag Generation: It is run by the data owner to generate the verification metadata, i.e., data block tags. It takes as inputs a public key $pk$, a secret key $sk$ and the file blocks $b$. It outputs the verifiable block tags $T_b$.

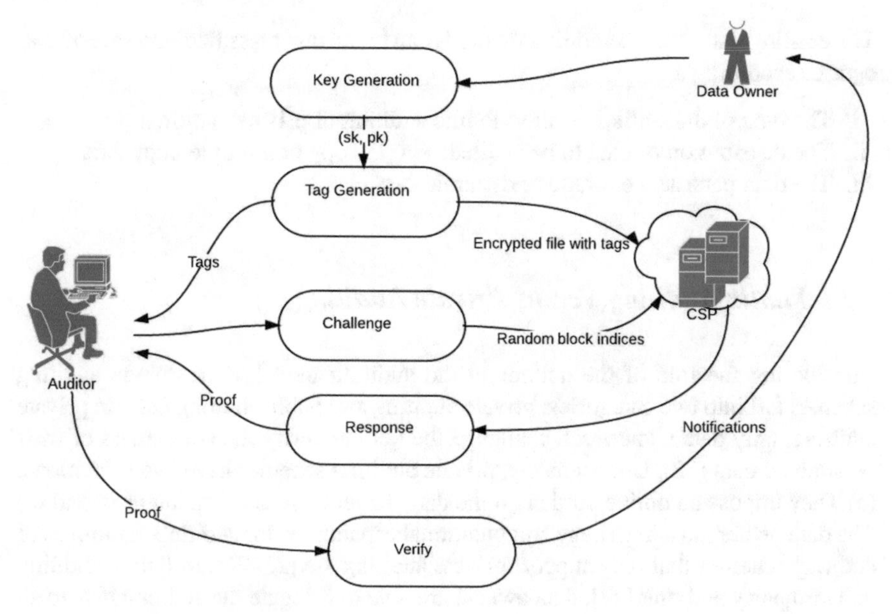


**Fig. 1** Structure of an auditing scheme

3. Challenge: It is run by the auditor in order to randomly generate a challenge that indicates the specific blocks. These random blocks are used as a request for a proof of possession.
4. Response/Proof: It is run by the CSP, upon receiving the challenge, to generate a proof that is used in the verification. In this process, the CSP proves that it is still correctly storing all file blocks.
5. Verify: It is run by the auditor in order to validate a proof of possession. It outputs TRUE if the verification equation passed, or FALSE otherwise.

The existing data storage auditing schemes can be basically classified into two main categories:

1. Provable Data Possession (PDP) methods: For verifying the integrity of data without sending it to untrusted servers, the auditor verifies probabilistic proofs of possession by sampling random sets of blocks from the cloud service provider [19].
2. Proof of Retrievability (PoR) methods: A set of randomly-valued check blocks called sentinels are embedded into the encrypted file. For auditing the data storage, the auditor challenges the server by specifying the positions of a subset of sentinels and asking the server to return the associated sentinel values [20].

The existing data storage auditing methods can be further classified into several categories according to:

i. The type of the auditor/verifier: Public auditing or private auditing.
ii. The distribution of data to be audited: Single-copy or multiple-copy data.
iii. The data persistence: Static or dynamic data.

## 2.1 Public Auditing Versus Private Auditing

Considering the role of the auditor in the auditing model, data storage auditing schemes fall into two categories; private auditing and pubic auditing [21]. In private auditing, only data owners can challenge the CSS to verify the correctness of their outsourced data [22]. Unfortunately, private auditing schemes have two limitations: (a) They impose an online burden on the data owner to verify data integrity and (b) The data owner must have huge computational capabilities for auditing. Examples of auditing schemes that only support private auditing are [23–26]. In Public auditing or Third party auditing [27], data owners are able to delegate the auditing task to an independent third party auditor (TPA), without the devotion of their computational resources. However, pubic auditing schemes should guarantee that the TPA keeps no private information about the verified data. Several variations of PDP schemes that support public auditing such as [17, 24, 25, 28], were proposed under different cryptographic primitives.

## 2.2 Auditing Single-Copy Versus Multiple-Copy Data

For verifying the integrity of outsourced data in the cloud storage, various auditing schemes have been proposed which can be categorized into:

i. Auditing schemes for single-copy data.
ii. Auditing schemes for multiple-copy data.

*i. Auditing Schemes for Single-Copy Data*

Shacham and Waters [29] proposed two fast PoR schemes based on an homomorphic authenticator that enables the storage server to reduce the complexity of the auditing by aggregating the authentication tags of individual file blocks. The first scheme is built from BLS signatures and allows public auditing. The second scheme is built on pseudorandom functions and allows only private auditing but its response messages are shorter than those of the first scheme, i.e., 80 bits only. Both schemes use Reed-Solomon erasure encoding method [30] to support an extra feature which allows the client to recover the data outsourced in the cloud. However, their encoding and decoding are slow for large files.

Yuan and Yu [31] managed to suppress the need for the tradeoff between communication costs and storage costs. They proposed a PoR scheme with public auditing and constant communication cost by combining techniques such as constant size polynomial commitment, BLS signatures and homomorphic linear authenticators. On the other hand, their data preparation process requires (s + 3) exponentiation operations where s is the block size. However, their scheme does not support data dynamics.

Xu and Chang [32] proposed a new PDP model called POS. POS requires no modular exponentiation in the setup phase and uses a smaller number, i.e., about 102 for a 1G file, of group exponentiation in the verification phase. POS only supports private auditing for static data and its communication cost is linear in relation to the number of encoded elements in the challenge query.

Ateniese et al. [33] introduced a PDP model supported with two advantages: Lightweight and robustness. Their challenge/response protocol transmits a small, and constant amount of data, which minimizes network communication. Furthermore, it incorporates mechanisms for mitigating arbitrary amounts of data corruption. On the other hand, It relies on Reed-Solomon encoding scheme in which the time required for encoding and decoding of n-block file is $O(n^2)$.

Cao et al. [34] proposed a public auditing scheme that is suitable for distributed storage systems with concurrent users access. In order to efficiently recover the exact form of any corrupted data, they utilized the exact repair method [35] where the newly generated blocks are the same as those previously stored blocks. So, no verification tags need to be generated on the fly for the repaired data. Consequently, it relieves the data owner from the online burden. However, their scheme increases the storage overheads at each server, uses an additional repair server to store the original packets besides the encoded packets, and does not support data dynamics.

*ii. Auditing Schemes for Multiple-Copy Data*

Barsoum and Hasan [36, 37] proposed two dynamic multi-copy PDP schemes: Tree-Based and Map-Based Dynamic Multi-Copy PDP (TB-DMCPDP and MB-DMCPDP, respectively). These schemes prevent the CSP from cheating and maintaining fewer copies, using the diffusion property of the AES encryption scheme. TB-DMCPDP scheme is based on Merkle hash tree (MHT) whereas MB-DMCPDP scheme is based on a map-version table to support outsourcing of dynamic data. The setup cost of the TB-DMCPDP scheme is higher than that of the MB-DMCPDP scheme. On the other hand, the storage overhead is independent of the number of copies for the MB-DMCPDP scheme, while the storage overhead is linear with the number of copies for the TB-DMCPDP scheme. However, the authorized users should know the replica number in order to generate the original file which may require the CSP reveal its internal structure to the users.

Zhu et al. [38] proposed a co-operative provable data possession scheme (CPDP) for multi-cloud storage integrity verification along with two fundamental techniques: Hash index hierarchy (HIH) and homomorphic verifiable response (HVR). Using the hash index hierarchy, multiple responses of the clients challenges that are computed from multiple CSPs, can be combined into a single response as the final result. Homomorphic verifiable response supports distributed cloud storage in a multi-cloud storage environment, and implements an efficient collision-resistant hash function.

Wang and Zhang [39] proved that the CPDP [38] is insecure because it does not satisfy the knowledge soundness, i.e., any malicious CSP or malicious organizer is able to pass the verification even if they have deleted all the stored data. Additionally, the CPDP does not support data dynamics.

Etemad and Kupcu [26] proposed a distributed and replicated DPDP (DR-DPDP) that provides transparent distribution and replication of the data over multiple servers where the CSP may hide its internal structure from the client. This scheme uses persistent rank-based authenticated skip lists to handle dynamic data operations more efficiently, such as insertion, deletion, and modification. On the other hand, DR-DPDP has three noteworthy disadvantages: First, it only supports private auditing. Second, it does not support recovery of corrupted data. Third, the organizer looks like a central entity that may get overloaded and may cause a bottleneck.

Mukundan et al. [24] proposed a Dynamic Multi-Replica Provable Data Possession scheme (DMR-PDP) that uses the Paillier probabilistic encryption for replica differentiation so that it prevents the CSP from cheating and maintaining fewer copies than what is paid for. DMR-PDP also supports efficient dynamic operations such as block modification, insertion and deletion on data replicas over cloud servers. However, it supports only private auditing and does not provide any security proofs.

Chen and Curtmola [25] proposed a remote data checking scheme for replication-based distributed storage systems, called RDC-SR. RDC-SR enables server-side repair and places a minimal load on the data owner who only has to act as a repair coordinator. In RDC-SR, each replica constitutes a masked/encrypted version of the original file in order to support replica differentiation. In order to overcome the replicate-on-the-fly (ROTF) attack, they make replica creation a more time-consuming process. However, RDC-SR has three remarkable limitations: First, the authorized users must know the random numbers used in the masking step in order to generate the original file. Second, it only supports private auditing. Third, it works only on static data.

## 2.3 *Static Versus Dynamic Data Auditing*

Considering the data persistence, existing auditing schemes can be categorized into: Auditing schemes that support only static archived data such as; [3, 5, 25] and auditing schemes that support data dynamics such as; insertion, deletion and modification. For enabling dynamic operations, existing data storage schemes utilize different authenticated data structures including: (i) Merkle hash tree [40], (ii) balanced update tree [41], (iii) skip-list [42–44] and (iv) map-version table (index table) [25, 45] that are illustrated in detail, as follows:

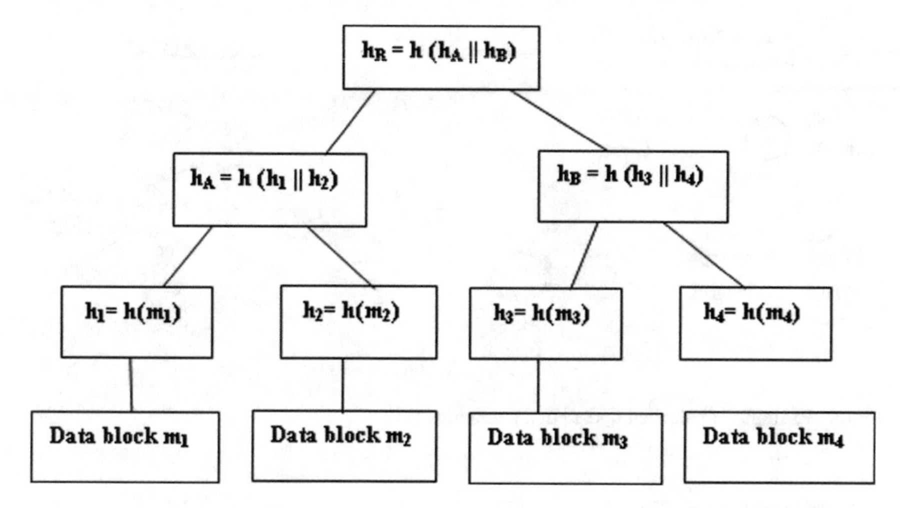


**Fig. 2** Merkle hash tree [40]

*i. Merkle Hash Tree*

Merkle Hash Tree (MHT) [40] is a binary tree structure used to efficiently verify the integrity of the data. As illustrated in Fig. 2, the MHT is a tree of hashes where the leaves of the tree are the hashes of the data blocks. Wang et al. [46] proposed a public auditing protocol that supports fully dynamic data operations by manipulating the classic Merkle Hash Tree construction for block tag authentication in order to achieve efficient data dynamics. They could also achieve batch auditing where different users can delegate multiple auditing tasks to be performed simultaneously by the third party auditor (TPA). Unfortunately, their protocol does not maintain the privacy of the data and TPA could derive users data from the information collected during the auditing process. Additionally, their protocol does not support data recovery in case of data corruption.

Lu et al. [47] addressed the security problem of the previous auditing protocol [46] in the signature generation phase that allows the CSP to deceive by using blocks from different files during verification. They presented a secure public auditing protocol based on the homomorphic hash function and the BLS short signature scheme which is publicly verifiable and supports data dynamics, it also preserves privacy. However, their protocol suffers from massive computational and communication costs.

Wang et al. [48] proposed a privacy-preserving public auditing scheme using random masking and homomorphic linear authenticators (HLAs) [49]. Their auditing scheme also supports data dynamics using Merkle Hash Tree (MHT) and it enables the auditor to perform audits for multiple users simultaneously and efficiently. Unfortunately, their scheme is vulnerable to TPA offline guessing attack.

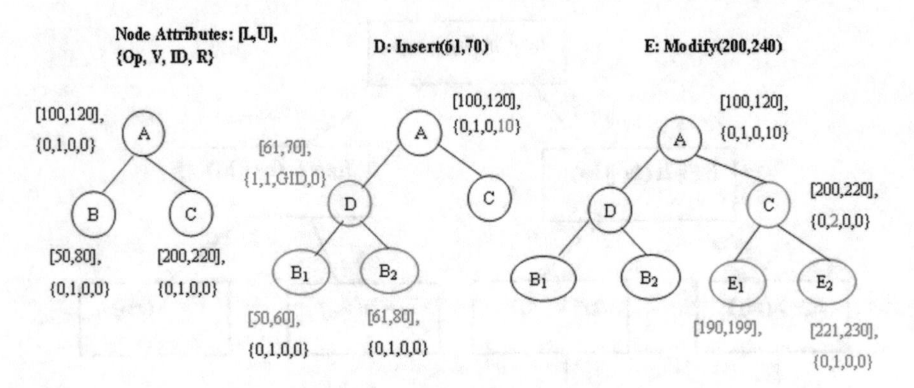


Fig. 3 Example of balanced update tree operations [41]

*ii. Balanced Update Tree*

Zhang and Blanton [41] proposed a new data structure called "balanced update tree," to support dynamic operations while verifying data integrity. In the update tree, each node corresponds to a range of data blocks on which an update (i.e., insertion, deletion, or modification) has been performed. The challenge with constructing such a tree was to ensure that: (i) a range of data blocks can be efficiently located within the tree and (ii) the tree is maintained to be balanced after applying necessary updates caused by clients queries, i.e., the size of this tree is independent of the overall file size as it depends on the number of updates. However, it introduces more storage overhead on the client. Besides, the auditing scheme requires the retrieval, i.e., downloading, of data blocks which leads to high communication costs. Figure 3 illustrates an example of balanced update tree operations.

*iii. Skip List*

A skip list [42] is a hierarchical structure of linked lists that is used to store an ordered set of items. An authenticated skip list [43] is constructed using a collision-resistant hash function and keeps a hash value in each node. Due to the collision resistance of the hash function, the hash value of the root can be used later for validating the integrity. Figure 4 [44] illustrates an example of a rank-based authenticated skip list, where the number inside the node represents its rank, i.e., the number of nodes at the bottom level that can be reached from that node. Lu et al. [47] used the skip-list structure to support data dynamics in their PDP model that reduces the computational and communication complexity from log(n) to constant. However, the use of a skip-list creates some additional storage overheads, i.e., about 3.7 % of the original file at the CSP-side and 0.05 % at the client-side. Additionally, it only supports private auditing.

*iv. Index Table*

A map-version table or an index table is a small data structure created by the owner and stored on the verifier side to validate the integrity and consistency of all file

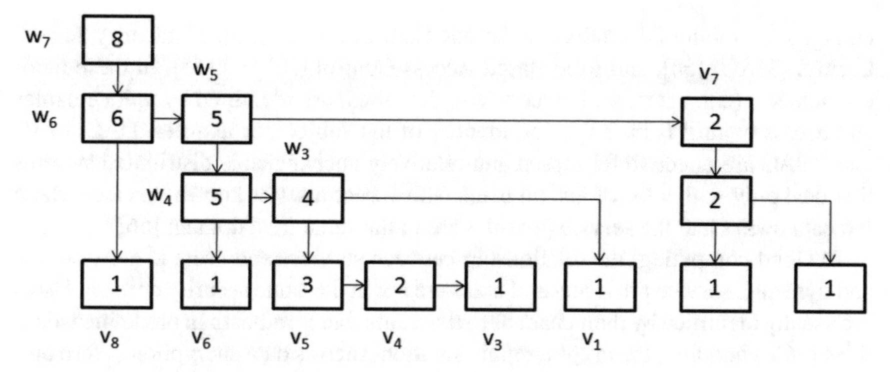


**Fig. 4** Rank-based authenticated skip list [44]

copies stored by the CSP [36]. The map-version table consists of three columns: Serial number (SN), block number (BN), and version number (VN). The SN represents an index to the file blocks that indicates the physical position of a block in a data file. The BN is a counter used to make logical numbering/indexing to the file blocks. The VN indicates the current version of file blocks.

Li et al. [50, 51] proposed a full data dynamic PDP scheme that uses a map-version table to support data block updates. Then, they discussed how to extend their scheme to support other features, including public auditing, privacy preservation, fairness, and multiple-replica checking.

Recently, Yang and Xiaohua [18] presented Third-party Storage Auditing Scheme (TSAS) which is a privacy-preserving auditing protocol. It also supports data dynamic operations using Index Table (ITable). Moreover, they add a new column to the index table, Ti, that is the timestamp used for generating the data tag to prevent the replay attack. They applied batch auditing for integrity verification for multiple data owners in order to reduce the computational cost on the auditor. However, this scheme moved the computational loads of the auditing from the TPA to the CSP, i.e., pairing operation, which introduced high computational cost on the CSP side.

## 3 Access Control Methods

Despite of the cost effectiveness and reliability of cloud storage services, data owners may consider them as an uncertain storage pool outside the enterprises. Data owners may worry that their data could be misused or accessed by unauthorized users as a reason of sharing the cloud infrastructure. An important aspect for the cloud service provider is to have in place an access control method to ensure the confidentiality and privacy of their data, i.e., their data are safe from unauthorized access and disclosure.

Generally, access control can be defined as restricting access to resources/data to privileged/authorized entities [52]. Various access control models have been emerged, including Discretionary Access Control (DAC) [53], Mandatory Access Control (MAC) [54], and Role Based Access Control (RBAC) [55]. In these models, subjects (e.g. users) and objects (e.g. data files) are identified by unique names and access control is based on the identity of the subject, or its roles. DAC, MAC and RBAC are effective for closed and relatively unchangeable distributed systems that deal only with a set of known users who access a set of known services where the data owner and the service provider are in the same trust domain [56].

In cloud computing, the relationship between services and users is more ad hoc and dynamic, service providers and users are not in the same security domain. Users are usually identified by their characteristics or attributes rather than predefined identities [56]. Therefore, the cryptographic solution, such as data encryption before outsourcing, can be the trivial solution to keep sensitive data confidential against unauthorized users and untrusted CSPs. Unfortunately, cryptographic solutions only are inefficient to encrypt a file to multiple recipients, and fail to support fine-grained access control, i.e., granting differential access rights to a set of users and allowing flexibility in specifying the access rights of individual users. Moreover, traditional ACL-based access control methods require attaching a list of authorized users to every data object. When ACLs are enforced with cryptographic methods, the complexity of each data object in terms of its ciphertext size and/or the corresponding data encryption operation is linear to the number of users in the system, and thus makes the system less scalable [57].

The following sections overview the existing access control methods and highlight their main advantages and drawbacks.

## 3.1 Traditional Encryption

One method for enforcing access control and assuring data confidentiality is to store sensitive data in encrypted form. Only users authorized to access the data have the required decryption key. There are two main classes of encryption schemes: (i) Symmetric key encryption and (ii) Public key encryption [45]. There are several schemes [58–60] proposed in the area of access control of outsourced data addressing the similar issue of data access control with conventional symmetric-key cryptography or public-key cryptography. Although these schemes are suitable for conventional file systems, most of them are less suitable for fine-grained data access control in large-scale data centers which may have a large number users and data files. Obviously, neither of them should be applied directly while sharing data on cloud servers, since they are inefficient in encrypting a file to multiple recipients in terms of key size, ciphertext length and computational cost for encryption. Moreover, they fail to support fine-grained attribute-based access control and key delegation.

## 3.2 Broadcast Encryption

In broadcast encryption (BE), a sender encrypts a message for some subset S of users who are listening on a broadcast channel, so that only the recipients in S can use their private keys to decrypt the message. The problem of practical broadcast encryption was first formally studied by Fiat and Naor in 1994 [61]. The BE system is secure against a collusion of k users, which means that it may be insecure if more than k users collude.

Since then, several solutions have been described in the literature such as schemes presented in Halevy and Shamir [62] and Boneh et al. [63] which are the best known BE schemes. However, the efficiency of both schemes depends on the size of the authorized user set. Additionally, they also require the broadcaster/sender to refer to its database of user authorizations.

Delerable et al. [64] proposed a dynamic public-key broadcast encryption that simultaneously benefit from the following properties: Receivers are stateless; encryption is collusion secure for arbitrarily large collisions of users and security is tight in the standard model; new users can join dynamically (i.e. without modification of user decryption keys and ciphertext size).

Recently, Kim et al. [65] proposed a semi-static secure broadcast encryption scheme with constant-sized private keys and ciphertexts that improves the scheme introduced by Gentry and Waters [66]. They reduce the private key and ciphertext size by half. In addition, the sizes of the public key and the private key do not depend on the total number of users. Unfortunately, it is only secure against adaptive chosen plaintext attacks (CPA).

Apparently, a BE system achieves an one-to-many encryption with general performance. However, it may not be applied directly while sharing data on cloud servers, since it fails to support attribute-based access control and key delegation.

## 3.3 Identity-Based Encryption

Identity-based encryption (IBE) was proposed by Boneh et al. in 1984 [63]. But, IBE remained an open problem for many years until a fully functional identity based encryption (IBE) scheme proposed by Halevy and Shamir [62]. It can be defined as a type of public-key cryptography (PKC), in which any arbitrary string corresponding to unique user information is a valid public key such as; an email address or a physical IP address. The corresponding private key is computed by a trusted third party (TTP) called the private key generator (PKG) as illustrated in Fig. 5 [67]. Compared with the traditional PKC, the IBE system eliminates online look-ups for the recipients authenticated public key. However, an IBE system introduces several problems: First, there is only one PKG to distribute private keys to each user which introduces the key escrow problem, i.e., the PKG knows the private keys of all users and may decrypt any message. Second, the PKG is a centralized entity; it may get overloaded and can

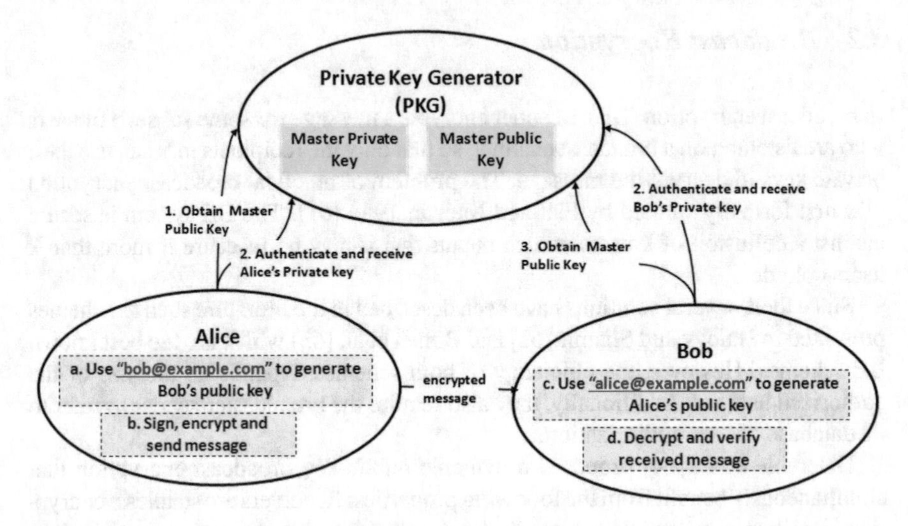


**Fig. 5** Example of an identity-based encryption [67]

cause a bottleneck. Third, if the PKG server is compromised, all messages used by that server are also compromised. Recently, Li et al. [68] proposed a revocable IBE scheme that handle the critical issue of overhead computation at the Private Key Generator (PKG) during user revocation. They employ a hybrid private key for each user, in which an AND gate is involved to connect and bound the identity component and the time component. At first, the user is able to obtain the identity component and a default time component, i.e., PKG can issue his/her private key for a current time period. Then, unrevoked users needs to periodically request on a key-update for the time component to a newly introduced entity named Key Update Cloud Service Provider (KU-CSP) which introduces further communication costs.

## 3.4 Hierarchical Identity-Based Encryption

Horwitz and Lynn [69] introduced the concept of a Hierarchical identity-based encryption (HIBE) system in order to reduce the workload on the root PKG. They proposed a two-level HIBE scheme, in which a root PKG needs only to generate private keys for domain-level PKGs that, in turn generates private keys for all the users in their domains at the next level. That scheme has a chosen ciphertext security in the random oracle model. In addition, it achieves total collusion resistance at the upper levels and partial collusion resistance at the lower levels.

Gentry and Halevi [70] proposed a HIBE scheme with total collusion resistance at an arbitrary number of levels, which has chosen ciphertext security in the random oracle model under the BDH assumption and key randomization. It is noteworthy that their scheme has a valuable feature which is one-to-many encryption, i.e., an

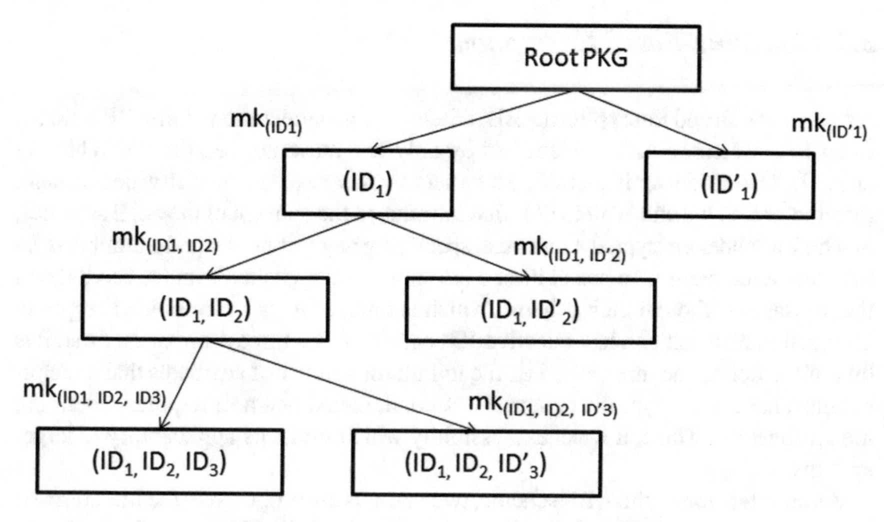


**Fig. 6** Hierarchical identity-based encryption [71]

encrypted file can be decrypted by a recipient and all his ancestors, using their own secret keys, respectively. However, the length of ciphertext and private keys, as well as the time of encryption and decryption, grows linearly with the depth of a recipient in the hierarchy.

Figure 6 [71] illustrates an example of HIBE in which the root PKG generates the system parameters for the HIBE system and the secret keys for the lower-level PKGs, which, in turn generate the secret keys for the entities in their domains at the bottom level. In other words, a user public key is an ID-tuple, which consists of the users identity (ID) and the IDs of the users ancestors. Each PKG uses its secret keys (including a master key and a private key) and a user public key to generate secret keys for each user in its domain. Liu et al. [72] utilized the 'one-to-many' encryption feature of [70] and proposed an efficient sharing of the secure cloud storage services scheme. In their scheme, a sender can specify several users as the recipients for an encrypted file by taking the number and public keys of the recipients as inputs of a HIBE system. Using their scheme, the sender needs to encrypt a file only once, and store only one copy of the corresponding ciphertext regardless of the number of intended recipients. The limitation of their scheme is that, the length of ciphertexts grows linearly with the number of recipients, so that it can only be used in the case of a confidential file involving a small set of recipients.

Recently, Mao et al. [73] presented a new HIBE system where the ciphertext sizes as well as the decryption costs are independent of the hierarchy depth, i.e., constant length of ciphertext and a constant number of bilinear map operations in decryption. Moreover, their scheme is fully secure in the standard model. The HIBE system obviously achieves key delegation and some HIBE schemes achieve one-to-many encryption with adequate performance. However, it may not be applied directly while sharing data on cloud servers because it fails to efficiently support fine-grained access control.

## 3.5 Attribute-Based Encryption

An Attribute-Based Encryption (ABE) scheme is a generalization of the IBE scheme. In an IBE system, a user is identified by only one attribute, i.e., the ID. While, in an ABE scheme, a user is identified by a set of attributes, e.g. specialty, department, location, etc. Sahai and Waters [74] first introduced the concept of the ABE schemes, in which a sender encrypted a message, specifying an attribute set and a number d, so that only a recipient who has at least d attributes of the given attributes can decrypt the message. Although their scheme, which is referred to as a threshold encryption, is collusion resistant and has selective-ID security, it has three drawbacks: First, it is difficult to define the threshold, i.e., the minimum number of attributes that a recipient must have to decrypt the ciphertext. Second, revoking a user requires redefining the attribute set. Third, it lacks expressibility which limits its applicability to larger systems.

As an extension of the ABE scheme, two variants are proposed in the literature: (i) the Key-Policy based ABE (KP-ABE) scheme and (ii) the Ciphertext-Policy based ABE (CP-ABE) scheme.

*i. Key-Policy Attribute-Based Encryption*

A key-policy attribute-based encryption (KP-ABE) scheme was first proposed by Goyal et al. [75] which supports any monotonic access formula consisting of AND, OR, or threshold gates. Their scheme is considered a fine-grained and expressive access control. KP-ABE is a scheme in which the access structure or policy is specified in the users' private keys, while ciphertexts are associated with sets of descriptive attributes.

As stated in [76], any monotonic access structure can be represented as an access tree over data attributes. For example, Fig. 7 presents an access structure and attribute sets that can be generated in a healthcare application [77]. The data owner encrypts the data file using a selected set of attributes (e.g., diabetes, A, asian, etc.) before

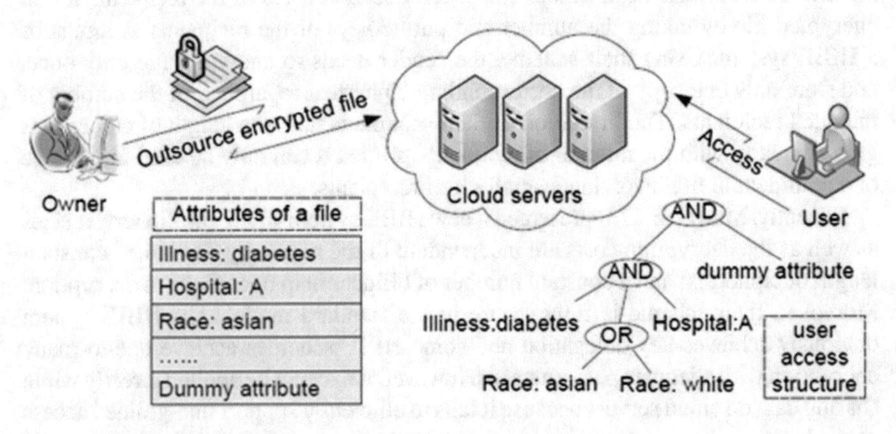


**Fig. 7** Key policy attribute-based encryption in a healthcare system [77]

uploading it to the cloud. Only users, whose access structure specified in their private keys matching the file attributes can decrypt the file, In other words, user with access structure as follows: diabetes (asian V white) A can decrypt the data file encrypted under the attributes diabetes, A, Asian. In recent times, [77, 78] proposed a user private key revocable Key Policy Attribute Based Encryption (KP-ABE) scheme with non-monotonic access structure, which can be combined with the XACML policy [79] to address the issue of moving the complex access control process to the cloud and constructing a security provable public verifiable cloud access control scheme.

All of the prior work described above assume the use of a single trusted authority (TA) that manages, e.g., add, issue, revoke, etc., all attributes in the system domain. This assumption not only may create a load bottleneck, but also suffers from the key escrow problem as the TA can decrypt all the files, causing a privacy disclosure. Thus, Chase [80] provided a construction for a multi-authority ABE scheme which supports many different authorities operating simultaneously, and each administering a different set of domain attributes, i.e., handing out secret keys for a different set of attributes [81]. However, his scheme is still not ideal because there are three main problems: first, there is a central authority that can decrypt all ciphertexts because it masters the system secret keys and thus a key-escrow problem is aroused; second, it is very easy for colluding authorities to build a complete profile of all of the attributes corresponding to each global identifier (GID); third, the set of authorities is pre-determined.

Chase and Chow [82] proposed a more practical multi-authority KP-ABE system, which removes the trusted central authority to preserve the user's privacy. Their scheme allows the users to communicate with AAs via pseudonyms instead of having to provide their GIDs. Moreover, they prevent the AAs from pooling their data and linking multiple attribute sets belonging to the same user.

Yu et al. [77] exploited the uniquely combined techniques of ABE, proxy re-encryption (PRE) [60], and lazy re-encryption (LRE) [83] to allow the data owner to delegate most of the computation tasks involved in user revocation to untrusted CSPs without disclosing the underlying data contents. PRE eliminates the need of direct interaction between the data owner and the users for decryption key distribution, whereas LRE allows the CSP to aggregate the computation tasks of multiple user revocation operations. For example, once a user is revoked, the CSP just record that. If only there is a file data access request from a user, the CSP then re-encrypts the requested files and updates the requesting users secret key.

Recently, Li et al. [84] propose an expressive decentralizing KP-ABE scheme with a constant ciphertext size, i.e., the ciphertext size is independent of the number of attributes used in the scheme. In their construction, there is no trusted central authority to conduct the system setup and the access policy can be expressed as any non-monotonic access structure. In addition, their scheme is semantically secure in so-called Selective-Set model based on the n-DBDHE assumption.

Hohenberger and Waters [85] proposed a KP-ABE scheme in which ciphertexts can be decrypted with a constant number of pairings without any restriction on the number of attributes. However, the size of the user's private key is increased by a factor of number of distinct attributes in the access policy. Furthermore, there is a trusted single authority that generates private keys for users which violates the user privacy and causes a key-escrow problem.

Unfortunately, in all KP-ABE schemes, the data owners have no control over who has access to the data they encrypt, except by their choice of the set of descriptive attributes for the data. Rather, they must trust that the key-issuer issues the appropriate keys to grant or deny access to the appropriate users. Additionally, the size of the user's private key and the computation costs in encryption and decryption operations depend linearly on the number of attributes involved in the access policy.

*ii. Ciphertext Policy Attribute-Based Encryption*

In the ciphertext policy ABE (CP-ABE) introduced by Bethencourt et al. [86], the roles of the ciphertexts and keys are reversed in contrast with the KP-ABE scheme. The data owner determines the policy under which the data can be decrypted, while the secret key is associated with a set of attributes.

Most proposed CP-ABE schemes incur large ciphertext sizes and computation costs in the encryption and decryption operations which depend at least linearly on the number of attributes involved in the access policy. Therefore, Chen et al. [87] proposed two CP-ABE schemes, both have constant-size ciphertext and constant computation costs for an access policy containing AND-Gate with wildcard. The first scheme is provably CPA-secure in standard model under the decision n-BDHE assumption while the second scheme is provably CCA-secure in a standard model under the decision n-BDHE assumption and the existence of collision-resistant hash functions.

Zhu et al. [88, 89] proposed a comparison-based encryption scheme that support a complete comparison relation, e.g., $<, >, \leq, \geq$, in policy specification to implement various range-constraints on integer attributes, such as temporal and level attributes. They combined proxy re-encryption (with CP-ABE to support key delegation and reduce computational overheads on lightweight devices by outsourcing the majority of decryption operations to the CSP. Their scheme provides O(1) size of private-key and ciphertext for each range attribute. Additionally, it is also provably secure under the RSA and CDH assumption. However, their scheme depends on a central single authority to conduct the system setup and manage all attributes. And, they do not provide a mechanism for efficient user revocation.

Zhang and Chen [90] proposed the idea of "Access control as a service" for public cloud storage, where the data owner controls the authorization, and the PDP (Policy Decision Point) and PEP (Policy Enforcement Point) can be securely delegated to the CSP by utilizing CP-ABE and proxy re-encryption. However, it incurs high communication and setup costs.

The main limitations of most traditional CP-ABE schemes are: First, compromising the users privacy since the access structure is embedded in the ciphertext that may reveal the scope of data file and the authorized users who have access. The obvious solution to this problem as proposed by Nishide et al. [91], is to hide ciphertext policy, i.e., hidden access structure. Subsequently, there are various efforts to improve the traditional CP-ABE scheme and to support privacy-preserving access policy such as in [92–94]. The other limitation of the traditional CP-ABE schemes is depending on a single central authority for monitoring and issuing user's secret keys. Recently, many CP-ABE schemes consider multi-authority environments [94–96].

To achieve fine-grained and scalable data access control for personal health records (PHRs), Li et al. [95] utilized CP-ABE techniques to encrypt each patients PHR file while focusing on the multiple data owner scenario. Moreover, they adopted proxy encryption and lazy-revocation to efficiently support attribute and user revocation. However, the time costs of the key generation, encryption, and decryption processes are all linear with the number of attributes.

## *3.6 Hierarchical Attribute-Based Encryption*

The Hierarchical Attribute-Based Encryption (HABE) model, as described in [97, 98], integrates properties in both a HIBE model [71] and an ABE model [74]. As illustrated in Fig. 8, it consists of a root master (RM) and multiple domains, where the RM functions as the TTP, and the domains are enterprise users. More precisely, a domain consists of many domain masters (DMs) corresponding to the internal trusted parties (ITPs) and numerous users corresponding to end users. The RM, whose role closely follows the root PKG in a HIBE system, is responsible for generation and distribution of system parameters and domain keys. The DM, whose role integrates both the properties of the domain PKG in a HIBE system and the AA in an ABE system, is responsible for delegating keys to the DMs at the next level and distributing secret keys to users.

Wang et al. [99] proposed fuzzy and precise identity-based encryption (FPIBE) scheme that supports the full key delegation and requires only a constant number of bilinear map operations during decryption. The FPIBE scheme is able to efficiently

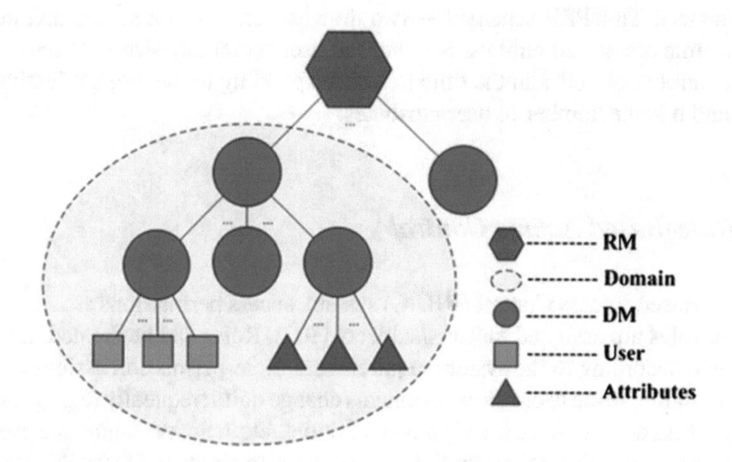


**Fig. 8** Hierarchical attribute-based encryption model [98]

achieve a flexible access control by combining the HIBE system and the CP-ABE system. Using the FPIBE scheme, a user can encrypt data by specifying a recipient ID set, or an access control policy over attributes, so that only the user whose ID belongs to the ID set or attributes satisfying the access control policy can decrypt the corresponding data. However, the ciphertext length is proportional to the number of authorized users and encryption time. As well as, the size of a user's secret key is proportional to the depth of the user in the hierarchy.

For supporting compound and multi-valued attributes in a scalable, flexible, and fine-grained access control, Wan et al. [100] proposed hierarchical attribute-set-based encryption (HASBE) by extending the ciphertext policy attribute-set-based encryption (CP-ASBE) with a hierarchical structure of users. HASBE employs multiple value assignments for access expiration time to deal with user revocation more efficiently. However, the granting access operation is proportional to the number of attributes in the key structure.

Chen et al. [101] proposed a new hierarchical key assignment scheme, called CloudHKA, that addresses a cryptographic key assignment problem for enforcing a hierarchical access control policy over cloud data. CloudHKA possesses many advantages: (1) Each user only needs to store one secret key, (2) Each user can be flexibly authorized the access rights of Write or Read, or both, (3) It supports a dynamic user set and access hierarchy, and (4) It is provably-secure against collusive attacks. However, the re-key cost in the case of user revocation is linear in the number of users in the same security class. CloudHKA does not consider the expressive user attributes, so it can be considered coarse-grained access control scheme.

Wang et al. [102] recently extended the hierarchical CP-ABE scheme [97, 98] by incorporating the concept of time to perform automatic proxy re-encryption. More specifically, they proposed a time-based proxy re-encryption (TimePRE) scheme to allow a user's access right to expire automatically after a predetermined period of time. In this case, the data owner can be offline in the process of user revocations. Unfortunately, TimePRE scheme has two drawbacks: first, It assumes that there is a global time among all entities. Second, the user secret key size is O(mn), where m is the number of nodes in the time tree corresponding to the user's effective time period and n is the number of user attributes.

## 3.7 Role-Based Access Control

In a Role-Based Access Control (RBAC) system, access permissions are assigned to roles and roles are assigned to users/subjects [103]. Roles can be created, modified or disabled according to the system requirements. Role-permission assignments are relatively stable, while user-role assignments change quite frequently (e.g., personnel moving across departments, reassignment of duties, etc.). So, managing the user-role permissions is significantly easier than managing user rights individually [104].

Zhou et al. [105] proposed a role-based encryption (RBE) scheme for secure cloud storage. This scheme specifies a set of roles assigned to the users, each role having a set of permissions. Roles can be defined in a hierarchy, which means a role can have sub-roles (successor roles). The owner of the data can encrypt the private data to a specific role. Only the users in the specified role or predecessor roles are able to decrypt that data. The decryption key size still remains constant regardless of the number of roles that the user has been assigned to. However, the decryption cost is proportional to the number of authorized users in the same role.

## 4 Comparative Analysis of Security Methods on Cloud Data Storage

This section evaluates the performance of auditing and access control methods, that were presented in the previous sections, against different performance criteria.

### *4.1 Performance Analysis of Data Storage Auditing Methods*

This section assesses the different characteristics of some existing data storage auditing schemes such as auditor type, supporting dynamic data, replication/multiple-copy and data recovery, as illustrated in Table 1. In addition, it evaluates their performance in terms of computational complexity at the CSP and the auditor, storage overheads at the CSP and the auditor, communication complexity between the auditor and the CSP, as illustrated in Table 3.

As shown in Table 1, most auditing schemes focus on a single copy of the file and provide no proof that the CSP stores multiple copies of the data owners file. Although data owners many need their critical data to be replicated on multiple servers across multiple data centers to guarantee the availability of their data. only few schemes [26, 37, 38] that support auditing for multiple replicas of data owners file.

Table 2 gives more notations for cryptographic operations used in the different auditing methods. Let r, n, k denote the number of replicas, the number of blocks per replica and the number of sectors per block (in case of block fragmentation), respectively. s denotes the block size. c denotes the number of challenged blocks. Let $\lambda$ be the security parameter which is usually the size of key. Let p denote the order of the groups and $\phi(N)$ denotes the Euler Function on the RSA modulus N.

As shown in Table 3, MAC-based schemes [29] require storage overheads of metadata, i.e., block tags, as long as each data block size. On the other hand, they have efficient computational complexity at the CSP and the auditor. The homomorphic tags based on BLS signatures [19, 29] are much shorter than the RSA-based homomorphic tags [29]. While the verification cost, i.e., the computational cost at the auditor in the BLS-based auditing schemes is higher than that of RSA-based auditing schemes due to the bilinear pairing operations which consume more time than other cryptographic operations.

**Table 1** Characteristics of data storage auditing schemes

| Scheme | Auditor type | Dynamic data | Replication | Data recovery |
|---|---|---|---|---|
| Liu et al. [17] | Public | Yes | No | No |
| Yang and Xiaohua [18] | Public | Yes | No | No |
| Ateniese et al. [19] | Public | No | No | No |
| Wang et al. [48] | Public | Yes | No | No |
| Shacham and Waters [29] (BLS) | Public | No | No | Yes |
| Shacham and Waters [29] (RSA) | Public | No | No | Yes |
| Shacham and Waters [29] (MAC) | Private | No | No | Yes |
| Yuan and Yu [31] | Public | No | No | Yes |
| Xu and Chang [32] | Public | No | No | Yes |
| Barsoum and Hasan [37] | Public | Yes | Yes | No |
| Zhu et al. [38] | Public | No | Yes | No |
| Etemad and Kupcu [26] | Private | Yes | Yes | No |

**Table 2** Notation of cryptographic operations

| Notation | Cryptographic operation |
|---|---|
| MUL | Multiplication in group G |
| ADD | Addition in group G |
| EXP | Exponentiation in group G |
| H | Hashing into group G |
| Pairing | Bilinear pairing; e(u, v) |
| SEncr | Stream Encryption |
| MOD | Modular operation in $z_N$ |

To reduce the storage overheads, the data owner can store the tags together with the data blocks only on the CSP. Upon challenging the CSP, the CSP generates the data proof and the tag to the auditor instead of only the data proof. But this solution will increase the communication cost between the CSP and the auditor. In fact, there is a tradeoff between the storage overheads at the auditor and the communication costs between the CSP and the auditor.

The scheme of Yuan and Yu [31] omits the tradeoff between communication costs and storage costs. The message exchanged between the CSP and the auditor during the auditing procedure consists of a constant number of group elements. However, it requires 4 Bilinear pairing operations for proof verification that result in high computational cost at the auditor. The communication cost can be reduced by using short homomorphic tags such as BLS tags [6, 24, 26] that enable the CSP to reduce the communication complexity of the auditing by aggregating the authentication tags of individual file blocks into a single tag. Batch auditing [18, 48] can further reduce the communication cost by allowing the CSP to send the linear combination of all the challenged data blocks whose size is equal to one data block instead of sending them sequentially.

Auditing schemes that support dynamic data [17, 18, 26, 37, 48] add more storage overheads at the auditor. MHT-based schemes [48] keep the metadata at the auditor side (i.e. the root of the MHT) smaller than schemes that utilize the index-table [18, 37], as the total number of index table entries is equal to the number of file blocks. On the contrary, the computational and communication costs of the MHT-based schemes are higher than those of the table-based schemes. During the dynamic operations of the MHT-based schemes, the data owner sends a modification request to the CSP and receives the authentication paths. The CSP updates the MHT according to the required dynamic operations, regenerates a new directory root and sends it to the auditor. On the other hand, for the table-based schemes, the data owner only sends a request to the CSP and updates the index table without usage of any cryptographic operations. However, auditing schemes based on the index table [18, 37] suffer a performance penalty during insertion or deletion operations, i.e. O(n) in the worst case, because the indexes of all the blocks after the insertion/deletion point are changed, and all the tags of these blocks should be recalculated. While the complexity of insertion or deletion operations when using skip lists [26] is O(log n).

## *4.2 Performance Analysis of Access Control Methods*

This section presents an evaluation of different access control methods in terms of the ciphertext size, user's secret key size, decryption cost, user revocation cost and the existence of multiple authorities. Table 4 evaluates the performance of different access control methods.

As illustrated in Table 4, user revocation is a very challenging issue in access control methods that requires re-encryption of data files accessible to the revoked user, and may need updates of secret keys for all the non-revoked users. So the user revocation requires a heavy computation overhead on the data owner and may also require him/her to be always online. Access control methods that utilize the proxy encryption can efficiently perform the user revocation operation such as [90, 95, 96, 101]. To further improvement on the complexity of user revocation, hierarchical access control methods such as [100, 105] use the concept of key delegation. Ciphertext and secret key sizes are other challenging issues in current access control methods that may cause high storage overhead and communication cost. Ciphertext and secret key sizes usually grow linearly with the number of attributes in the system domain. Methods of [90, 92, 100, 101, 105] have constant ciphertext sizes while only the methods of [92, 105] achieve constant secret key size.

In most of the access control methods, decryption cost scales with the complexity of the access policy or the number of attributes which is infeasible as in case of

**Table 3** Performance analysis of different auditing schemes

| Scheme | Computational complexity | | Storage overhead | | Communication complexity |
|---|---|---|---|---|---|
| | CSP | Auditor | CSP | Auditor | |
| Liu et al. [17] | 1 SEncr + O(c) [MUL + EXP + ADD] | O(c) [MUL + H + EXP] + 2 *pairing* | (l + l/k)n | O(1) | O(1) |
| Yang and Xiaohua [18] | O(c) [ADD + MUL + EXP] + O(k) [EXP + *pairing*] | O(c) [H + EXP + MUL] + 2 *pairing* | s · n/k | O(1) | O(c) |
| Ateniese et al. [19] | O(c) [H + MUL + EXP] | O(c) [H + MUL + EXP] | n.\|N\| | O($\lambda$) | 3\|N\| |
| Wang et al. [48] | O(c) [MUL + EXP + ADD] + H + 1 *pairing* | O(c) [MUL + H + EXP] + 2 *pairing* | (l + l/k)n | O(l + l/k) | O(kc) |
| Shacham and Waters [29] **(BLS)** | O(c) [ADD + MUL + EXP] | O(k) [MUL + EXP] + O(c) [H + EXP] + 2 *pairing* | n − \|p\| | O($\lambda$) | O(c) + \|p\| |
| Shacham and Waters [29] **(RSA)** | O(c) [ADD + MUL + EXP] | O(k) [MUL + EXP] + O(c) [H + EXP] | n − \|N\| | O($\lambda$) | O(c) + \|N\| + s |
| Shacham and Waters [29] **(MAC)** | O(c) | O(c)H | n − \|p\| | N/A | O(c) + \|p\| + s |
| Yuan and Yu [31] | O(c + s) [MUL + EXP) | O(c) [MUL + EXP] + 4 *pairing* | (l + l/s)n | O(1) | O(1) |
| Xu and Chang [32] | O(s) EXP | 3 EXP + O(c) [ADD + MUL + PRF] | (l + l/s)n | O(1) | O($\lambda$) |
| Barsoum and Hasan [37] | O(c) [EXP + MUL + ADD] | O(c) [H + MUL + EXP] + O(k) [MUL + EXP] + O(r)ADD + 2 *pairing* | 2n | O(2n) | O(kr) |
| Zhu et al. [38] | O(c) *pairing* + O(ck)EXP | 3 *pairing* + O(k) EXP | n − \|p\| | O($\lambda$) | O(c) + O(k) |
| Etemad and Kupcu [26] | O(1 + log nr) [EXP + MUL] | O(1 + log nr) [EXP + MUL] | log nr | O(1) | O(log nr) |

**Table 4** Performance analysis of different access control methods

| Access control method | Approach | Ciphertext size | User key size | Decryption cost | User revocation cost | Multiple authority |
|---|---|---|---|---|---|---|
| Hohenberger and Waters [85] | KP-ABE | Linear with no. of attributes | Linear with no. of attributes | Constant no. of pairings | N/A | No |
| Doshi and Jinwala [92] | CP-ABE | Constant | Constant | Constant no. of pairings | N/A | No |
| Qian et al. [93] | CP-ABE fully hidden access structure | Linear with no. of attributes | Linear with total no. of attributes | Linear with no. of attribute authorities | N/A | Yes |
| Li et al. [95] | CP-ABE with proxy encryption and lazy re-encryption | Linear with no. of attributes and no. of AA | Linear with no. of attributes in the secret key | Linear with total no. of attributes | Linear with no. of attributes in the secret key | Yes |
| Yang et al. [96] | CP-ABE with proxy encryption | Linear with no. of attributes in the policy | Linear with the total no. of attributes | Constant | Linear with no. of non-revoked users who hold the revoked attribute | Yes |
| Wang et al. [99] | HIBE + CP-ABE | Linear with no. of users and the max. depth of hierarchy | Linear with no. of attributes of a user | Constant | N/A | Yes |
| Wan et al. [100] | CP-ABE + HIBE | Constant | Linear with no. of attributes of a user | Linear with no. of attributes in the key | Constant | Yes |
| Chen et al. [101] | HIBE with proxy encryption | Constant | Read and write keys independent of number of ciphertexts | Linear with the depth of the user in hierarchy | Linear with no. of authorities and no. of ciphertext accessed by revoked user | Yes |
| Zhou et al. [105] | Hierarchical RBAC | Constant | Constant | Linear with no. of users in the same role | Linear with no. of roles | Yes |
| Zhang and Chen [90] | CP-ABE + proxy encryption | Constant | Linear with no. of attributes | Constant no. of pairings | Constant | No |

lightweight devices such as mobiles. Some access control methods such as [85, 90, 92, 96, 99] have constant decryption costs. Some access control methods, e.g. [85, 90, 92], assume the use of a single trusted authority (TA) that manages (e.g., add, issue, revoke, etc.,) all attributes in the system domain. This assumption not only may create a load bottleneck, but also suffers from the key escrow problem as the TA can decrypt all the files, causing a privacy disclosure. Instead, recent access control methods, such as [99–101], consider the existence of different entities, called attribute authorities (AAs), responsible for managing different attributes of a person, e.g. the Department of Motor Vehicle Tests whether you can drive, or a university can certify that you are a student, etc. Each AA manages a disjoint subset of attributes, while none of them alone is able to control the security of the whole system.

## 5 Discussions and Concluding Remarks

From the extensive survey in the previous sections, we conclude that data security in cloud is one of the major issues that is considered as a barrier in the adoption of cloud storage services. The most critical security concern is about data integrity, availability, privacy and confidentiality.

For validating data integrity and availability in cloud computing, many auditing schemes have been proposed under different security levels and cryptographic assumptions. Most auditing schemes are provably secure in the random oracle model. The main limitations of existing auditing schemes can be summarized as follows: (1) Dealing only with archival static data files and do not consider dynamic operations such as insert, delete and update, (2) Relying on spot checking that can detect if a long fraction of the data stored at the CSP has been corrupted but it cannot detect corruption of small parts of the data, (3) Relying on Reed-Solomon encoding scheme to support data recovery feature in case the data is lost or corrupted. It leads to inefficient encoding and decoding for large files, (4) Supporting only private auditing that impose computational and online burdens on the data owners for periodically auditing their data, (5) verifying only single copy data files and not considering replicated data files, and (6) Incurring high computational costs and storage overheads.

Therefore, to overcome the prior limitations, an ideal auditing scheme should have the following features:

1. **Public auditing**: To enable the data owners to delegate the auditing process to a TPA in order to verify the correctness of the outsourced data on demand.
2. **Privacy-preserving assurance**: To prevent the leakage of the verified data during the auditing process.
3. **Data dynamics**: To efficiently allow the clients to perform block-level operations on the data files such as: insertion, deletion and modification while maintaining the same level of data correctness assuring and guaranteeing data freshness.
4. **Robustness**: To efficiently recover an arbitrary amount of data corruptions.
5. **Availability and reliability**: To support auditing for distinguishable multi-replica data files to make sure that the CSP is storing all the data replicas agreed upon.

6. **Blockless verification**: To allow TPA to verify the correctness of the cloud data on demand without possessing or retrieving a copy of challenged data blocks.
7. **Stateless verification**: Where the TPA is not needed to maintain a state between audits.
8. **Efficiency**: To achieve the following aspects:
   (a) Minimum computation complexity at the CSP and the TPA.
   (b) Minimum communication complexity between the CSP and the TPA.
   (c) Minimum storage overheads at the CSP and the TPA.

For ensuring data confidentiality in the cloud computing, various access control methods were proposed in order to restrict access to data and guarantee that the outsourced data is safe from unauthorized access. However, these methods suffer many penalties such as: (1) Heavy computation overheads and a cumbersome online burden towards the data owner because of the operation of user revocation, (2) Large ciphertext and secret key sizes as well as high computation costs in the encryption and decryption operations which depend at least linearly on the number of attributes involved in the access policy, (3) Compromising the users privacy by revealing some information about the data file and the authorized users in the access policy, and (4) Relying on single authority for managing different attributes in the access policy which may cause a bottleneck and a key escrow problem.

Consequently, we propose that the ideal access control methods should satisfy the following requirements:

1. **Fine-grained**: Granting different access rights to a set of users and allowing flexibility and expressibility in specifying the access rights of individual users.
2. **Privacy-preserving**: Access policy does not reveal any information to the CSP about the scope of data file and the kind or the attributes of users authorized to access.
3. **Scalability**: the number of authorized users cannot affect the performance of the system.
4. **Efficiency**: The method should be efficient in terms of: ciphertext size, user's secret key size, and the costs of encryption, decryption, and user revocation.
5. **Forward and backward security**: The revoked user should not be able to decrypt new ciphertexts encrypted with the new public key. And, the newly joined users should be able to decrypt the previously published ciphertexts encrypted with the previous public key if they satisfy the access policy.
6. **Collusion resistant**: Different users cannot collude each other and combine their attributes to decrypt the encrypted data.
7. **Multiple authority**: To overcome the problems of load bottleneck and key escrow problems, there should be multiple authorities that manage the user attributes and issue the secret keys, rather than a central trusted authority.

## 6 Conclusion

Cloud storage services have become extremely promising due to its cost effectiveness and reliability. However, this service also brings many new challenges for data security and privacy. Therefore, cloud service providers have to adopt security practices to ensure that the clients' data is safe. In this paper, the different security challenges within a cloud storage service are introduced and a survey of the different methods to assure data integrity, availability and confidentiality is presented. Comparative evaluations of these methods are also conducted according to various pre-defined performance criteria. Finally, we suggest the following research directions for data security within cloud computing environments: (1) Integrate attribute-based encryption with role-based access control models such that the user-role and role-permission assignments can be separately constructed using access policies applied on the attributes of users, roles, the objects and the environment. (2) Develop a context-aware role-based control model and incorporate it to the Policy Enforcement Point of a cloud, and enable/activate the role only when the user is located within the logical positions time intervals and under certain platforms in order to prevent the malicious insiders from disclosing the authorized user's identity. (3) Integrate efficient access control and auditing methods with new hardware architectural and virtualization features that can help protect the confidentiality and integrity of the data and resources. (4) Incorporate the relationship between auditing and access control for guaranteeing secure cloud storage services. (5) Extend current auditing methods with data recovery features.

## References


1. Attebury, R., George, J., Judd, C., Marcum, B.: Google docs: a review. Against Grain **20**(2), 14–17 (2008)
2. Tim, M., Subra, K., Shahed, L.: Cloud Security and Privacy. O'Reilly and Associates, USA (2009)
3. Chambers, J.: Windows Azure Web Sites. Wiley (2013)
4. Pandey, U.S., Anjali, J.: Google app engine and performance of the web application. Int. J. **2**(2) (2013)
5. Gonzalez, C., Border, C., Oh, T.: Teaching in amazon EC2. In: The 13th Annual ACM SIGITE Conference on Information Technology Education. ACM (2013)
6. Srinivasan, S.: Cloud computing providers. In: Cloud Computing Basics. Springer, New York (2014)
7. Bhadauria, R., Sanyal, S.: Survey on security issues in cloud computing and associated mitigation techniques. Int. J. Comput. Appl. **47**(18), 47–66 (2012)
8. Borgmann, M., Hahn, T., Herfert, M., Kunz, T., Richter M., Viebeg, U., Vowe, S.: On the Security of Cloud Storage Services. Fraunhofer-Verlag (2012)
9. Berriman, G.B., Deelman, E., Good, J., Juve, G., Kinney, J., Merrihew, A., Rynge, M.: Creating A Galactic Plane Atlas With Amazon Web Services (2013). arXiv:1312.6723
10. Garg, S.K., Versteeg, S., Buyya, R.: A framework for ranking of cloud computing services. Future Gener. Comput. Syst. **29**(4), 1012–1023 (2013)

11. Miller, R.: Amazon Addresses EC2 Power Outages. Data Center Knowledge (2010). http://www.datacenterknowledge.com/archives/2010/05/10/amazon-addresses-ec2-power-outages/
12. Aboalian, A., Badr, N.L., Tolba, M.F.: Keystroke dynamics based user authentication service for cloud computing. In: Practice and Experience: Concurrency and Computation (2015)
13. Cong, W., Ren, K., Lou, W., Li, J.: Toward publicly auditable secure cloud data storage services. IEEE Netw. **24**(4), 19–24 (2010)
14. Shalabi, S.M., Doll, C.L., Reilly, J.D., Shore, M.: Access Control List. U.S. Patent Application 13/311, 278 (2011)
15. Abo-alian, A., Badr, N.L., Tolba, M.F.: Hierarchical attribute-role based access control for cloud computing. In: The 1st International Conference on Advanced Intelligent System and Informatics (AISI2015). Springer (2016)
16. Blum, M., Evans, W., Gemmell, P., Kannan, S., Naor, M.: Checking the correctness of memories. In: The 32nd Annual Symposium on Foundations of Computer Science. IEEE Computer Society, Washington, DC, USA (1991)
17. Liu, H., Zhang, P., Lun, J.: Public data integrity verification for secure cloud storage. J. Netw. **8**(2), 373–380 (2013)
18. Yang, K., Xiaohua, J.: TSAS: third-party storage auditing service. In: Security for Cloud Storage Systems. Springer Briefs in Computer Science (2014)
19. Ateniese, G., Burns, R.C., Curtmola, R., Herring, J., Kissner, L., Peterson, Z.N.J., Song, D.X.: Provable data possession at untrusted stores. In: The 2007 ACM Conference on Computer and Communications Security. ACM (2007)
20. Juels, A., Kaliski, B.S.: Pors: proofs of retrievability for large files. In: The 2007 ACM Conference on Computer and Communications Security. ACM (2007)
21. Zheng, Q., Xu, S.: Secure and efficient proof of storage with deduplication. In: The Second ACM Conference on Data and Application Security and Privacy. ACM (2012)
22. Yang, K., Jia, X.: Data storage auditing service in cloud computing: challenges, methods and opportunities. World Wide Web **15**(4), 409–428 (2012)
23. Chen, B., Curtmola, R.: Robust dynamic provable data possession. In: The 32nd International IEEE Conference on Distributed Computing Systems Workshops. IEEE (2012)
24. Mukundan, R., Madria, S., Linderman, M.: Replicated data integrity verification in cloud. IEEE Data Eng. Bull. **35**(4), 55–64 (2012)
25. Chen, B., Curtmola, R.: Towards self-repairing replication-based storage systems using untrusted clouds. In: The 3rd ACM Conference on Data and Application Security and Privacy (CODASPY '13). ACM (2013)
26. Etemad, M., Kupcu, A.: Transparent distributed and replicated dynamic provable data possession. In: The 11th International Conference on Applied Cryptography and Network. Springer, Berlin (2013)
27. Zhu, Y., Ahn, G., Hu, H., Yau, S.S., An, H.G., Hu, C.: Dynamic audit services for outsourced storages in clouds. IEEE Trans. Serv. Comput. **6**(2), 227–238 (2013)
28. Abo-alian, A., Badr, N.L., Tolba, M.F.: Auditing-as-a-service for cloud storage. In: Intelligent Systems' 2014. Springer (2015)
29. Shacham, H., Waters, B.: Compact proofs of retrievability. J. Cryptol. **26**(3), 442–483 (2013)
30. Plank, J.S.: A tutorial on Reed-Solomon coding for fault-tolerance in RAID-like systems. Softw. Pract. Exp. **27**(9), 995–1012 (1997)
31. Yuan, J., Yu, S.: Proof of retrievability with public verifiability and constant communication cost in cloud. In: The 2013 International ACM Workshop on Security in Cloud Computing. ACM (2013)
32. Xu, J., Chang, E.C.: Towards efficient provable data possession. In: IACR Cryptology ePrint Archive 574. ASIACCS (2011)
33. Ateniese, G., Burns, R., Curtmola, R., Herring, J., Khan, O., Kissner, L., Peterson, Z., Song, D.: Remote data checking using provable data possession. ACM Trans. Inf. Syst. Secur. **14**(1), 121–155 (2011)

34. Cao, N., Yu, S., Yang, Z., Lou, W., Hou, Y.T.: LT codes-based secure and reliable cloud storage service. In: The 2012 INFOCOM. IEEE (2012)
35. Rashmi, K.V., Shah, N.B., Kumar, P.V., Ramchandran, K.: Exact regenerating codes for distributed storage. In: Allerton Conference on Control Computing and Communication (2009)
36. Barsoum, A.F., Hasan, M.A.: On verifying dynamic multiple data copies over cloud servers. IACR Cryptol. ePrint Arch. **447** (2011)
37. Barsoum, A.F., Hasan, M.A.: Integrity verification of multiple data copies over untrusted cloud servers. In: The 12th IEEE/ACM International Symposium on Cluster, Cloud and Grid Computing (2012)
38. Zhu, Y., Hu, H., Ahn, G.J., Yu, M.: Cooperative provable data possession for integrity verification in multicloud storage. IEEE Trans. Parallel Distrib. Syst. **23**(12), 2231–2244 (2012)
39. Wang, H., Zhang, Y.: On the knowledge soundness of a cooperative provable data possession scheme in multicloud storage. IEEE Trans. Parallel Distrib. Syst. **25**(1), 264–267 (2014)
40. Merkle, R.C.: Protocols for public key cryptosystems. In: IEEE Symposium on Security and Privacy. IEEE Computer Society (1980)
41. Zhang, Y., Blanton, M.: Efficient dynamic provable possession of remote data via balanced update trees. In: The 8th ACM SIGSAC Symposium on Information, Computer and Communications Security (2013)
42. Pugh, W.: Skip lists: a probabilistic alternative to balanced trees. Commun. ACM **33**(6), 668–676 (1990)
43. Goodrich, M.T., Tamassia, R., Schwerin, A.: Implementation of an authenticated dictionary with skip lists and commutative hashing. In: DARPA Information Survivability Conference (2001)
44. Erway, C., Kp, A., Papamanthou, C., Tamassia, R.: Dynamic provable data possession. In: The 16th ACM Conference on Computer and Communications Security. ACM (2009)
45. Fujisaki, E., Okamoto, T.: Secure integration of asymmetric and symmetric encryption schemes. In: Advances in Cryptology CRYPTO99. Springer, Heidelberg (1999)
46. Wang, Q., Wang, C., Ren, K., Lou, W., Li, J.: Enabling public auditability and data dynamics for storage security in cloud computing. IEEE Trans. Parallel Distrib. Syst. **22**(5), 847–859 (2011)
47. Liu, F., Gu, D., Lu, H.: An improved dynamic provable data possession model. In: The IEEE International Conference on Cloud Computing and Intelligence Systems (CCIS). IEEE (2011)
48. Wang, C., Chow, S.S., Wang, Q., Ren, K., Lou, W.: Privacy-preserving public auditing for secure cloud storage. IEEE Trans. Comput. **62**(2), 362–375 (2013)
49. Ateniese, G., Kamara, S., Katz, J.: Proofs of Storage from homomorphic identification protocols. In: The 15th International Conference on Theory and Application of Cryptology and Information Security: Advances in Cryptology (ASIACRYPT). Springer, Heidelberg (2009)
50. Li, C., Chen, Y., Tan, P., Yang, G.: An efficient provable data possession scheme with data dynamics. In: Tthe International Conference on Computer Science and Service System (CSSS). IEEE (2012)
51. Li, C., Chen, Y., Tan, P., Yang, G.: Towards comprehensive provable data possession in cloud computing. Wuhan Univ. J. Nat. Sci. **18**(3), 265–271 (2013)
52. Menezes, A.J., Van Oorschot, P.C., Vanstone, S.A.: Handbook of Applied Cryptography. CRC Press (1996)
53. Li, N.: Discretionary access control. In: Encyclopedia of Cryptography and Security. Springer (2011)
54. Lindqvist, H.: Mandatory access control. Master's Thesis in Computing Science, Umea University, Department of Computing Science (2006)
55. Ferraiolo, D., Kuhn, D.R., Chandramouli, R.: Role-Based Access Control. Artech House (2003)
56. Cha, B., Seo, J., Kim, J.: Design of attribute-based access control in cloud computing environment. In: The International Conference on IT Convergence and Security. Springer, Netherlands (2012)

57. Yu, S.: Data sharing on untrusted storage with attribute-based encryption. PhD diss, Worcester Polytechnic Institute (2010)
58. Kallahalla, M., Riedel, E., Swaminathan, R., Wang, Q., Fu, K.: Plutus: scalable secure file sharing on untrusted storage. In: FAST03 Berkeley, California, USA (2003)
59. Vimercati, S.D.C. di, Foresti, S., Jajodia, S., Paraboschi, S., Samarati, P.: Over-encryption: management of access control evolution on outsourced data. In: The 33rd International Conference on Very Large Data Bases, VLDB Endowment (2007)
60. Goh, E., Shacham, H., Modadugu, N., Boneh, D.: Sirius: securing remote untrusted storage. In: NDSS 03, San Diego, CA, USA (2003)
61. Fiat, A., Naor, M.: Broadcast encryption. In: CRYPTO 93 (Lecture Notes in Computer Science), Santa Barbara, CA, USA (1993)
62. Halevy, D., Shamir, A.: The LSD broadcast encryption scheme. In: CRYPTO 02 (Lecture Notes in Computer Science), Santa Barbara, CA, USA (2002)
63. Boneh, D., Gentry, C., Waters, B.: Collusion resistant broadcast encryption with short ciphertexts and private keys. In: Proceedings of CRYPTO 05 (Lecture Notes in Computer Science), Santa Barbara, CA, USA (2005)
64. Delerable, C., Paillier, P., Pointcheval, D.: Fully collusion secure dynamic broadcast encryption with constant-size ciphertexts or decryption keys. In: Pairing-Based Cryptography Pairing 2007. Springer, Heidelberg (2007)
65. Kim, J., Susilo, W., Au, M.H., Seberry, J.: Efficient semi-static secure broadcast encryption scheme. In: Pairing-Based Cryptography Pairing 2013. Springer (2014)
66. Gentry, C., Waters, B.: Adaptive security in broadcast encryption systems (with short ciphertexts). In: Advances in Cryptology-EUROCRYPT 2009. Springer, Heidelberg (2009)
67. Wikipedia: ID-based encryption (2014). http://en.wikipedia.org/wiki/ID-based_encryption
68. Li, J., Chen, X., Jia, C., Lou, W.: Identity-based encryption with outsourced revocation in cloud computing. IEEE Trans. Comput. 1–12 (2013)
69. Horwitz, J., Lynn, B.: Toward hierarchical identity-based encryption. In: EUROCRYPT 02 (Lecture Notes in Computer Science), Amsterdam, The Netherlands (2002)
70. Gentry, C., Halevi, S.: Hierarchical identity based encryption with polynomially many levels. In: TCC 09 (Lecture Notes in Computer Science), San Francisco, CA, USA (2009)
71. Gagn, M.: Identity-based encryption. In: Encyclopedia of Cryptography and Security. Springer Science Business Media, LLC (2011)
72. Liu, Q., Wang, G., Wu, J.: Efficient sharing of secure cloud storage services. In: IEEE TSP 10 in Conjunction with IEEE CIT 10, Bradford, UK (2010)
73. Mao, Y., Zhang, X., Chen, M., Zhan, Y.: Constant size hierarchical identity-based encryption tightly secure in the full model without random oracles. In: The 2013 Fourth International Conference on Emerging Intelligent Data and Web Technologies (EIDWT). IEEE (2013)
74. Sahai, A., Waters, B.: Fuzzy identity-based encryption. In: EUROCRYPT 05 (Lecture Notes in Computer Science), Aarhus, Denmark (2005)
75. Goyal, V., Pandey, O., Sahai, A., Waters, B.: Attribute-based encryption for fine-grained access control of encrypted data. In: ACM CCS 06, Alexandria, VA, USA (2006)
76. Waters, B.: Ciphertext-policy attribute-based encryption: an expressive, efficient, and provably secure realization. In: Public Key Cryptography|PKC, LNCS. Springer (2011)
77. Yu, S., Wang, C., Ren, K., Lou, W.: Achieving secure, scalable, and fine-grained data access control in cloud computing. In: The 2010 IEEE INFOCOM. IEEE (2010)
78. Si, X., Wang, P., Zhang, L.: KP-ABE based verifiable cloud access control scheme. In: 2013 12th IEEE International Conference on Trust, Security and Privacy in Computing and Communications (TrustCom). IEEE (2013)
79. Moses, T.: Extensible access control markup language (xacml) version 2.0. Oasis Standard 200502 (2005)
80. Chase, M.: Multi-authority attribute based encryption. In: TCC 07 (Lecture Notes in Computer Science), Amsterdam, The Netherlands (2007)
81. Li, M., Yu, S., Ren, K., Lou, W.: Securing personal health records in cloud computing: patient-centric and fine-grained data access control in multi-owner settings. In: Proceedings of Security and Privacy in Communication Networks. Springer, Heidelberg (2010)

82. Chase, M., Chow, S.: Improving privacy and security in multi-authority attribute-based encryption. In: ACM CCS 09, Chicago, IL, USA (2009)
83. Blaze, M., Bleumer, G., Strauss, M.: Divertible protocols and atomic proxy cryptography. In: EUROCRYPT 98, Espoo, Finland (1998)
84. Li, Q., Xiong, H., Zhang, F., Zeng, S.: An expressive decentralizing kp-abe scheme with constant-size ciphertext. Int. J. Netw. Secur. **15**(3), 161–170 (2013)
85. Hohenberger, S., Waters, B.: Attribute-based encryption with fast decryption. In: Public-Key Cryptography PKC 2013. Springer, Heidelberg (2013)
86. Bethencourt, J., Sahai, A., Waters, B.: Ciphertext-policy attribute-based encryption. In: IEEE Symposium on Security and Privacy. IEEE Computer Society (2007)
87. Chen, C., Zhang, Z., Feng, D.: Efficient ciphertext policy attribute-based encryption with constant-size ciphertext and constant computation-cost. In: Proceedings of Provable Security. Springer, Heidelberg (2011)
88. Zhu, Y., Hu, H., Ahn, G., Huang, D., Wang, S.: Towards temporal access control in cloud computing. In: The 2012 IEEE INFOCOM. IEEE (2012)
89. Zhu, Y., Hu, H., Ahn, G., Yu, M., Zhao, H.: Comparison-based encryption for fine-grained access control in clouds. In: The Second ACM Conference on Data and Application Security and Privacy. ACM (2012)
90. Zhang, Y., Chen, J.: Access control as a service for public cloud storage. In: Distributed Computing Systems Workshops (ICDCSW). IEEE (2012)
91. Nishide, T., Yoneyama, K., Ohta, K.: Attribute-based encryption with partially hidden encryptor-specified access structures. In: Applied Cryptography and Network Security. Springer, Heidelberg (2008)
92. Doshi, N., Jinwala, D.: Hidden access structure ciphertext policy attribute based encryption with constant length ciphertext. In: Advanced Computing, Networking and Security. Springer, Heidelberg (2012)
93. Qian, H., Li, J., Zhang, Y.: Privacy-preserving decentralized ciphertext-policy attribute-based encryption with fully hidden access structure. In: Information and Communications Security. Springer (2013)
94. Jung, T., Li, X., Wan, Z., Wan, M.: Privacy preserving cloud data access with multi-authorities. In: The 2013 IEEE INFOCOM. IEEE (2013)
95. Li, M., Yu, S., Zheng, Y., Ren, K., Lou, W.: Scalable and secure sharing of personal health records in cloud computing using attribute-based encryption. IEEE Trans. Parallel Distrib. Syst. **24**(1), 131–143 (2013)
96. Yang, K., Jia, X., Ren, K., Zhang, B.: Dac-macs: effective data access control for multi-authority cloud storage systems. In: The 2013 IEEE INFOCOM. IEEE (2013)
97. Wang, G., Liu, Q., Wu, J.: Hierarchical attribute-based encryption for fine-grained access control in cloud storage services. In: The 17th ACM Conference on Computer and Communications Security. ACM (2010)
98. Wang, G., Liu, Q., Wu, J., Guo, M.: Hierarchical attribute-based encryption and scalable user revocation for sharing data in cloud servers. Comput. Secur. **30**(5), 320–331 (2011)
99. Wang, G., Liu, Q., Wu, J.: Achieving finegrained access control for secure data sharing on cloud servers. Concurr. Comput. Pract. Exp. **23**(12), 1443–1464 (2011)
100. Wan, Z., Liu, J., Deng, R.H.: HASBE: a hierarchical attribute-based solution for flexible and scalable access control in cloud computing. IEEE Trans. Inf. Forensics Secur. **7**(2), 743–754 (2012)
101. Chen, Y., Chu, C., Tzeng, W., Zhou, J.: Cloudhka: A cryptographic approach for hierarchical access control in cloud computing. In: Applied Cryptography and Network Security. Springer, Heidelberg (2013)
102. Wang, G., Liu, Q., Wu, J.: Time-based proxy re-encryption scheme for secure data sharing in a cloud environment. Inf. Sci. **258**, 355–370 (2014)
103. Wikipedia: Role-based access control (2014). http://en.wikipedia.org/wiki/Role-based_access_control

104. Ferrara, A.L., Madhusudan, P., Parlato, G.: Policy analysis for self-administrated role-based access control. In: Tools and Algorithms for the Construction and Analysis of Systems. Springer, Heidelberg (2013)
105. Zhou, L., Varadharajan, V., Hitchens, M.: Enforcing role-based access control for secure data storage in the cloud. Comput. J. **54**(10), 1675–1687 (2011)

# Homomorphic Cryptosystems for Securing Data in Public Cloud Computing

**Nihel Msilini, Lamri Laouamer, Bechir Alaya and Chaffa Hamrouni**

**Abstract** No one could deny that Cloud Computing has always been the most amazing innovation over the last decade. It is a platform that hosts computing resources, either hardware or software, offering data storage in remote sites. Due to its numerous advantages, it is doubtless a promising perspective. It is available for all time over the internet, offers a low cost service, easy to maintain, environmentally friendly and has a tremendous performance. However, users and companies relying on this technique have always struggled with the lack of security and privacy in the platform, especially when dealing with critical data storage and transfer. Moreover, even if users encrypt their messages in advance, the cloud service provider ought to decrypt it first in order to carry out different operations which can be a considerable risk, since clouds are not fully trusted. So, to solve this problem, a new encryption manner came along, it is the homomorphic encryption. Using this technique, a cloud service provider can

N. Msilini
MAC'S, National Engineering School of Gabes, University of Gabes,
Avenue Omar Ibn El Khattab, 6029 Zerig, Gabes, Tunisia
e-mail: nihel.msilini@gmail.com

L. Laouamer (✉) · B. Alaya
Department of Management Information Systems, CBE Qassim University,
6633, Buraidah 51452, Saudi Arabia
e-mail: laoamr@qu.edu.sa

B. Alaya
e-mail: b.alaya@qu.edu.sa; bechir.alaya@litislab.fr

L. Laouamer
Lab-STICC (UMR CNRS 6285), University of Western Brittany,
20 Avenue Victor Le Gorgeu, BP 817 - CS 93837, 29238 Brest Cedex, France

C. Hamrouni
Higher Institute of Technological Studies, University of Gabes, Gabes, Tunisia
e-mail: chafaa.hamrouni.tn@ieee.org

B. Alaya
LITIS-Lab, University of Le Havre, UFR Sciences et Techniques,
25 rue Philippe Lebon, BP 540, 76058 Le Havre Cedex, France

C. Hamrouni
MAC'S-Lab, National Engineering School of Gabes, Avenue Omar Ibn El Khattab,
6029 Zerig, Gabes, Tunisia



A.E. Hassanien et al. (eds.), *Multimedia Forensics and Security*,
Intelligent Systems Reference Library 115, DOI 10.1007/978-3-319-44270-9_3

perform different operations on the cipher text without the need to decrypt it. Thus security and confidentiality are guaranteed. The main purpose of this chapter is to discuss the concepts and significance of homomorphic cryptosystems along with the different techniques such as the Fully Homomorphic Encryption (FHE) and the Somewhat Homomorphic Encryption (SHE) and also the related works based on these techniques.

**Keywords** Cloud computing · Security · Homomorphic Encryption · FHE · SHE

## 1 Introduction

With today's fast pace of the world's business, relying on new technologies is now compulsory to maintain a company development. Cloud Computing is considered to be one of the revolutionary advances of the century. Data can be stored remotely on servers across the world. Different companies can rent hardware other than buying a whole infrastructure which could be really costly. Furthermore, stored data can be remotely accessed anywhere and anytime. However, a major hurdle to the adoption of Cloud-based services is security. Cloud users are immensely concerned with losing control of their most sensitive data. Moreover, it is difficult for public cloud service user to feel comfortable that their files are well protected. Yet, encryption could lighten this matter, in that case the user has to share his secret key with the cloud provider in order to decrypt the data, thus, sharing the secret key allows the cloud provider to access to your data. The answer to this problem is Homomorphic Encryption. The reminder of this chapter is structured as described below: in Sect. 2, we present the Cloud Computing. Section 3 highlights the security issues faced in cloud. Section 4 deals with the current security techniques compared to homomorphic technique. Then, in Sect. 5, we present the homomorphic encryption technique as a solution to provide security in the Cloud. Next, in Sect. 6, we describe several cryptographic techniques and the current cryptosystems essentially deployed to achieve these security goals. We present the result and discussions in Sect. 7 and finally, Sect. 8, presents some headlines for near researches in the cloud security field.

## 2 Cloud Computing

Cloud Computing [1] became the most highly used platform due to the services it guarantees and according to the NIST [2]. Cloud computing is defined as a platform offering on demand services based on an internet connection, data remote storage and data access anywhere and anytime. Besides, the keystone elements of cloud computing is the different deployment models, we could find as illustrated in Fig. 1.

- **Public Cloud**: This type of deployment is shared with the public. A Cloud service provider is the main administrator who rents out different services on demand to a cloud service consumer. It is intended to take in web applications and is only accessible via Internet. It is an infrastructure that offers an optimal use of resources. For every end of month the cloud service consumer has to pay for its consumption.
- **Private Cloud**: This type of clouds is dedicated to a specific company. The infrastructure hosts different services and resources on demand to a limited number of people through virtualization technologies. Or else it could be taken in by a Cloud service provider based on a VPN connection.
- **Community Cloud**: This model of deployment hosts different companies with common interests. Just like the private Cloud, the community cloud can be administrated by the companies themselves or by a foreign cloud service provider.
- **Hybrid Cloud**: It is the combination of two deployment models, the public cloud and the private cloud. The infrastructure can hold either two or more cloud model. It is mostly used to guarantee direct access through internet based on the private cloud. The private and the public infrastructure communicate with each other, then it is a way to gather the benefits of both platforms.

Cloud computing platform offers different services on demand such as:

- **Software as a Service (SaaS)**: This service is being manipulated by the service cloud provider running on his computers and not the user computers. It is accessed through public Internet and generally lent on a monthly or yearly subscription.
- **Infrastructure as a Service (IaaS)**: An IaaS provider can offer computing, repository, networking through public Internet, a VPN connection or a dedicated network connection. Service consumers own and manage operating systems, applications, and every single file existing on the infrastructure. The payment is by usage.
- **Platform as a Service (PaaS)**: With the aim to build applications based on cloud, either software or hardware resources, are given by the provider, via public Internet, a VPN, or a dedicated network connection. Service consumers pay by use of the platform. They also control how applications can be used throughout their lifecycle.

**Fig. 1** Cloud Computing deployment models

## 3 Security Issues in Cloud Computing

Cloud computing has a major weakness when it comes to security. This weakness became a important risk to critical data stored and it widely boosted from the urge to store sensitive data in a non-fully trusted cloud service provider. Different parts in the cloud infrastructure can be supervised by different non-trusted parts, and could be vulnerable to internal as well as external attacks, either from other cloud roomers or malicious insiders. Furthermore, a Cloud is a multitenant infrastructure, in such a way that there is a lack of trust among the users which can lead to data leakage thus, there is a tremendous lack in security assurance. Even if a Cloud Computing technology offers advanced services and practical deployment models, it doesn't offer, in any way, the security feature of the data stored or being transferred to the Cloud. Hence, Cloud Computing platforms struggle with the lack of security, confidentiality and visibility. Other than providing different types of services, like Infrastructure, Platform or Software as a Service, a cloud service provider should better security or privacy to a cloud service consumer. Thus, security and confidentiality must be as important as the need to protect the data stored in remote sites. In a public cloud, security is ensured by encrypting the data stored. Cryptography, a technique to convert a plain text to a cipher text, is then, a probable solution. It is the most common form to secure data communication. However, performing different operations on the data stored in requires decrypting it beforehand, which makes sensitive data open to the cloud service provider. Moreover, there are no manipulations that could be carried out on the cipher text. It is very easy to have a secure transmission from a local device to a cloud infrastructure, in case where, the data stored is encrypted and the channel used in transmission is well secured. The proposal here is, before delivering the messages to the cloud service providers, they should be encrypted in advance. In order to achieve this and to allow the cloud party to carry out the operations on cipher texts without decrypting them first, it is necessary to adopt the cryptosystems based on Homomorphic Encryption.

## 4 Why Homomorphic Encryption (HE)?

Cloud service providers have always struggled with the data security challenges within their platforms. Moreover, a cloud service consumer can't be fully confident that his most sensitive data stored in a remote site is secure. And even if, your Cloud service providers holds efficient encryption techniques, they are still able to access to your data with or without your knowledge. The main risk seems to be internal, and out of your control. Data manipulation and updating is way too easy to carry out by a user, however, this have no meaning without secure storage. Thus, providers must strengthen their defenses to guarantee a better security. In fact, many service providers employ different techniques to ensure that their Cloud platforms can meet the users requirements when it comes to data protection. Some Cloud providers choose to create dedicated infrastructure and divide their client's data in order to conduct frequent auditing of their platforms, others are leaning on the use of Cloud encryption techniques to enhance their security strategies. Data encryption solves many of the control problems found in the cloud. Additionally, placing the interest on the data rather than on the platform, can ensure that the data will remain protected, even if the hardware vulnerabilities can be exploited. As far as Cloud Computing Encryption is concerned, we consider three cryptographic techniques that are applicable to achieve security, within three scenarios: Verifiable Computation (VC) [3] for a trusted private Cloud, Secure Multi-Party Computation (MPC) [4] for a semi-trusted Cloud and Homomorphic Encryption (HE) [5] for an non-trusted public Cloud.

In order to secure its own cloud, a cloud service provider tend to apply the verifiable computation technique, in other words, it is efficient when used on private clouds, a fully trusted cloud. This technique ensures the transfer of computations between machines in the same cloud, while checking if the operations done are correct. As a first scenario, a cloud provider deploys his own resources while completely isolating it from the outside, not letting any information out. This procedure guarantees the data confidentiality without the need to rely on other encryption techniques. However, all these setups don't necessarily offer security since we are never sure that all parties are honest or trustworthy. An internal attacker could easily attempt to corrupt machines through malwares in order to compromise the integrity of operations. VC seem to be, hence, the greatest tool to apply in this kind of scenarios, when it comes to the owner to verify the integrity of the computations. But still, for big data analytic this technique is too slow.

For the second scenario, we consider the case of a semi trusted cloud, when a Cloud provider tends to build his own cloud while creating algorithms for nodes in order to carry out an operation over their most confidential data, side by side. In this case, the cloud won't been considered as entirely adversarial. However, as for the first scenario, there is always internal danger. In fact, it is dif?cult to prevent the risk of malwares. Thus we apply the secure multiparty computation. This technique strengthens the fact that many honest nodes are present in order to achieve integrity. But, it is still a weaker technique when it comes to security concept, compared to the fully homomorphic encryption. Using MPC, no one of the nodes has any information about the data, but if an attacker corrupts many nodes at once, the confidentiality will be broken.

Finally, for the last scenario, we will consider the case where we have a completely non-trusted public Cloud. A cloud service consumer chooses to store his data remotely in a public cloud in order to lower his costs. The main idea was to encrypt data in advance and then send it to the cloud, where the cloud service provider to carry out his different operations on encrypted data. Homomorphic encryption is a technique offering that feature. A cloud service provider can perform different operations on cipher texts without the need to decrypt them beforehand.

In conclusion, the last scenario based on Homomorphic Encryption, seems to be the closest and the most efficient scenario to enhance security in a public Cloud platform. This technique encrypts data and then allows the cloud service provider to perform different operations on it without having to decrypt it first. In other words, if the scheme is thought to be homomorphic, the cloud can still carry out pretty good evaluations on the cipher text. In the following section, we are going to present the Homomorphic encryption technique [6].

## 5 Homomorphic Encryption (HE)

Lately, Homomorphic encryption schemes have become the most studied techniques in the majority of works related to cloud computing encryption. This technique offers different complex mathematical operations to be carried out on cipher texts. Homomorphism is the transformation of a data set to another with maintaining the relationship between the elements of both sets. The concept comes from a Greek definition for "same structure". Strictly speaking, a Homomorphic cryptosystem is a cryptosystem characterized by a powerful algorithm that operates an encryption of an addition or a multiplication of two texts while providing the public key and the cipher texts but not the raw messages themselves, it's like performing operations on a black box. More formally, homomorphic encryption is defined as an encryption method that allows definite operations, carried out on a message corresponds to another algebraic operation performed on its encryption. Mathematically, these operations are called mappings or functions. For example, if we consider two groups **A** and **B** where each group is characterized by a specific function, which are equivalent, a mapping **F** from **A** to **B** is called a homomorphism. **F** preserves the structure from one set of data to the other. Basically, the main idea is to make sure whether there exists any homomorphic scheme that can be set. But, there wasn't a remarkable development in finding if such schemes exist offering efficiency and security, until 2009, when Craig Gentry [16], theoretically, proved the possibility of the construction of such system. In order to preserve the homomorphic property and the efficiency of the scheme, it is essential to verify that the size of the cipher texts are still polynomially bounded in the security parameters while performing continuous computations. Let's consider the following example, we have a message $m = abc$, $Enc$ is the encryption function where $Enc(m)$ = @!$. The algorithm is given as follows (Fig. 2):

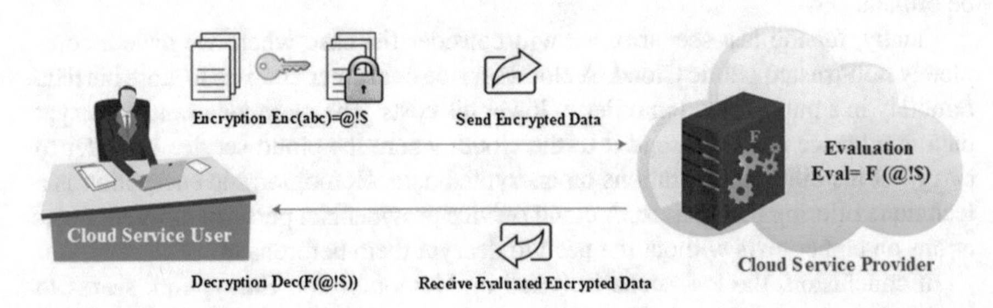


**Fig. 2** HE algorithm

1. **Encryption**: Using the secret key $s_k$, the client encrypt the plain text **m** and generate $Enc_s k$(m, the encrypted message @!$ will be sent to the cloud server.
2. **Evaluation**: The cipher text @!$ is evaluated by a function **F** where Eval = F(@!$).
3. **Decryption**: The generated **Eval** will be decrypted by the client using its $s_k$.

# 6 Different Encryption Cryptosystem

The encryption scheme's main goal is to retain confidentiality. The security feature of a scheme must not be built on the complexity of the adopted codes, but rather on the secrecy of the key. For homomorphic cryptosystems, we could find either a Somewhat Homomorphic Encryption (SHE) or a Fully Homomorphic Encryption (FHE). In the following section, we are going to describe the different Homomorphic cryptosystem techniques, the related works based on these techniques, the advantages and the drawbacks of each cryptosystem.

## *6.1 Somewhat Homomorphic Encryption (SHE)*

A Somewhat Homomorphic cryptosystem is a given cryptosystem that it relies either on additive or on multiplicative homomorphism. We distinguish the multiplicative homomorphic encryption is the RSA and El Gamal and the additive Homomorphic encryption is Pailler and Goldwasser-Micalli. Below, we will describe the most known Somewhat Homomorphic cryptosystems.

### 6.1.1 RSA Algorithm

RSA [7] encryption scheme is an asymmetric technique and is multiplicatively homomorphic. It offers security mainly built on the complexity of factoring, where n is product of two large primes. An encryption of a given message m is: $E(m) = m^e mod\, n$. The homomorphic operation provided by RSA is as follows: given two messages $m_1$ and $m_2$ and their encryptions $E(m_1)$ and $E(m_2)$, it is very easy to prove that $E(m_1) * E(m_2) = E(m_1 * m_2)$. The homomorphism property is then:

$$E(m_1) * E(m_2) = m_1^e * m_2^e \; mod\, n = (m_1 * m_2)^e \; mod\, n = E(m_1 * m_2) \qquad (6.1)$$

The RSA cryptosystem operates as follows:

| | Key generation | Encryption | Decryption |
|---|---|---|---|
| Input | p, q : large primes | message m | C |
| Computer | $n = p * q$<br>$\Phi(n) = (p-1)(q-1)$<br>$gcd(e, \Phi(n)) = 1$<br>$e * d \equiv 1 \bmod \Phi(n)$ | $C = m^e \bmod n$ | $m = c^d \bmod n$ |
| Output | Public key $p_k = (e, n)$<br>Secret key $S_k = d$ | C | M |

**Problem**: this algorithm is not randomized and is not semantically secure and this is a major disadvantage, since trying to make this procedure semantically secure destroys the Homomorphic properties. Besides, in the case of relying on a deterministic scheme, it seems to be effortless to know if a user has sent the same message two times while operated with the same key. This cryptosystem can evaluate only multiplication and a limited number of multiplications and other operations on the encrypted message will break the decryption. Furthermore, when the output is larger than **n** the decryption fails. Besides, when it comes to hardware and software implementation RSA, seems to be not very efficient.

### 6.1.2 ELGamal Algorithm

ElGamal [8] encryption is an asymmetric encryption algorithm. It is basically built on the Diffie Helman exchange. Same as the RSA technique, ElGamal is also multiplicatively homomorphic. The encryption of a message $m$ is given as $E(m) = (g^r, m * h^r)$, for some random $r \in \{0, \ldots, q-1\}$. The homomorphic property is as follows:

$$E(m_1) * E(m_2) = (g^{r_1}, m_1 * h^{r_1}) * (g^{r_2}, m_2 * h^{r_2}) = (g^{r_1+r_2}, (m_1 * m_2)h^{r_1+r_2}) = E(m_1 * m_2)$$

ElGamal algorithm is shown below:

| | Key generation | Encryption | Decryption |
|---|---|---|---|
| Input | Cyclic group G of order q and generator g<br>$x \in 1, \ldots, q-1$ | message m<br>$y \in 1, \ldots, q-1$ | (C1, C2) |
| Computer | $h := g^x$ | $c_1 := g^y$<br>Shared secret<br>$s := h^y$<br>Mapping in G: $m \rightarrow m'$<br>then $C_2 := m*s$ | $s := G_1^x$<br>$m' := C_2 * S^{-1}$ |
| Output | Public key $p_k = (G, q, g, h)$<br>Secret key $S_k = x$ | $(C_1, C_2) = (g^y, m' * h^y)$<br>$= (g^y, m' * (g^x)^y)$ | M |

**Problem**: ElGamal encryption is a probable victim a well-known attack, the Meet-in-the-Middle Attacks [9]. Given an encryption $(C_1, C_2)$ of a message *m*, one can easily construct a an encryption $(C_1, 2 * C_2)$ of the message $2 * m$. This technique is probabilistic, which means that a single message can be encrypted to many possible encrypted forms, while producing a *2:1* expansion in size from plaintext to cipher text.

### 6.1.3 Paillier Algorithm

Paillier [10] is a known to be a probabilistic scheme. It is essentially based on composite residuosity and has an additive homomorphic property. Therefore, this technique has various applications for instance we could mention the e-voting systems. The scheme is very efficient and is considered to be a high-leveled security cryptosystem. Furthermore, characterized by a smaller expansion and lower cost, this scheme has found great acceptance. In the Paillier cryptosystem the encryption of a message $m$ is $E(m) = g^x * r^n \bmod n^2$, for some random $r \in \{0, \ldots, q-1\}$. The homomorphic property is:

$$
\begin{aligned}
E(m_1) * E(m_2) &= (g^{m_1} * r_1{}^n) * (g^{m_2} * r_2{}^n) \; mod \; n^2 \\
&= (g^{m_1+m_2} * (r_1 r_2)^n \; mod \; n^2 \\
&= E(m_1 + m_2 \; mod \; n^2)
\end{aligned}
$$

Basic ideas of the algorithm is given as follows :

| | Key generation | Encryption | Decryption |
|---|---|---|---|
| Input | p, q : large primes | message m | C |
| Computer | $n = p * q$<br>$gcd(L(g^{\lambda} \bmod n^2), n) = 1$<br>With $L(u) = \frac{(u-1)}{n}$ | $C = g^m * r^n \bmod n^2$ | $z = \frac{L(C^{\lambda} mod\, n^2}{L(g^{\lambda} mod n^2}$<br>$m = z \bmod n$ |
| Output | Public key $p_k = (e,n)$<br>Secret key $S_k = d$ | C | M |

**Problem**: The encryption process has a low cost however, the decryption is a bit heavyweight since it needs an exponentiation modulo $n^2$(to $\lambda$(n)) and a multiplication modulo *n*. Besides, the main problem resides on the complexity of computing *n-th* residue classes.

### 6.1.4 Goldwasser-Micali Algorithm

Goldwasser-Micali (GM) [11] cryptosystem is an asymmetric algorithm. The latter has the feature of being the first probabilistic scheme that is proved to be secure under standard cryptographic assumptions. The encryption of a bit $b$ is $E(b) = x^b r^2 \ mod \ n$, for some random r ∈ {0, …, n − 1}. The encryption algorithm is simple, it uses a product and a square but the decryption is thought to be more complex and heavy and involves exponentiation. The homomorphic property is:

$$E(b_1) * E(b_2) = (x^{b_1} * {r_1}^2) * (x^{b_2} * {r_2}^2) \ mod \ n$$
$$= (x^{b_1+b_2} * (r_1 r_2)^2 \ mod \ n$$
$$= E(b_1 \oplus b_2)$$

where ⊕ denotes addition modulo *2*. Key generation and message encryption and decryption processes are given as follows:

| | Key generation | Encryption | Decryption |
|---|---|---|---|
| Input | p, q : large primes | $m = (m_1 \ldots m_n)$ | $C = (C_1 \ldots C_n)$ |
| Computer | n = p * q<br>non-residue x satisfying<br>$\frac{x}{p} = \frac{x}{q} = -1$ | $C_i = y_i^2 \, x^{m_i}$ mod n | If $C_i$ quadratic residue<br>$m_i = 0$<br>Else $m_i = 1$ |
| Output | Public key $p_k = (x, n)$<br>Secret key $S_k = (p, q)$ | C | $m = (m_1 \ldots m_n)$ |

**Problem**: The main weakness in this cryptosystem is that the cipher text may be way too larger than the plaintext. Furthermore, the decryption process has a complexity of $O(k * l(p)^2)$ where $l(p)$ stand for the number of bits in p. Moreover, the input of this scheme consist of a single bit. Besides, years later, Benaloh [12] has worked on the GM cryptosystem and the main improvement resides on the ability of the new cryptosystem to encrypt longer blocks of data at once rather than one bit individually. In the Benaloh cryptosystem, the encryption of a message $m$ is $E(m) = g^x r^c$ mod n, for some random r ∈ {0, …, n − 1}, where the homomorphic property is:

$$E(m_1) * E(m_2) \ mod \ n = (g^{m_1} * {r_1}^c) * (g^{m_2} * {r_2}^c) \ mod \ n$$
$$= (x^{m_1+m_2} * (r_1 r_2)^c \ mod \ n$$
$$= E(m_1 + m_2 \ mod \ n)$$

| | Key generation | Encryption | Decryption |
|---|---|---|---|
| Input | p, q : large primes<br>block size r | m, random u | C |
| Computer | n = p * q<br>Φ(p − 1)(q − 1)<br>$y^{\Phi/r} \neq 1$ mod | $E_r(m) = y^m \, u^r$ mod n | $a = C^{\Phi/r} \ mod \ n$ |
| Output | Public key $p_k = (y, n)$<br>Secret key $S_k = (\Phi, x)$ | C | $m = log_x(a)$ |

## 6.2 *Fully Homomorphic Encryption (FHE)*

In order to enhance security feature public cloud computing, an innovative solution is proposed, we are talking about the fully homomorphic encryption. This algorithm is fully adequate for the processing and recovery of the encrypted data, and strongly leads to the data storage and transmission. Different works were carried out to enhance the performance of such algorithm which thought to be impossible till its discovery. Below, we will present the most admitted works related to FHE.

### 6.2.1 First FHE BGN 2005

In 2005, there was a small but also highly significant advance in the homomorphic encryption schemes; it was made by Boneh, Goh and Nissim [13]. This cryptosystem was the first FHE cryptosystem to introduce using pairings on elliptic curves [14]. It is based on both additions and multiplications, while the encrypted data has a constant size. The additive property is the same as for the El Gamal algorithm and only one multiplication is permitted. The multiplication is thought to be possible because of the feature that pairings can be set for elliptic curves. This scheme uses different and multiple presentations of its ring $R$, and so comes its importance. The scheme is presented below: Let's consider $G_1$ and $G_2$ as two additive groups and $G_m$ as a multiplicative group of an order $p$. P $\in G_1$, Q $\in G_2$ are generators of $G_1$ and $G_2$, respectively. A pairing is a map as follows:

$$\mathrm{e} : G_1 * G_2 \rightarrow G_m$$

For which holds:

- $\mathrm{e}(P^a, Q^b) = \mathrm{e}(P, Q)^{ab}$
- Non-degeneracy: $\mathrm{e}(P, Q) \neq 1$
- $e$ has to be computable in an efficient way. If $G_1 = G_2 = G$, the pairing is symmetric. Even more, if $G$ is cyclic then e will be commutative for any P, Q $\in$ G, then: $e(P, Q) = e(Q, P)$. This is due to the existence of integers $p$, $q$ for a generator $g$ where $P = g^p$ and $Q = g^q$. So, $\mathrm{e}(P, Q) = \mathrm{e}(g^p, g^q) = \mathrm{e}(g, g)^{pq} = \mathrm{e}(g^q, g^p) = \mathrm{e}(Q, P)$

The algorithms are given as follows:

- **Key Generation**
  Having a security parameter $\lambda$, generate a tuple ($q_1$, $q_2$, G, $G_1$, e), such that $q_1$ and $q_2$ are two large primes, G a cyclic group of order $q_1\ q_2$, and $e$ the pairing map e:G * G $\rightarrow G_1$. Let's consider n = p * q. Choose two random generators $g$ and $u$ from $G$ and compute $h = u_2^q$. $h$ is a random generator for the subgroup of $G$ of order $q_1$. The public key is then $P_k = \{n, G, G_1, e, g, h\}$ and the private key $S_k = q_1$.

- **Encryption**
  Given a message $m$ and the public key $P_k$, pick up a random $r$ from $\{1, 2, \ldots, n\}$:

$$\mathrm{C} = g^m h^r \in \mathrm{G}$$

- **Decryption**
  To decrypt $C$ using $S_k = q_1$, we have:

$$C^{q_1} = (g^m h^r)^{q_1} = (g^{q_1})^m$$

  To get m, it is enough to calculate the discrete logarithm of $C^{q_1}$ to the base $g^{q_1}$. But, the message space is limited due to the complexity of this decryption.
- **The homomorphic property**
  It is clear that the *BGN* algorithm is homomorphically additive. Given $P_k =$ n, G, $\mathrm{G}_1$, e, g, h, two encrypted messages $m_1$ and $m_2$, $C_1 = g^{m_1} h^{r_1} \in \mathrm{G}$, $C_2 = g^{m_2} h^{r_2} \in \mathrm{G}$. We can set a distributed encryption of $m_1 + m_2$ mod $n$ by calculating: $\mathrm{C} = C_1 C_2 h^r$ for a random $r \in \{0, \ldots, \mathrm{n} - 1\}$ since $C_1 C_2 h^r = (g^{m_1} h^{r_1})(g^{m_2} h^{r_2}) h^r = g^{m_1+m_2} h^{r_1+r_2+r}$ is the encryption of $m_1 + m_2$. Moreover, we can calculate the multiplication of two cipher texts using the map $e : g_1 = \mathrm{e(g, g)}$ and $h_1 = \mathrm{e(g, h)}$ while $g_1$ is of order $n$ and $h_1$ of order $q_1$. We choose an integer $\alpha$ such that h $= g^{\alpha q_2}$. Given $C_1 = g^{m_1} h^{r_1} \in \mathrm{G}$, $C_2 = g^{m_2} h^{r_2} \in \mathrm{G}$ G we want to get an encryption of $m_1 + m_2$ mod n: $\mathrm{C} = \mathrm{e}\ (C_1, C_2)\ h_1^r \in G_1$, We have:

$$\begin{aligned} C = e(C_1, C_2)h_1^r &= e(g^{m_1}h^{r_1}, g^{m_2}h^{r_2})h_1^r \\ &= e(g^{m_1+\alpha q_2 r_1}, g^{m_2+\alpha q_2 r_2})h_1^r \\ &= e(g, g)^{(m_1+\alpha q_2 r_1)(m_2+\alpha q_2 r_2)}h_1^r \\ &= e(g, g)^{m_1 m_2+\alpha q_2(m_1 r_2+m_2 r_1+\alpha q_2 r_1 r_2)}h_1^r \\ &= e(g, g)^{m_1 m_2}h_1^{r+m_1 r_2+m_2 r_1+\alpha q_2 r_1 r_2} \end{aligned}$$

  where $\mathrm{r} + m_1\ r_2 + m_2\ r_1 +\ \alpha q_2\ r_1\ r_2$ is uniformly distributed. $C$ is then uniformly distributed encryption of $m_1 + m_2$ *mod n* not in $G$ but in $G_1$. BGN is still homomorphic in $G_1$ with respect to addition. Boneh, Goh and Nissim have presented the subgroup decision problem [15] of a pair of group (G, $G_1$) of composite order $n = pq$ where we can set a non-degenerate map e:G * G → $G_1$. The main issue is to find out if a given element y ∈ G is in the subgroup of order $p$. Moreover, if $g$ generates $G$ we get e(g, y) is an important keyword for the same problem in $G_1$, so, if the *SD* problem in unpractical in $G$, then it is the same in $G_1$.

Finally, for all a and $b$ in $P$, and $k$ in $K$, we have:

$$E_k(\mathrm{a}) \oplus E_k(\mathrm{b}) = E_k(\mathrm{a} \oplus \mathrm{b})$$
$$E_k(\mathrm{a}) \otimes E_k(\mathrm{b}) = E_k(\mathrm{a} \otimes \mathrm{b})$$

The encryption scheme is fully homomorphic.
The BGN cryptosystem was the first introduction to the fully homomorphic encryption and made its mark in the data encryption field. As the ubiquity of the encryption schemes continue to extend its reach into every facet of our lives, performance becomes paramount since this cryptosystem, operating with one multiplication and many additions, had to be improved, until 2009, when Craig Gentry, made a revolutionary cryptographic achievement in the fully homomorphic encryption.

### 6.2.2 Gentry 2009 and Gentry and van Dijk Simplify

In 2009, a revolutionary advance came along. It was Craig Gentry's [16]. This achievement made a greater difference later on and helped to realize the FHE which, previously thought to be impossible. The Gentry's scheme holds processes of arbitrary depth circuits. The scheme starts from an SHE based on ideal lattices [17], only limited to evaluating low-degree polynomials. However, each cipher message is noisy in some way, and the latter is boosted with every operation done (multiplication multiplies it and addition adds it) till making the decryption hard to compute. But then, he introduced the bootstappable feature which allows lowering the noise when Gentry made a slight modification on every SHE scheme in such a way to evaluate its own decryption circuit. And so on, every bootstrappable SHE scheme could be converted to be a FHE in a recursive self-embedding. The bootstrapping feature refreshes the encrypted message by lowering its related noise in order to be used again in other addition and multiplication operations without causing a complex decryption thereafter. Gentry's schemes are based on the complexity of two main criteria: worst-case problem over ideal lattices and low-weight subset sum. However, this cryptosystem is impractical, since the cipher text's size and operations time increase rapidly as the security level increases. The algorithm goes as follows:
**Encrypt**: A bit b $\in \{0,1\}$ is encrypted with a public key (d, r). First let's randomly generate a noise vector u $= (u_0, u_1, \ldots u_{n-1})$, with every entry selected as 0 with some probability p and as $(+/-)1$ with probability $(1-$ p)/2 each. He proved that u can hold a big number of zeros without influencing the security level.

**Eval**: operations can be directly operated on the cipher text:
$c_1 = \mathrm{Encrypt}(m_1)$ and $c_2 = \mathrm{Encrypt}(m_2)$, we will get:
$\mathrm{Encrypt}(m_1 + m_2) = (c_1 + c_2) \bmod \mathrm{d}$
$\mathrm{Encrypt}(m_1 * m_2) = (c_1 * c_2) \bmod \mathrm{d}$

**Decrypt**: m $= [\mathrm{c} * \omega]_d \bmod 2$, where $\omega$ is the private key and is part of the public key.

**Recrypt**: Process that homomorphically evaluates the decryption circuit on the cipher text. Given one bit and only operate certain fixed number of evaluations. To make this possible we need to restructure the decryption algorithm. The private key satisfies $\Sigma_s \omega_i = \omega$. Each $\omega_i$, where $\omega_i = x_i\ R^{l_i}$ mod d where $R$ is constant, $x_i$ and $l_i$ $\in$ {1, 2, 3,..., s} are random. The process can be operated in two steps: Compute the sum of $cx_i\ R^{l_i}$ for each block $i$. For further optimization this process, encode $l_i$ to a 0–1 vector { $\eta_1(i), \eta_2(i), \ldots, \eta_n(i)$} where only two elements are set to 1 and the rest to 0. The two positions are labeled as a and b, l(a, b) is referred to as the corresponding value of l. We can obtain $cx_i\ R^{l_i}$ from only when $\eta_a(i)$ and $\eta_b(i)$ are both 1, and the matching $cx_i\ R^{l(a,b)}$ is selected.

Further on, in the same year, Martenvan Dijl, Craig Gentry, Shai Halvei and Vaikuntanathan [18, 19] proposed a second FHE scheme, based on the improvements done by Gentry [16]. However, this time, the construction drops the ideal lattices and as an alternative with a simpler SHE homomorphic scheme that is based on integers. This scheme is simpler than the based-ideal lattice and applying the squash decryption and the cipher text refresh both proposed by Gentry.

Let's consider a set of integers $x_i = pq_i + r_i$, where 0 < i < r and p is secret. **Key generation**: Generate an integer p with a size $n_p$ bits. Let $p_k = (x_0, x_1 \ldots, x_r)$ and $s_k = \text{p}$.
**Encrypt**: Given m$\in${0,1} and $p_k$. Choose a random r$\in$ $(-2^{\rho'}, 2^{\rho'})$, a bit length for the noise and $\rho$' is a secondary noise parameter $r_i$:

$$\text{C} = 2\text{m} + 2\text{r} + 2\Sigma x_i$$

**Evaluate**: $(p_k, \text{C}, c_0, c_1 \ldots, c_t)$, with t input bits, and t cipher text $c_i$, perform the addition and the multiplication over the integers.
**Decrypt**: $(s_k, \text{C})$, output m = (c mod p) mod 2
However, the public key is too large for any practical use.

### 6.2.3 Smart and Vercauteren Reduce Key Size Algorithm

In 2010, Nigel P. Smart and Frederik Vercauteren proposed a FHE with a reduced key and cipher text sizes. The Smart-Vercauteren [20] scheme follows the lattice-based scheme proposed by Gentry. But it is also based on the SHE. For the latter, the keys (public and private) include two large integers (one is shared) and the encrypted text is one large integer. However, the proposed cryptosystem has reduced cipher text and key sizes than the lattice-based Gentry's scheme. Furthermore, has an efficient feature when it comes to any field of characteristic two. But, the main problem resides on the heaviness of the key generation algorithm, which prevents the practical implementation of this scheme.

### 6.2.4 Gentry and Halevi Present Working FHE Implementation

In 2011, at the rump session of Eurocrypt, Gentry and Shai Halevi gave a possible implementation [21] (bootstrapping feature) with performance numbers. They described an outstanding array of optimizations aiming to lower the size of the public key as well as the latencies of primitives. But the encryption is a bit slow, taking more than a second to be done, same for the recrypt procedure. Besides, after every some AND operations, another recrypt operation is carried out to lower noise in the cipher text. After this improvement, many schemes came along, such as the BGV leveled [22] scheme in 2012. But more recently, Coron, Naccache and Tibouchi presented a procedure aiming to reduce the size of the public key of the van Dijk et al. scheme to 600 Kb.

In 2013, via GitHub, the software HElib written in C++, was released to the open source community. This library holds the BGV HE [23] cryptosystem along with numerous optimizations to run the homomorphic operations faster and with a great interest in the use of the Smart-Vercauteren cipher text techniques and the Gentry, Halevi and Smart optimizations. In December 2014, the software even included the bootstrapping feature. And so on, in March 2015, it supported multithreading.

Recently, in 2015 a new homomorphic encryption cryptosystem of the AES circuit [24] was proposed by Gentry, Halevi and Smart. It is an updated implementation of the leveled FHE, either with or without bootstrapping, evaluating AES-128 circuit. The implementation was held by the HElib library presented beforehand.

## 7 Result and Discussion

Homomorphic encryption gave a whole other perspective one it came to cryptography. This new technique offers data confidentiality but one put together with the Verifiable Computation, we can tend for integrity as a plus. These techniques can guarantee secure computation in a totally non-trusted public cloud. Further works are carried out to improve the scheme's efficiency and to look for a practical implementation. Since BGN in 2005, going through Gentry's innovation in 2009 till 2015 and so on, we can hope for a practical, efficient secure system implementation in the cloud computing field. For example, the works held in 2015 by the HElib community, are thought to be great steps to an efficient software library of the FHE. Besides, homomorphic techniques could be used in different applications with great performance, below some applications:

- E-voting [25]: Homomorphic can be really useful in this case where, compared to traditional paper-based elections, it guarantees that elections results can be delivered quickly while the chance of the human error in lower which will reduce the costs. For instance, we can mention the Chaum's first e-voting scheme.

- Protection of mobile agent [26]: When using HE, the host is able to perform an encrypted function with no need to decrypt it. The resulting changes can be used as a mobile program that can be run on a remote host.
- Private Information Retrieval (PIR) [27]: We can mention Chor et al. scheme with communication overhead, where HE lowers the user's communication overhead as well as the total communication, in order to minimize the computed complexity needed to decode the server's response.
- Hybrid Wireless Network [28]: With no need to participate in verification, data can be directly aggregated. But these different schemes is characterized by a high and a complex operations to cipher texts and to aggregate it but still, sensor nodes are able to offer enough CPU for such functions.

## 8 Conclusion

The security feature based on homomorphic encryption is a new technique that enables providing results of operations carried out cipher texts without knowing the raw data, with respect of the data confidentiality. This paper presents the basic concept of the homomorphic encryption and the various encryption algorithms as per the properties of the encryption cryptosystems: Paillier can be used for preserving the additive property of homomorphic encryption while El Gamal and RSA can be used for multiplicative property. This paper can be useful for those who are wishing to carry out research in the direction of the cryptographic algorithms used for fully homomorphic encryption. It can be helpful to know which and how various cryptographic algorithms are being used for applying this kind of encryption for privacy preservation. There are various homomorphic encryption schemes described in this paper which may help to extend current research techniques.

## References


1. Patidar, S., Rane, D., Jain, P.: A Survey Paper on Cloud Computing. Adv. Comput. Commun. Technol. (ACCT) 394–398 (2012)
2. Mell, P., Grance, T.: The NIST Definition of Cloud Computing. NIST Special Publication, pp. 145–800 (2011)
3. Blumberg, A., Thaler, J., Vu, V., Walsh, M.: Veriable computation using multiple provers. ePrint Cryptology archive, pp. 1–37 (2014)
4. Cramer, R., Damgard, I.: Multiparty Computation, an Introduction. In: Lecture Notes, pp. 1–83 (2009)
5. Parmar, P., Padhar, S., Patel, S., Bhatt, N., Jhaveri, R.: Survey of various homomorphic encryption algorithms and schemes. Int. J. Comput. Appl. **91**(8), 26–32 (2014)
6. Yakoubov, S., Gadepally, V., Schear, N., Shen, E., Yerukhimovich, A.: A survey of cryptographic approaches to securing big-data analytics in the cloud. In: IEEE Conference on High Performance Extreme Computing (HPEC), pp. 1–6 (2014)
7. Padmavathi, B., Ranjitha Kumari, S.: A survey on performance analysis of DES, AES and RSA algorithm along with LSB substitution. Int. J. Sci. Res. **2**(4), 170–174 (2013)

8. Shetty, A., Shravya, K., Krithika, K.: A review on asymmetric cryptography RSA and ElGamal algorithm. Int. J. Innovative Res. Comput. Commun. Eng. **2**(5), 98–105 (2014)
9. Fouque, P., Karpman, P.: Security amplication against meet-in-the-middle attacks using whitening. In: 14th International Conference on Cryptography and Coding, pp. 1–18 (2013)
10. Paillier, P.: Public-key cryptosystems based on composite degree residuosity classes. In: Advances in Cryptology, vol. 1592, pp. 223–238 (1999)
11. Goldwasser, S., Micalli, S.: Probabilistic encryption. J. Comput. Syst. Sci. **28**(2), 270–299 (1984)
12. Fousse, L., Lafourcade, P., Alnuaimi, M.: Benaloh's dense probabilistic encryption revisited. In: Progress in Cryptology, vol. 6737, pp. 348–362 (2011)
13. Freeman, D.: Homomorphic Ecryption and the BGN Cryptosystem, pp. 1–7 (2011)
14. Silverman, J.: A survey of local and global pairings on elliptic curves and abelian varieties. In: International Conference on Pairing-Based Cryptography, vol. 6487, pp. 377–396 (2010)
15. Miller, C.: Decision problems for groups - survey and reflections. In: Algorithms and Classification in Combinatorial Group Theory, vol. 23, pp. 1–59 (1992)
16. Gentry, C.: A fully homomorphic encryption scheme. A Dissertation (2009)
17. Gentry, C.: Fully homomorphic encryption using ideal lattices. In: Proceedings of the forty-first annual ACM symposium on Theory of computing, pp. 169–178 (2009)
18. Dijk, M., Gentry, C., Halevi, S., Vaikuntanathan, V.: Fully homomorphic encryption over the integers. In: Advances in Cryptology, vol. 6110, pp. 24–43 (2010)
19. Coron, J., Mandal, A., Naccache, D., Tibouchi, M.: Fully homomorphic encryption over the integers with shorter public keys. In: Advances in Cryptology, vol. 6841, pp. 487–504 (2011)
20. Smart, N., Vercauteren, F.: Fully Homomorphic encryption with relatively small key and ciphertext sizes. In: Public Key Cryptography, vol. 6056, pp. 420–443 (2010)
21. Gentry, C., Halevi, S.: Implementing Gentry's fully-homomorphic encryption Scheme. In: Advances in Cryptology, vol. 6632, pp. 129–148 (2011)
22. Gentry, C., Brakerski, Z., Vaikuntanathan, V.: Fully homomorphic encryption without bootstrapping. In: Proceedings of the 3rd Innovations in Theoretical Computer Science Conference, pp. 309–325 (2012)
23. Moore, C., O'Neill, M., O'Sulivan, E., Doroz, Y., Sunar, B.: Practical homomorphic encryption: a survey. In: IEEE international Symposum, pp. 2792–2795 (2014)
24. Gentry, C., Halevi, S., Smart, N.: Homomorphic evaluation of the AES circuit. In: Advances in Cryptology, vol. 7417, pp. 850–867 (2012)
25. Huszri (Debrecen), A.: A homomorphic encryption-based secure electronic voting shceme. Publ. Math Debrecen **79**(3–4), 479–496 (2011)
26. Batarfi, O., Metro, A.: Protecting mobile agents against malicious hosts using dynamic programming homomorphic encryption. Int. J. Sci. Emerging Technol. **1**(1), 1–6 (2011)
27. Kreuter, B.: Private information retrieval based on fully homomorphic encryption, pp. 1–9 (2010)
28. Bahi, J., Guyeux, C., Makboul, A.: Secure data aggregation in wireless sensor networks, homomorphism versus watermarking approach. In: International Conference on Ad Hoe Networks, pp. 344-358 (2010)

# An Enhanced Cloud Based View Materialization Approach for Peer-to-Peer Architecture

**M.E. Megahed, Rasha M. Ismail, Nagwa L. Badr and Mohamed Fahmy Tolba**

**Abstract** Cloud computing is considered as a technology paradigm shift as it enables users to save both development and deployment time. It also reduces the operational costs of using and maintaining systems and applications by using only what you want. Moreover, it allows usage of any resources with elasticity instead of predicting workloads. There are many technologies that can be merged with Cloud computing to gain more benefits. One of these technologies is data warehousing, which can benefit from this trend when it's used to save large amounts of data with unpredictable sizes and if used in distributed environments. In this paper, a Cloud based view allocation algorithm is presented to enhance the performance of the data warehousing system over a Peer-to-Peer architecture. The proposed approach improves the allocation of the materialized views on cloud peers. It also reduces the cost of the dematerialization process and furthermore, the proposed algorithm saves the transfer cost by distributing the free space based on the required space to store the views and on the placement technique.

**Keywords** Data warehouse · Cloud computing · View materialization · Peer-to-peer cloud · Materialized views allocation · Peers allocation · Storage · Query processing

## 1 Introduction

Cloud computing is the term of using any computer resource (processing, storage or network bandwidth) as a service over the internet based on a virtualization concept and is a very successful service oriented computing paradigm [1–3]. The definition of the National Institute of Standards and Technology (NIST) for Cloud computing is, "a model for enabling ubiquitous, convenient, on-demand network access to a shared pool of configurable computing resources (e.g., networks, servers, storage,

M.E. Megahed (✉) · R.M. Ismail · N.L. Badr · M.F. Tolba
Faculty of Computer and Information Sciences, Ain Shams University, Cairo, Egypt
e-mail: mohammed_ezzat@live.com



A.E. Hassanien et al. (eds.), *Multimedia Forensics and Security*,
Intelligent Systems Reference Library 115, DOI 10.1007/978-3-319-44270-9_4

applications and services) that can be rapidly provisioned and released with minimal management effort or service provider interactions" [4]. The most used cloud computing types based on how the service is provided are: Infrastructure (IaaS), Platform (PaaS) and Software as a Service (SaaS) [5, 6]. Cloud computing has many architectures; one of these architectures is the Peer-to-Peer cloud architecture which is most suitable for distributed environments as it enables data travelling between the peers, data communication directly from each peer to any other peer in the architecture, it also allows data marts distribution and communication which allows query processing over all the peers [4, 7, 8].

View materialization is the process of computing and saving the views which contain the results of queries in the data warehouse in order to save processing costs. Every time the query is posed, the processing cost is saved, if the view is materialized, which means that the result of the query is saved, it will just need to be updated with the new data instead of computing the results from the tables [9, 10].

Data warehouse is considered as one of the solutions for storing large amounts of data to be used in decision making or supporting analysis and business managerial operations. A data warehouse can store a very large size and be geographically distributed leading to the occurrence of some problems in performance and storage. So, the large amount of data saved on a data warehouse can be distributed to data marts, which store smaller amounts of data and allow users to pose queries to these data marts directly instead of the data warehouse. This distribution may be based on the kind of data or time of data storage that is to be distributed to the workload, making the search easier, faster and more accurate, as it is a smaller amount of data and makes balance, saves time and avoids single points of failure [11–13].

The best solution to save a lot of time is to materialize all the views, which is not an applicable solution; it is not possible to materialize all the views, because saving these materialized views costs a lot of storage. So some views will be materialized and others not, but which set of views? These should be the most important ones which will save more time when being materialized. One of the selection criteria aspects is the frequency of usage; the views which contain the results of the highest number of queries.

There are two ways of view selection based on frequency: Static and dynamic. Static view selection is a traditional way to select views to be materialized based on prediction before the query answering process, but it has two main weaknesses. First, the workload of the queries is not easily predicted, especially in some kind of data warehouses. The second problem is the periodical changes of the workload [14], data and queries [15–17]. The dynamic view selection is based on data stored and the usage patterns, but it periodically changes over time based on the data usage patterns during running. Thus, the dynamic view selection will select the views based on the frequencies, their changes and also the changes on the relations and could be considered better than the static view selection, especially in our case (Cloud computing environment) [18].

In a peer-to-peer cloud view, the materialization process has two challenges. The first challenge is the evaluation of the views and selecting some of them to be materialized, because all views cannot be materialized as it will be space wasting.

Secondly, as we are working in a peer-to-peer cloud [7, 19], is to decide where the materialized views should be allocated on the peers of the cloud to get the most enhanced performance, considering the transfer cost between the peers and the processing cost of the queries [15].

This work presents a view materialization selection technique and a cloud dynamic view allocation algorithm to select the views to be materialized and allocates the materialized views at the suitable peers, in order to reduce the storage costs. (Utilizing the elasticity of the cloud [20]), transfer and query processing costs including the dematerialization of the already materialized views.

## 2 Related Work

Many researches applied an approach to materialize views in peer-to-peer data warehousing systems [12, 15, 21] and others working in cloud workload distribution and storage over peer-to-peer architectures, like; [22–26].

The authors of [15] provided two placement policies: The Isolated Policy: This policy places each materialized view in the peer with the highest frequency of queries with answers based on that view. In the case of no storage available, to save the materialized view, the replacement policy makes dematerialization, which frees up enough space to store the new materialized views. However this policy ignores the peer-to-peer architecture benefit of direct connection between the peers and also ignores the cloud computing advantage of elasticity and usage utilization. The isolated policy also wastes time in the materialization process, when there is no space, it will remove materialized views.

The second policy is the Voluntary policy: This policy is based on placing the materialized view in the peer that has enough space to store the materialized view and then selects the peer which has the highest frequency of queries with answers based on this view. If there is no peer having enough free space to store the materialized view, then some materialized views will be dematerialized to free up enough space to store the new materialized view which is more beneficial. However the voluntary policy also wastes time in the materialization process when there is no space, as it will remove materialized views and also wastes network transfer time when locating the materialized view in the peer which has enough space regardless of the frequency of queries.

The work of [27] proposed the DynaMat system that creates dynamic materialization. It aims to unify the problems of view selection and maintenance by taking into consideration maintenance restrictions such as; view updating and available space. This system is working in 2 modes, the first of which is the online mode, at this mode the system will try to answer the queries and at same time saves the new patterns. The second mode is the offline mode; the system in this mode will update and maintain the views. The main approach of the system is to monitor the incoming queries to get the frequencies and materialize the views of the more frequent queries. In the DynaMat system, there is no dynamic detection of the redundancies (common views between queries); this process is done by the data warehouse administrator and the authors did not provide an allocation technique for the views after materialization, they also did not apply their work over a peer-to-peer architecture or cloud computing and the proposed system was compared against the static queries and static view selection system.

Furthermore, others proposed a cloud view distributed data processing system which allows users to define the required views to be maintained [28]. The authors proposed a system called cloud view, which the users can define the orientation of their applications based on their XML based scheme descriptions called RQL. The system generates the required jobs and distributes them to a set of nodes to be executed. The queries from the clients are served and executed by using the materialized and time versioned views in less time than the traditional queries. However, the authors did not provide a data storage technique or allocation approach and there were no scheduling techniques for the queries or data sets, they also did not work on view materialization dynamically, they just introduced a static materialization process which depends on the users experience, therefore the system performance may be affected negatively in the case of any new or unexpected data patterns that are added.

However, in [21] the authors presented a PeerOLAP system which supports the Online Analytical Processing queries based on a Peer-to-Peer or client server network, each end client contains a cache with the most asked queries results. The proposed system is based on 2 main sub-systems; a chunk based system providing chunk management techniques and a caching management system which manages the cached data at each peer to find if it contains the answer of the query or not. If a query cannot find the result on its peer, it will distribute through the network until finding the result at the peer which contains it. The query answer may be located in more than one peer and this answer will be constructed from these peers. This research presents an approach of OLAP but it works in a traditional data warehouse distributed on traditional peer-to-peer network distributions without any central processing or management and also without query answering at the peer side. Finally the proposed PeerOlap does not provide any information about the peers to each other.

The proposed work of [29] is called a Relational cloud and utilizes the cloud scalability and provides database privacy over the cloud, the authors also mention that the queries are posted over encrypted data. They also provide a resources placement technique to allocate the workload and assign the required resources. The authors do not provide any enhancements.

In [10] the authors presented a view selection algorithm called the View Recommendation Greedy algorithm (VRGA) which is based on the traditional greedy algorithm but with recommended views. The algorithm work is based on two phases; the recommendation phase, then the greedy selection phase. This work is applied over a traditional data warehouse.

The work of [30] presented a PeerDW system, which applies the data warehouse upon peer-to-peer architectures, it utilizes the benefits this architecture and overcomes some of the distributed systems problems. The authors also presented a stable hybrid peer-to-peer network structure for the PeerDW to adapt the data warehouse and the peer-to-peer nature. The dynamic caching peer materialized views (PMV) algorithm is also presented as materialized views distributed and cached dynamically on the peer side but not on the data warehouse side, in order to reduce the query response time and increase the system throughput but it was not on a cloud environment so the authors did not provide any utilization to the cloud features.

The work of [31] presented an analysis study for scalable multi-tenant database management systems and they provided their architecture which utilizes cloud features like scalability and elasticity, the proposed system was applied over a client server cloud architecture and it provided dynamic expanding, dynamic partition over nodes and/or sharing nodes and the system also considered the workload balance between the clients. Their proposed approach did not provide database query answering and they did not consider the time factor.

In the work of [32, 33] they presented an ASM based approach to data warehouses dynamic design and OLAP systems with a foundational approach to data intensive software services. They proposed the abstract state services or (AS2) and they applied their proposed system on a database and then on a data warehouse. Technically, data warehouse and OLAP features can be considered as an instantiation of an abstract state service. They explained this view and provided examples to clarify their idea.

The work of [34] proposed a cloud based database operation support system. Their system basically is about storage and is called: The "Elastic Cloud data storage system" or ES2. The proposed system was designed to support both OLTP and OLAP in workload storage from different sources, data access and query processing. They also provided an indexing mechanism and their system was integrated with an epic cloud system. However the authors did not provide any work regarding allocation or view materialization.

In [35, 36] the authors presented frameworks to support indexing for DBMS in the peer-to-peer cloud. They support multiple types of indexes (specially the trees) and the proposed work enhanced the scalability by supporting the indexing operation and also the maintenance costs were reduced. The work was applied over a database, they did not use view materialization and also they did not provide any placement approach.

In the work of [15, 21, 30] they did not work in cloud computing, they just concentrated on data warehousing techniques, the work of [27] was in view materialization but not on the cloud, also in the work of [32, 33] their work was in the core of the data warehousing not the materialization. Therefore our dynamic allocation methodology addresses a different algorithm, as it considers view materialization based on cloud computing and not a traditional network. Besides, it utilizes the cloud features to improve the placement of materialized views.

Based on this research, survey's in the mentioned topics and researches produced the following issues. First, no previous research works handled a data warehouse with its data marts distributed over a peer-to-peer cloud architecture.

Secondly, some of the previous work proposed some views on materialization selection techniques but not one of these works was considering or utilizing any of the cloud computing features like; elasticity which is very effective in this process.

Thirdly, there were no research works on the allocation process for materialized views based on a peer-to-peer cloud computing architecture, especially when these peers store data marts with materialized views to be ready to answer the queries.

## 3 Solutions and Recommendations

Our proposed methodology proposes a new architecture called Dynamic View Allocation Architecture as shown in Fig. 1, to improve the placement of the materialized views. Moreover, our placement policy keeps the materialized views and reduces the cost of the dematerialization process (dematerialization process wastes processing costs as the views may need to be materialized again with a new cost) more than any other policies. It also enhances the views selection to be materialized and these processes save transfer costs. The architecture consists of the following modules:

- **Module One: The Materialization module**:

The main challenge of view materialization is based on a peer-to-peer cloud and it is to select the best combination set of views to be materialized and to consider query performance over the cloud, the overall processing time and the maintenance cost of views. Thus, each view should be evaluated before the selection process; this evaluation is based on the calculation of the saved cost if this view is materialized. The cost saved $CS_{(V_J)}$ could be calculated as follows:

$$CS_{(V_J)} = \sum f_{(q_j)} * Ct_{(V_j)} \quad (1)$$

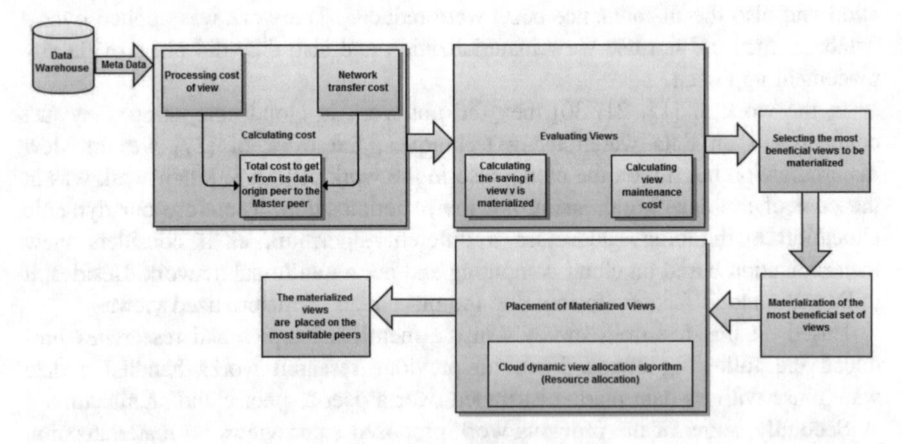


**Fig. 1** The proposed Dynamic View Allocation Architect

where $f_{(q_j)}$ is the frequency of Query q and $Ct_{(V_j)}$ is the total cost of materializing view V. The total cost $Ct_{(V_j)}$ is calculated as follows:

$$Ct_{(V_j)} = Ce(vj) + PeerCost(pi \rightarrow cp) * Size(vj) \tag{2}$$

where Ce(vj) is the processing cost of view v and PeerCost(pi → cp) the transfer cost from peer pi to the cloud peer. The view maintenance cost of the view could be calculated using the following equation:

$$CM_{(V_j)} = \sum F_{Br_{Pi}} * Cm_{(V_j)} \tag{3}$$

where $F_{Br_{Pi}}$ is the relation update frequency of the view relations (tables) and $Cm_{(V_j)}$ is the maintenance cost of each update (i). The views will be ranked based on their goodness,

$$Goodness(Vj) = CS_{(V_j)} - CM_{(V_j)} \tag{4}$$

Then the system can make combinations of these views and calculate the benefits of these combinations. Finally, it materializes the set of views which achieve the most benefits of the materialization process.

The second challenge is the view placement on the peers, by considering the utilization of the materialized views, the query processing and transfer costs between the peers.

- **Module Two: The Placement Module**:

In this module, a candidate peer is selected to store the materialized view. This selection is done according to the predefined equations. The placement phase is based on utilizing some of the cloud computing features; the elasticity feature and resources distribution over the peers. Utilizing these features in this module leads to avoiding waste in already materialized views and reduces the transfer costs between the peers in the cloud.

The policy is based on placing the new materialized view on the peer which has the highest frequency of accessed queries which require this materialized view to be answered. This reduces both the transfer and query processing costs.

- **Module Three: The Resource Allocation Module**:

Resource allocation is one of the main challenges within the cloud environment. If there is not enough space to store the view at the peer, the free space of the other peers can be used. The materialized view will not be transferred to these peers but a storage re-allocation process will be done (by doing resource allocation) and the required free spaces are located in the peer which has the highest use of this view, utilizing the elasticity feature of the cloud. So, the peer can have enough free space to store the view to be materialized and placed on it. Therefore, the network transfer costs will be saved.

```
Calculate the total cost of each view at the cloud
Begin                For i=1 to P
                         For j= 1 to Pi
                             Ct(vj)=Ce(vj) + PeerCost(pi → cp) * Size (vj)
                             Next j
                             End For
     Next i
     End For;   End;
The Evaluation phase: Calculate both cost saved of each view and the total maintenance cost
     Begin
        For i=1  to P
                         For j= 1 to Pi

                             CM(vj)= ∑  F_BrPi * Cm(vj)
                             CS(vj)= ∑  f(qj) * Ct(vj)
                             Next j
                            End For ;
                            Next i
        End For;      End;
Determine the final candidate views to be materialized
     Begin
                    For each View V ∈ EvArrOfViews
                            Goodness(Vj) = CS(vj) − CM(vj)
                    End For
For i=1 to Cnt Do
                            CVi=Max of ∑ Goodness (vj) in AllCombArray[i]
                            Next i
                                   For each CVi  in CV
                                          CV_Final[i]= Min Number of Views in CV[i]
                            Next
                                   End For;        End For ;       End;
View placement phase
       Begin
      For each v_k ∈ {vc_final } Do;
                                   ViewPeerNo(Vk)={Ø}; X=0;
                                   For i 1 To P;
                                   CntQuery(Vk) = 0;
                                   For j=1 To P(i);
                                   CntQuery(Vk) = CntQuery(Vk)+ Count(Qj for Vk);
                                   Next J;
   End For;
                                    If X > CntQuery(Vk);
                                    ViewPeerNo = [i];
                                    Next i;
                                    End if;
                                   Return ViewPeerNo(Vk);
                                   Append Both Vk and ViewPeerNo(Vk)in MaxPeerView Array;
Resource allocation phase
For Each View (Vk) in MaxPeerView;
                    Check the Free Space [Pi,FreeSpace(Pi)],
                            If FreeSpace in Pi > Size (Vk);
                            Locate Vk At Pi;
         Else Enter Cloud De-fragmentation Phase Defragment The Cloud Adding extra space for Pi
End For;     End;
```

**Fig. 2** Cloud based dynamic view allocation algorithm

The main contribution of the proposed dynamic view allocation approach is to allocate the materialized views to suitable peers in order to reduce:

1. Storage costs. (utilizing the elasticity feature which is one of the cloud computing main features)
2. Transfer costs between the peers of the cloud architecture.
3. Query processing costs.
4. Save time and resources used to materialize the views by decreasing the dematerialization process of the already materialized views.

The algorithm is based on our proposed dynamic view allocation architecture as shown in the following pseudo code Fig. 2.

# 4 The Experimental Results

A simulated cloud environment has been developed with the help of a VMW workstation [37] to test the proposed cloud dynamic view allocation algorithm. It simulates the cloud peers who contain the data marts and the cloud super peer that contains the data warehouse and where the system is run.

The cloud simulation contains 3 peers and the cloud super peer. TPC-H database [38] is used as a dataset (scale factor 1) and 6 sets of queries have been run. These queries come from each one that contains a number of queries (20, 100, 200, 500, 1000 and 2000). These sets of queries are posed to the cloud on each peer randomly.

In order to evaluate the performance of our policy, the proposed cloud dynamic view allocation algorithm is compared to the two algorithms of [15]; the isolated policy and the voluntary policy. The saving transfer cost, processing time and space usage are calculated and compared for the three polices through three simulation experiments. The algorithms are tested using the work of [39].

**Experiment 1** Measures the transfer cost. Table 1 and Fig. 3 show the comparison between the cloud view allocation algorithm, the isolated policy and the voluntary policy against the transfer cost based on four sets of queries (20, 100, 200 and 500) and it shows that the cloud view allocation algorithm saves the transfer cost, especially in the two cases of 200 and 500 query sets.

Figure 4 shows the transfer cost savings of the three policies for all 6 queries sets (20, 100, 200, 500, 1000 and 2000 queries). It shows that our policy saved the transfer cost by about 79.3 %, while the voluntary policy saved it by about 63.2 %

**Table 1** Comparisons between the three policies against the transfer cost (time in milliseconds)

| | Isolated | Voluntary | Cloud view allocation |
|---|---|---|---|
| 20 queries | 411.86 | 1626.02 | 24.7 |
| 100 queries | 4888.74 | 3760.16 | 6040.18 |
| 200 queries | 16852.34 | 24977.13 | 9133.03 |
| 500 queries | 44432.36 | 44228.71 | 27488.94 |

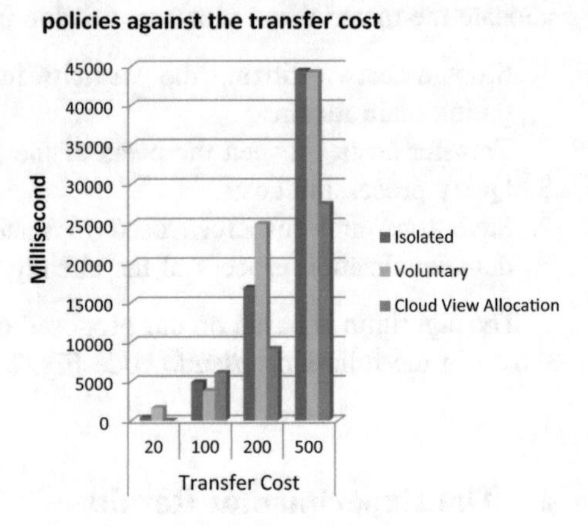


**Fig. 3** Comparisons between the three policies against the transfer cost

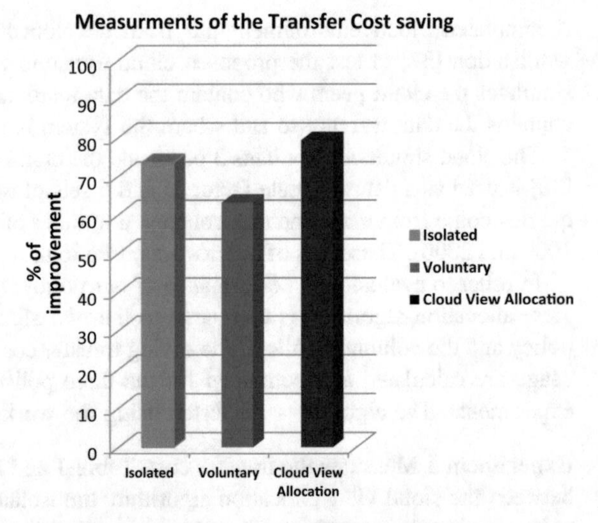


**Fig. 4** Measurements of the saved transfer cost

and the isolated policy saved it by about 73.6 %. This result is based on our cloud dynamic view allocation algorithm that saves transfer costs by allocating the views in the peer with the highest frequency of queries. In the case of no free space, additional space is added to the peer from the other peers on the cloud by utilizing the elasticity of the cloud.

**Experiment 2** Measures the processing cost. Table 2 and Fig. 5 show the comparison between the Cloud view allocation algorithm, the isolated policy and the voluntary policy against the processing cost, based on four sets of queries (20, 100, 200 and 500) and it shows that the cloud view allocation algorithm saves more processing time than the two other policies.

Figure 6 shows the processing cost savings of the three policies for all the 6 query sets (20, 100, 200, 500, 1000 and 2000 queries). It shows that our dynamic

Table 2 Comparisons between the three policies against the processing cost (time in milliseconds)

| | Isolated | Voluntary | Cloud view allocation |
|---|---|---|---|
| 20 queries | 650 | 946 | 320 |
| 100 queries | 3993 | 1533 | 1757 |
| 200 queries | 9896 | 7828 | 6911 |
| 500 queries | 16580 | 13301 | 12807 |

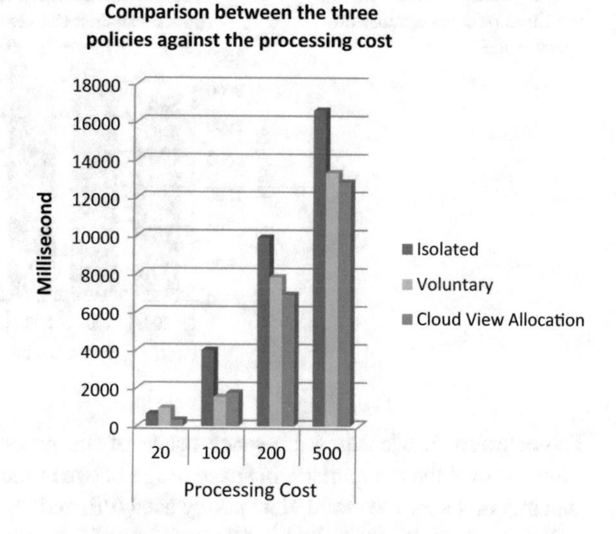


Fig. 5 Comparisons between the three policies against the processing cost

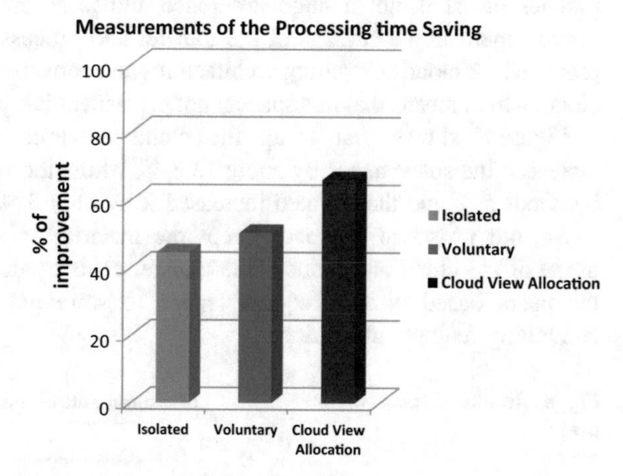


Fig. 6 Measurements of the saved processing time

allocation policy enhanced the processing cost by about 66 %, while the voluntary policy enhanced it by about 50.2 % and the isolated was enhanced by about 44.4 %.

This result was expected because the processing cost is based on the transfer cost between the peers to get the data from the relations (the tables which the views are based on) and the transfer cost of the views themselves.

**Table 3** Comparisons between the three policies against the space usage (space in kilobytes)

| | Isolated | Voluntary | Cloud view allocation |
|---|---|---|---|
| 20 queries | 2940 | 2790 | 2590 |
| 100 queries | 2930 | 2890 | 3330 |
| 200 queries | 2840 | 2930 | 3280 |
| 500 queries | 2730 | 2930 | 2990 |

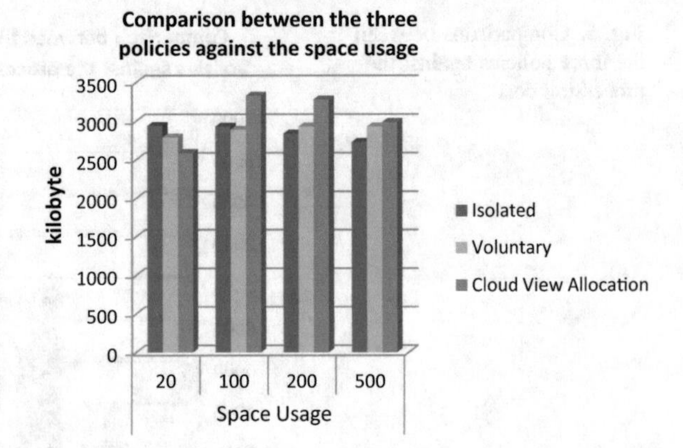


**Fig. 7** Comparisons between the three policies against the space usage

**Experiment 3**: Measures the space usage of the peers of the cloud. Table 3 and Fig. 7 shows the comparison in space usage between the three policies and it shows that in 3 of 4 sets, the cloud view policy used (utilized) space more than the other two policies but it is not a huge difference and is accepted when compared to the improvement in the savings of the transfer and processing costs. This system was proposed for cloud computing architectures and considers the elasticity feature of the cloud, which means that this space it not permanent it is just used when it is required.

Figure 8 shows that using the cloud dynamic view allocation algorithm increased the space usage by about 13.8 %, while, the voluntary policy increased it by about 5 % and the isolated increased it by only 3 %.

So, our proposed approach keeps the materialized views, increases the space usage of the cloud and reduces the transfer costs by distributing the free space on the peers based on the required space to store the views. This means cloud computing features utilization.

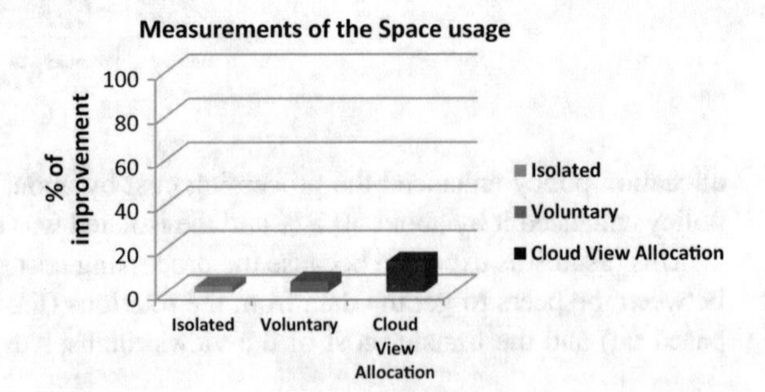


**Fig. 8** Results of space usage

## 5 Future Research Directions

According to the wide use of information technology, database management systems and other types of information systems, there are huge amounts of data generated and used daily. Big data is the new term of storage, access and analysis of these huge terabytes of data, in order to be used in business aspects, predictions and any other data analysis issues. We think that data warehouse and no sql databases [40] which contain big data still have a lot of benefits. especially when deployed on the cloud computing environment [41], so we propose working in the same cloud architecture, peer-to-peer architecture with data warehouse or no sql databases and OLAP technology by converting the data to cubes and applying the proposed approach of cloud peer-to-peer allocation on cubes to be allocated at the peers instead of the materialized views and we may use techniques like MapReduce in query answering which is used to search in big data. The future proposed work will be based on applying our new research over big data formed as cubes and allocated at the cloud peers using our cloud allocation approach.

Our proposed work is supposed to make enhancements on cloud resource utilization; storage and processing resources. The proposed work will also save time in query processing and answering as well as the huge storage and processing resources needed and as we will work on a peer-to-peer architecture the workload balance between the peers will be considered [42].

## 6 Conclusion

View materialization in a peer-to-peer cloud has some challenges such as; the evaluation of the views that will be selected to be materialized. Moreover, it decides where the materialized views should be allocated on the peers of the cloud to get the most enhanced performance, considering the transfer cost between the peers and the processing cost of the queries. Our proposed methodology overcomes the previously mentioned challenges.

In this paper, the Cloud dynamic view allocation algorithm is introduced to improve the data warehouse system over peer-to-peer cloud computing architectures. The main idea of the proposed approach is based on the view materialization of the most accessed queries on its peer by utilizing cloud computing features, especially the elasticity. The performance of the data warehouse system is based on the query processing time as well as the transfer time between the peers and the space utilized of the views. Our proposed methodology improves them by about 66 %, 79 % and 14 % respectively.

# Appendix

The full Pseudo Code of the proposed algorithm:

```
Calculate the total cost of each view at the cloud
Define P int "Number of Peers in the Cloud";
Define PeerCost (pi→cp) Cost between any peer and
cloud Peer;
Define FBrpi The update frequency of base relation
Br at Peer p;
Define Q = {q1,q2,.....qT} query Workload , T is the
total number of queries;
Define fpi(qj)The access frequency of query j at
peer i;
define Cm(vi) The maintenance cost of query vi;
Define Cp(vi) The processing cost of view vi;
Define Ct(vi) The total cost of view vi;
Define Goodness(vi) "A value Reflects the
efficiency of materializing vi  ;
Define Vc Set of candidate Views;
Define Cs(vi) "The saving cost if View vi is
materialized";
Define CM(vi) The Maintenance cost of view VI
Define I,J int;
Define Size[vi , Size] " Array of views and their
sizes";
Define TotalCostArray [vi , Ct(vi)]"Array of views
and their total cost";
       Begin
   Begin
       For i=1 to P
                  For j= 1 to Pi
                   Ct(Vj)=Ce(vj) + PeerCost(pi → cp) * Size (vj)
                   Next j
                   End For
// Append vj & ct(vj) in TotalCostArray
       Next i
      End For
End
The evaluation phase: calculate both cost saved of
each view and the total maintenance cost
Define EvArrOfViews [V,CSVJ,CMVj,Goodness] array of
the Views and their Saving cost , maintenance cost
and
goodness : initially empty;
```

```
Define AllCombOfView: A set of all combinations of
views to be materialized.
Define CvFinal a set of the final candidate view
that will be materialized.
Define i int; Define j int
   Begin
       For i=1  to P
                  For j= 1 to Pi
```

$CM_{(V_j)} = \sum F_{Br_{Pi}} * Cm_{(V_j)}$

// Append $CM_{(V_j)}$ to EvArrOfViews

$CS_{(V_j)} = \sum f_{(q_j)} * Ct_{(V_j)}$

// Append $CM_{(V_j)}$ to EvArrOfViews

```
                   Next j
                  End For
       Next i
       End For
   End
```

**Determine the final candidate views to be materialized**

```
Define EvArrOfViews [V,CSVJ,CMVj,Goodness] array of
the Views and their Saving cost , maintenance cost
and goodness : initially
Define AllCombArray: Array of sets of all
combinations of views to be materialized: initially
empty;
Define Cnt int : total number of the available
combinations of views in the AllCombArray array
Define CvFinal a set of the final candidate view
that will be materialized;
       Begin
            For each View V ∈ EvArrOfViews
```

Goodness(Vj) = $CS_{(V_j)} - CM_{(V_j)}$

//AppendGoodness$_{(V_j)}$ to the EvArrOfViews

```
            End For
               //Create all combinations of views
            to be materialized
               // Append each of the combinations
    to the AllCombArray
              For i=1 to Cnt Do
                  CVi=Max of ∑ Goodness(vj) in
            AllCombArray[i]
                  //append CVi to the CV
```

```
                    Next i
                          For each CVi  in CV
                                CVFinal[i]= Min
                          Number of Views in CV[i]
                                //Append CVi to
                          CVFinal
                                            Next
                          End For
                End For
                       End
```

**View placement phase**

```
Define CvFinal [{V1 ,Size1},{V2,Size2},......} array of
the final candidate views that will be materialized
and its sizes;
Define Qpi={q1,q2,......} a set of quires at peer (i);
Define [Pi,FreeSpace(Pi)] array of Peers in the
cloud and the free space for each;
Define CntQuery(Vj) int number of quires that
access view (j);
Define MaxPeerView[{Vj,Pi}] Array of peers id in
which view(j) is mostly accessed by query initially
empty;
                Begin
                       For each vk ∈ {vcfinal} Do;
                        ViewPeerNo(Vk)={Ø}; X=0;
                        For i 1 To P;
                        CntQuery(Vk) = 0;
                        For j=1 To P(i);
                        CntQuery(Vk) = CntQuery(Vk)+
Count(Qj for Vk);
                        Next J;
                            End For;
                         If X > CntQuery(Vk);
                         ViewPeerNo = [i];
                         Next i;
                         End if;
                        Return ViewPeerNo(Vk);
                        Append Both Vk and
ViewPeerNo(Vk)in MaxPeerView Array;
```

**Resource allocation phase**

```
           For Each View (Vk) in MaxPeerView;
```

```
    Check the Free Space
    [Pi,FreeSpace(Pi)],
          If FreeSpace in Pi > Size (Vk);
          Locate Vk At Pi;
          Else //Enter Cloud De-
          fragmentation Phase Defragment
          The Cloud Adding extra space for
          Pi
   End For;
```

## References


1. Abadi, D.J.: Data management in the cloud: limitations and opportunities. IEEE Data Eng. Bull. **32**(1), 3–12 (2009)
2. Panzieri, F., Babaoglu, O., Ferretti, S., Ghini, V., Marzolla, M.: Distributed computing in the 21st century: some aspects of cloud computing, pp. 393–412. Springer, Berlin Heidelberg (2011)
3. Agrawal, D., Das, S., El Abbadi, A.: Big data and cloud computing: new wine or just new bottles? Proc. VLDB Endow. **3**(1–2), 1647–1648 (2010)
4. Mell, P., Grance, T.: The NIST definition of cloud computing (Draft)—Recommendations of the National Institute of Standards and Technology. Gaithersburg (MD), Jan 2011
5. Buyya, R., Yeo, C.S., Venugopal, S., Broberg, J., Brandic, I.: Cloud computing and emerging IT platforms: Vision, hype, and reality for delivering computing as the 5th utility. Future Gener. Comput. Syst. **25**(6), 599–616 (2009)
6. Rimal, B.P., Choi, E., Lumb, I.: A taxonomy and survey of cloud computing systems. In: Fifth International Joint Conference on INC, IMS and IDC, 2009. NCM'09, pp. 44–51. IEEE (2009)
7. Babaoglu, O., Marzolla, M., Tamburini, M.: Design and implementation of a P2P Cloud system. In: Proceedings of the 27th Annual ACM Symposium on Applied Computing, pp. 412–417, Mar 2012. ACM (2012)
8. Fox, A., Griffith, R., Joseph, A., Katz, R., Konwinski, A., Lee, G., Stoica, I., et al.: Above the clouds: a Berkeley view of cloud computing. Dept. Electrical Eng. and Computer. Sciences, University of California, Berkeley (2009). Rep. UCB/EECS, 28, 13
9. Nguyen, T.V.A., Bimonte, S., d'Orazio, L., Darmont, J.: Cost models for view materialization in the cloud. In: Proceedings of the 2012 Joint EDBT/ICDT Workshops, Mar 2012, pp. 47–54. ACM (2012)
10. Kumar, T.V., Haider, M., Kumar, S.: A view recommendation greedy algorithm for materialized views selection. In: Information Intelligence, Systems, Technology and Management, pp. 61–70. Springer Berlin Heidelberg (2011)
11. Ismail, R.M.: Maintenance of materialized views over peer-to-peer data warehouse architecture. In: 2011 International Conference on Computer Engineering & Systems (ICCES), Nov 2011, pp. 312–318. IEEE (2011)
12. Ismail, R.M., Karam, O. H., El-Sharkawy, M., Wahaab, M.S.: Streaming real time data in data warehouses for just-in-time views maintenance. In: IADIS AC, vol. 1, pp. 101–108 (2009)
13. Zhao, J., Schewe, K.D., Koehler, H.: Dynamic data warehouse design with abstract state machines. J. UCS **15**(1), 355–397 (2009)
14. Quiroz, A., Kim, H., Parashar, M., Gnanasambandam, N., Sharma, N.: Towards autonomic workload provisioning for enterprise grids and clouds. In: 2009 10th IEEE/ACM International Conference on Grid Computing, Oct 2009, pp. 50–57. IEEE (2009)

15. Bellahsene, Z., Cart, M., Kadi, N.: A cooperative approach to view selection and placement in P2P systems. In: On the Move to Meaningful Internet Systems: OTM 2010, pp. 515–522. Springer, Berlin, Heidelberg (2010)
16. Krompass, S., Kuno, H., Dayal, U., Kemper, A.: Dynamic workload management for very large data warehouses: Juggling feathers and bowling balls. In: Proceedings of the 33rd International Conference on Very large data bases, pp. 1105–1115. VLDB Endowment, Sept 2007
17. Chen, G., Wu, Y., Liu, J., Yang, G., Zheng, W.: Optimization of sub-query processing in distributed data integration systems. J. Netw. Comput. Appl. **34**(4), 1035–1042 (2011)
18. Xiao, L.Q.Z.: Research survey of cloud computing. Comput. Sci. **4**, 008 (2011)
19. Ranjan, R., Harwood, A., Buyya, R.: Peer-to-peer-based resource discovery in global grids: a tutorial. Commun. Surv. Tutor. IEEE **10**(2), 6–33 (2008)
20. Agrawal, D., El Abbadi, A., Das, S., Elmore, A.J.: Database scalability, elasticity, and autonomy in the cloud. In: Database Systems for Advanced Applications, pp. 2–15. Springer, Berlin, Heidelberg (2011)
21. Kalnis, P., Ng, W.S., Ooi, B.C., Papadias, D., Tan, K.L.: An adaptive peer-to-peer network for distributed caching of OLAP results. In: Proceedings of the 2002 ACM SIGMOD International conference on Management of data, Jun 2002, pp. 25–36. ACM (2002)
22. Cunsolo, V.D., Distefano, S., Puliafito, A., Scarpa, M.: Volunteer computing and desktop cloud: The cloud@ home paradigm. In: Eighth IEEE International Symposium on Network Computing and Applications, NCA, Jul 2009, pp. 134–139. IEEE (2009)
23. Aversa, R., Avvenuti, M., Cuomo, A., Di Martino, B., Di Modica, G., Distefano, S., Villano, U.: The Cloud@ Home project: towards a new enhanced computing paradigm. In: Euro-Par 2010 Parallel Processing Workshops, Jan 2011, pp. 555–562. Springer, Berlin, Heidelberg (2011)
24. Ramachandran, K., Lutfiyya, H., Perry, M.: Decentralized approach to resource availability prediction using group availability in a P2P desktop grid. Future Gener. Comput. Syst. **28**(6), 854–860 (2012)
25. Tung, T.S., Richter, O.E., Savjani, V.: U.S. Patent Application 12/642,656 (2009)
26. Lakshman, A., Malik, P.: Cassandra: structured storage system on a p 2p network. In: Proceedings of the 28th ACM symposium on Principles of distributed computing, Aug 2009, p. 5. ACM (2009)
27. Kotidis, Y., Roussopoulos, N.: DynaMat: a dynamic view management system for data warehouses. In: ACM SIGMOD Record, vol. 28, no. 2, pp. 371–382. ACM (2009)
28. Zhou, D., Zhong, L., Wo, T., Kang, J.: CloudView: describe and maintain resource view in cloud. In: 2010 IEEE Second International Conference on Cloud Computing Technology and Science (CloudCom), Nov 2010, pp. 151–158. IEEE (2010)
29. Curino, C., Jones, E.P., Popa, R.A., Malviya, N., Wu, E., Madden, S., Zeldovich, N., et al.: Relational cloud: a database-as-a-service for the cloud. In: 5th Biennial Conference on Innovative Data Systems Research, CIDR, Jan 2011
30. Ismail, R.M., Karam, O.H., El-Sharkawy, M.A., Abdel-Wahab, M.S.: An adaptive peer-to-peer network for distributed data warehouse views. Int. J. Intell. Comput. Inf. Sci. (IJICIS) **8**(9) (2009)
31. Elmore, A. J., Das, S., Agrawal, D., Abbadi, A.E.: Towards an elastic and autonomic multitenant database. In: Proceedings of NetDB Workshop (2011)
32. Ma, H., Schewe, K.D., Thalheim, B., Wang, Q.: Cloud warehousing. J. Universal Comput. Sci. **16**(8), 1183–1201 (2010)
33. Mian, R., Martin, P., Vazquez-Poletti, J.L.: Provisioning data analytic workloads in a cloud. Future Gener. Comput. Syst. **29**(6), 1452–1458 (2012)
34. Cao, Y., Chen, C., Guo, F., Jiang, D., Lin, Y., Ooi, B. C., Xu, Q., et al.: Es 2: a cloud data storage system for supporting both OLTP and OLAP. In: 2011 IEEE 27th International Conference on Data Engineering (ICDE), Apr 2011, pp. 291–302. IEEE (2011)
35. Chen, G., Vo, H.T., Wu, S., Ooi, B.C., Özsu, M.T.: A framework for supporting dbms-like indexes in the cloud. Proc. VLDB Endow. **4**(11), 702–713 (2011)

36. Papadopoulos, A., Katsaros, D.: A-tree: distributed indexing of multidimensional data for cloud computing environments. In: 2011 IEEE Third International Conference on Cloud Computing Technology and Science (CloudCom), Nov 2011, pp. 407–414. IEEE (2011)
37. VMware, http://www.vmware.com/
38. Transaction processing and database benchmark, http://www.tpc.org/tpch/
39. Boghdady, P. N., Badr, N. L., Hashim, M.A., Tolba, M.F.: An enhanced test case generation technique based on activity diagrams. In: 2011 International Conference on Computer Engineering & Systems (ICCES), Nov 2011, pp. 289–294. IEEE (2011)
40. Han, J., Haihong, E., Le, G., Du, J.: Survey on NoSQL database. In: 2011 6th international conference on Pervasive computing and applications (ICPCA), Oct 2011, pp. 363–366. IEEE (2011)
41. Agrawal, D., Das, S., El Abbadi, A.: Big data and cloud computing: current state and future opportunities. In: 14th International Conference on Extending Database Technology, Mar 2011, pp. 530–533. ACM (2011)
42. Rao, A., Lakshminarayanan, K., Surana, S., Karp, R., Stoica, I.: Load balancing in structured P2P systems. In: Peer-to-Peer Systems II, pp. 68–79. Springer, Berlin, Heidelberg (2003)

# Distributed Database System (DSS) Design Over a Cloud Environment

**Ahmed E. Abdel Raouf, Nagwa L. Badr and Mohamed Fahmy Tolba**

**Abstract** An efficient way to improve the performance of database systems is the distributed processing. Therefore, the functionality of any distributed database system is highly dependent on its proper design in terms of adopted fragmentation, allocation, and replication methods. As a result, fragmentation including its allocation and replication is considered as a key research area in the distributed environment. Cloud computing is an emerging distributed environment that uses central remote servers and the internet to maintain data and applications. This research presents an enhanced dynamic distributed database system over a cloud environment. The proposed system allows fragmentation, allocation and replication decisions to be taken dynamically at run time. It also allows users to access the distributed database from anywhere. Moreover, this research presents an enhanced allocation and replication technique that can be applied at the initial stage of the distributed database design when no information about the query execution is available. It also presents different clustering techniques and their advantages and disadvantages.

## 1 Introduction

Distributed database systems typically consist of a number of distinct database fragments located at different geographic sites, which can communicate through a network and are managed by a distributed database management system (DDBMS) [15].

A.E. Abdel Raouf (✉) · N.L. Badr · M.F. Tolba
Faculty of Computer and Information Sciences, Ain Shams University, Cairo, Egypt
e-mail: ahmed_ezzat991@yahoo.com

N.L. Badr
e-mail: dr.nagwabadr@gmail.com

M.F. Tolba
e-mail: fahmytolba@gmail.com

 
A.E. Hassanien et al. (eds.), *Multimedia Forensics and Security*,
Intelligent Systems Reference Library 115, DOI 10.1007/978-3-319-44270-9_5

An efficient support is needed to databases that consist of very large amounts of data which are used by applications at different physical locations. Telecom databases, scientific databases, and large distributed enterprise databases are examples of application areas [18]. The main problem of many of these applications is the delay of accessing remote databases. As a result, it is necessary to use a distributed database that employs fragmentation, allocation and replication [18].

Fragmentation is the process of dividing a single database into two or more pieces; known as database fragment, the combinations of the pieces yielded from the original database are without any loss of information [15].

The process of placing data fragments at the sites are in order to, minimize the overall data transmission cost required to answer the query known as the allocation process [15].

To enhance the system performance and increase the availability, copies of the fragments are allocated to other locations within the distributed database. This process is known as the replication process. Too many copies of a fragment will enhance the performance of read only queries and increase the availability; however, it will also slow down updates. While too fewer copies of a fragment will decrease the performance of read only queries and the availability.

The design of a distributed database is one of the major research issues within the distributed database system area. The main challenges facing the DDBS design are: How to fragment database tables and which type of fragmentation will be used, when to replicate fragments, what is the optimal number of replications that can be taken for each fragment to enhance the system performance and increase availability, how to allocate fragments to sites where they are frequently accessed, do we group distributed database sites into disjoint clusters and which type of clustering will be used. These issues were previously solved either by static and dynamic solutions or based on a priori query analysis.

Fragmentation, replication and allocation are considered the most important design issues that lead to optimal solutions particularly in a dynamic distributed environment. They also have a great impact on the Distributed Database Systems (DDBS) performance. In distributed databases, the communication costs can be reduced by partitioning database tables into fragments. The fragments are then allocated to the sites where they are most frequently accessed, aiming at maximizing the number of local accesses compared to accesses from remote sites. The cost of the read operation can be further reduced by the replication of fragments when beneficial. Fragmentation, allocation and replication will be referred to as FAR in the rest of the proposed research.

Many applications of DDBS generate very dynamic workloads with frequent changes in access patterns from different sites. Consequently, static/manual FAR may not always be optimal. As a result, FAR should be automatic and completely dynamic. Any change in access patterns should result in re-fragmentation of existing fragments or tables and reallocation of fragments to different sites, as well as creation or removal of fragment replicas [18].

This research presents an enhanced dynamic DDBS over the cloud environment that allows dynamic FAR decisions to be taken dynamically over a clustered distributed database sites. Dynamic FAR decisions are based on the access pattern and the load of the sites after allocation or migration of fragments or a replica to it. Moreover, this research presents an optimal allocation and replication technique, which can be applied at the initial stage of the distributed database design, when no information about the query execution is available. It also presents different clustering techniques and their advantages and disadvantages and which one we will use in our system.

The rest of this paper is organized as follows; Sect. 2 reviews the related works of dynamic fragmentation, allocation and replication. The research issues are presented in Sect. 3. The solutions and recommendations are given in Sect. 4 and its subsections. Section 5 presents the enhanced allocation and replication technique. The experimental results are given in Sect. 6. Section 7 presents the previous works done in clustering distributed database sites. Finally, the future research directions and the conclusions are given in Sects. 8 and 9.

## 2 Background

The authors of [18] present a decentralized approach for dynamic table FAR in distributed database systems. It performs FAR based on recent access history, aiming at maximizing the number of local accesses compared to accesses from remote sites. In this approach FAR decisions are fully decentralized. Each site decides over its own fragments to split, migrate and/or replicate independently of other sites. The FAR decisions are made on the fly based on current operations and recent history of local reads and writes. This makes it possible to use this approach without communication overhead. This approach has two main components first, detecting replica access patterns. Second, based on these statistics to decide on re-fragmentation and reallocation. This approach ensures that decisions taken are not in conflict with each other by handling master replica and read replicas differently. It use cost functions to take decision by estimate the difference in future communication costs between a given replica change and keeping it as is. However, this algorithm doesn't consider site constraints, the load of the site after the allocation or migration of replica or fragments to it and the optimal number of replicas that can be taken for each fragment to enhance the system performance and increase availability.

The author of [3] presents a synchronized horizontal FAR model. It adopts a new approach to perform horizontal fragmentation of database relation based on attribute retrieval and update frequency. It proposes a new heuristic technique to satisfy horizontal fragmentation and allocation using a cost model to minimize the total cost of distribution. This technique performs the allocation process based on the fragment access pattern and the cost of moving the data fragments from one site to the other. It provides an optimal allocation that leads to best fragments distribution which avoids frequent remote accesses. In this technique first, performance and cost implications of different allocation choices are evaluated. Second, the replication decision is taken based on the computed threshold values for the average retrieval and update costs of all fragments individually. This technique use site constraints to improve the efficiency which has been confirmed by the produced empirical results.

The process of placing data fragments at the sites in order to minimize the overall data transmission costs that are required to answer the query is known as fragment allocation. The fragment allocation has two types: Non-redundant fragment allocation and redundant fragment allocation. In non-redundant fragment allocation, each fragment of each global relation is allocated to exactly one site. As a result, the optimum allocation of the fragments is the only way, which can be exploited to increase the performance, efficiency, reliability and availability of the distributed database. In redundant fragment allocation, the fragments of each global relation are allocated to one or more sites introducing replication of the fragments.

The work of [1] proposes an approach that contributes in determining best possible allocation of a data fragment in a distributed environment. This approach is based on the fragment access patterns and the cost of moving data fragments from one site to the other. Different SAGA methods for data allocation were employed. However, the implementation confirmed that SAGA 100 outperformed all other SAGA.

The authors of [30] give the brief overview of dynamic fragment allocation in non-replicated distributed database system algorithms. They also propose new algorithm called region based fragment allocation (RFA). The proposed algorithm considers the frequency of fragment accessed by region as well as individual nodes to move fragment from source node to target node. The RFA algorithm is designed to address the issues in existing approaches where fragments movement depends only on the frequency of access to the fragments. The RFA algorithm decreases the migration of fragments using knowledge of the network topology in comparison to optimal [14] and threshold [29] algorithms. In comparison to the BGBR algorithm [11], the RFA algorithm reduces the amount of topological data required in decision making. The proposed algorithm first, chooses the region that has high fragment access. Second, chooses node in that region that has high fragment access. Third, allocates the fragment to that node. However, this solution will not be the optimal solution in the case of a node in a region having a high fragment access while the other nodes in that region have low fragment access. As a result, the fragment will allocate to it, in this case, the problem is solved for only one node and is not solved for the other nodes in other regions.

The author of [2] presents a new data reallocation model for replicated and non-replicated constrained DDBSs. The proposed model takes site constraints into account in the process of reallocation. It reallocates data fragments across sites based on communication and update cost for each fragment individually. The reallocation process performed by selecting the site that has the highest query updates cost for fragment Fi to be chosen as the candidate site to store fragment Fi in order to, minimize communication cost. If the candidate site that chosen to store fragment violate site constraints then, the fragment will be migrate to the site with the next highest update cost value. Moreover, if more than one site has the same update cost for a certain fragment. In this case, this model will use fragment priority

(FP) procedure to allocate fragment to the site with the highest FP value in order to, avoid fragment duplication over sites. The main advantages of this model are that any change in the site queries and their frequency will have an effect on the reallocation process. In addition, this model guarantees that no fragment duplication at post allocation. However, this model will be more complicated when queries information continuously changes in faster way or when the number of fragments and sites largely increase.

The work of [16] proposes a new dynamic data allocation algorithm for non-replicated distributed database system. The proposed algorithm named Near Neighborhood Allocation (NNA). The NNA algorithm reallocates data with respect to the changing in data access pattern with time constraint. This algorithm is based on optimal algorithm, but with different strategy for selecting nodes for data movement. It moves data to a node which is the neighborhood and also placed in the path to the node with the maximum access counter. The experiments of this algorithm find that the NNA algorithm performs better for large fragment size and query production. However, the optimal algorithm performs better for small query production and fragment size. It also finds that the threshold for the fragment size is almost 8000 byte. The NNA can be used for larger networks to decrease the delay of response to a fragment regarding to optimal algorithm.

The work of [22] proposes a new integer programming formulations for the non-redundant version of the fragment allocation problem. This work assumes that the fragments already determined, and focuses on the problem of allocating them in such a way as to minimize the total cost resulting from transmissions generated by user queries.

The authors of [10] present and analyze a new approach for dynamic data allocation algorithm named Fuzzy Neighborhood Allocation (FNA) algorithm. This algorithm is based on NNA in [27]. It is different with NNA in selecting nodes for data movements and different strategy in migrating fragments. The proposed algorithm uses fuzzy method to prevent redundant data fragment migration and avoid oscillation condition. The experiments results indicated that, the proposed algorithm performs better for larger fragment size. However, the NNA and optimal algorithm performs better for small fragment size. The future of this work is to test FNA, NNA and optimal algorithms in replicated distributed database systems.

The author of [23] proposes a new dynamic fragment allocation algorithm in non-replicated allocation scenario. The proposed algorithm takes into account the time constraints of database accesses, volume threshold, and the volume of data transmitted in successive time intervals in order to, dynamically reallocate fragments to sites at runtime in accordance with the changing access patterns. The proposed algorithm migrates a fragment located at a certain site to another site, which not only makes number of accesses to that Fragment greater than the access threshold for reallocation in the specific period of time, but also results in transmission of maximum volume of data from or to that fragment in that specific period of time. The proposed algorithm improves the overall performance of the distributed database by imposing a more strict condition for fragment reallocation in distributed database. It results in fewer migrations of fragments from one site to other sites over the network. However, the volume threshold, the number of time intervals and duration are the most important factors that regulate the frequency of fragment reallocations.

The authors of [5] propose a new dynamic data allocation algorithm for non-replicated DDBS named Performance Optimality Enhancement Algorithm (POEA). This works explores and improves some concepts used in previously developed algorithms to reallocates fragments to different sites given the changing data access patterns, time, and sites constraints of the DDBS. When the migration decisions are made it adopts the shortest path between the old location and the new anticipated location for the transferred fragments. The POEA algorithm is the most efficient one among all previous algorithms for dynamic data allocation as, it has certainly improves the DDBS performance by further minimizing network traffic. It also reduces data transmission cost compared to the previous methods, since it adopts the shortest path algorithm once data movement decisions are taken. The only drawback of the POEA algorithm is that it requires more storage compared to some previous algorithms. However, this is compensated by DDBS performance enhancement and is considered as a very trivial drawback due to the dramatic fall down in the storage hardware prices.

The authors in [28] solve the fragment allocation problem by using the well-known Quadratic Assignment Problem solution algorithms. The authors of [21] propose a new technique for horizontal fragmentations of the relations of distributed databases. This technique can be applied at the initial stage as well as in later stages of DDBS for partitioning the relations. The authors of [20] address some important scalability issues. They provide some algorithms to ensure generality of the technique developed in [21].

A new vertical fragmentation, allocation and replication scheme of a distributed database called (VFAR) was proposed by our previous work in [7]. The proposed scheme partitions the distributed database relations vertically at the initial stage of the database design by using the enhanced minimum spanning tree (MST) Prim's algorithm. In addition, it allocates and replicates the resulted fragments to the sites that require it.

The authors of [4] propose a heuristic technique to satisfy horizontal fragmentations and allocations using a cost model to minimize the total cost of distribution. Furthermore, the authors of [9, 19] present a new framework for dynamic fragment allocation and replication. The authors of [19] consider replication and fragment correlation under a flexible network topology. This framework tackles multiple issues in this system, including lazy replication strategy, fragments' correlation, on the-fly fragment allocation in the face of changing query access patterns, and non-uniform distances between network sites. In this framework the correlation between fragments is modeled and an algorithm to find near optimal dynamic allocation is presented. It also provides a simple methodology to update the allocation when access patterns change. The experiments demonstrate that this algorithm provides efficient solutions for the fragment allocation problem in distributed database systems. The future of this work is firstly, leveraging fragment ordering to further improve the performance of the algorithms. Secondly, generalize reallocation by employing a mechanism that does data mining of the access patterns to detect and decide on a reallocation schedule. Finally, incorporate into the algorithms the ability to handle extra characteristics, such as bounds on the capacity of sites, constraints on the number of replicas for each fragment, and constraints on fragments that are not allowed to be replicated, e.g., due to access control or security constraints.

The author of [33] proposes two algorithms for dynamic data redistribution: Part of the redistribution (Partial Reallocate) and full redistribution (Full Reallocate) algorithm. The two algorithms are linear complexity and can be used in a variety of different sizes distributed database system. The authors of [25] propose a new dynamic data allocation algorithm for non-replicated distributed database system. This algorithm is called Threshold and Time Constraint Algorithm (TTCA), which is an extension of the optimal algorithm [12] and threshold algorithm [29]. The TTCA algorithm removes problems of threshold algorithm by adding time constraint to the existing threshold technique.

The authors of [8] introduced cluster based peer to peer architecture for distributed databases. The proposed architecture is named flexipeer. This work uses predicate based fragmentation of previous work done in [21]. It is used to address the fragmentation and allocation of database designs. The proposed work introduces a clustering approach for partitioning database sites. The clustering process is done based on the unique numbers of region sites. It also allocates fragments across the sites of each cluster. This paper tries to implement the concept of chord in peer to peer based data management. The sites of each cluster are managed by the local cluster administrator LCA and the whole architecture is managed by a global cluster administrator GCA. However, it doesn't address the replication phase of database design. It also wastes a lot of time between node, LCA, LCA Validator, Resource Checker and GCA until it reaches the required data.

The work of [17] presents a novel algorithm for grouping distributed database network sites into disjoint clusters based on communication time. The authors see that performing clustering algorithms after fragmentation will speed up the process of data allocation by eliminating extra communication costs between sites. The proposed algorithm creates disjoint clusters according to the least average communication cost between network sites. It also distributes the DDBS sites over the clusters. In addition, it generates near optimal numbers of clusters required to achieve high network system performance. The results obtained from the simulation demonstrated the significant network server's load balance and network delay. The experimental outcomes confirmed that this approach can be implemented in different DDBS environments even if the network sites are enormous.

Another method of clustering the sites in which low communication cost sites are grouped in one cluster was proposed in [15]. Furthermore it allows the fragmentation of structured data, fragmentation of unstructured data and it describes the allocation of fragments to the cluster of sites in order to reduce communication cost. However, this work doesn't mention which sites in the cluster will hold the fragment when the fragment is allocated to it.

The authors of [31] proposed a dynamic data replication strategy using historical access records and proactive deletions called Closest Access Greatest Weight with Proactive Deletion (CAGW_PD). The authors of [13, 24, 26, 32] believe that a replication method has three important issues to consider: When the new replica should be replicated, which file should be replicated, and where the new replica should be placed. However, the authors of [31] see that there is still one additional issue which should be resolved, i.e., how to control the number of replicas.

## 3 Research Issues

Based on the above survey, the following key findings are highlighted. First, no previous works handle dynamic fragmentation, allocation, and replication decisions of distributed database fragments and replica at cloud environment.

Second, no existing work takes the fragmentation, allocation, and replication decisions dynamically over a clustered distributed database sites. In spite of, performing clustering algorithm after fragmentation will speed up the process of data allocation by eliminate extra communication costs between sites.

Third, few existing works handle FAR decisions dynamically at run time. However, these works does not consider site constraints, load of the sites after allocate or migrate a fragment or replica to it, and the limit of the replica numbers into account in order to take the dynamic decisions.

## 4 Solutions and Recommendations

To overcome the limitations of existing literature highlighted by the above survey, this research proposes an enhanced DDBS design over a cloud environment. In our previous work in [6], a DDBS design over the cloud was introduced.

To extend this work, contributions in this proposed research includes adding new layers and modules to enhance the DDBS design and allows users to access a database from anywhere in the world without owning any technology infrastructure. It can be accessed through: A web browser, mobile application or desktop application while the database is stored on servers at a remote site. It also allows FAR decisions to be taken dynamically at run time. These decisions are based on access patterns and the load of sites after the allocation and migration of fragments as well as its replicas.

The proposed architecture is shown in Fig. 1. It consists of three layers: Application client layer, distributed database system manager layer and distributed database clusters layer.

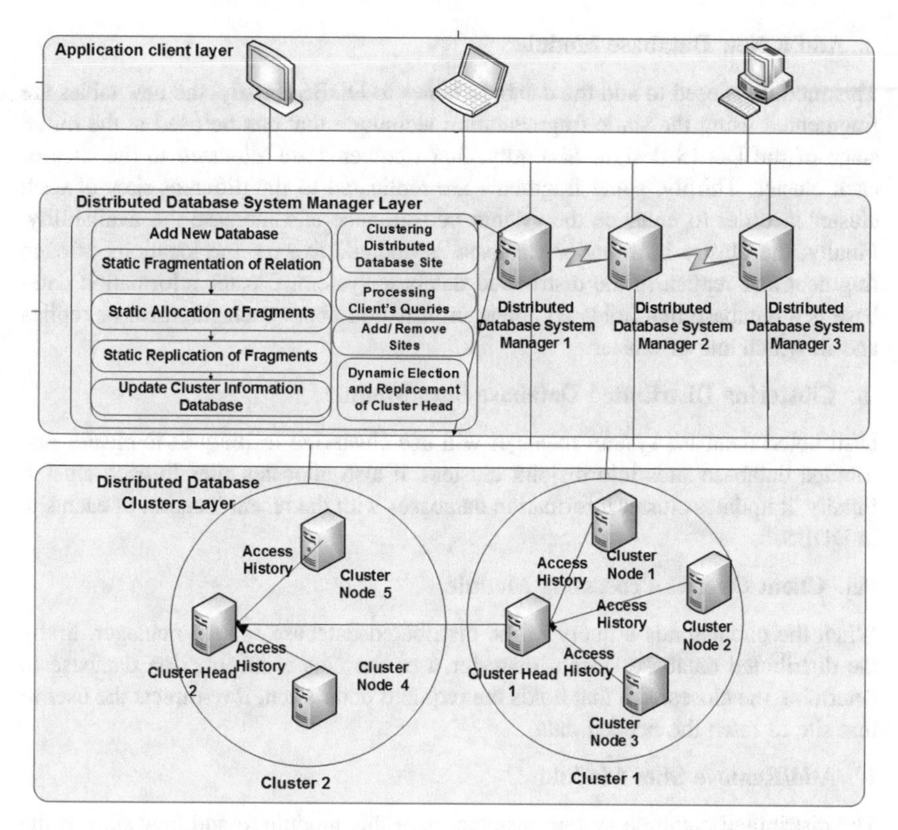


**Fig. 1** A dynamic distributed database system over a cloud environment architecture

## *4.1 Application Client Layer*

This layer consists of a set of interfaces which allow the users of distributed database systems to access the databases from anywhere. In order to join the distributed database system, first the client calls the distributed database system manager to get all the information about the structure and the location of the sites of the distributed system. Second, the client joins the distributed system by sending a query to the distributed database system manager.

## *4.2 Distributed Database System Manager Layer*

This layer is composed of five modules: Add a new database, clustering distributed database sites, client queries processing, add/remove sites and dynamic election and replacement of the cluster head.

i. **Add a New Database Module**

This module is used to add the database tables to DDBS. Firstly, the new tables are fragmented using the static fragmentation technique that can be used at the initial stage of the DDBS design. Secondly, that fragments are allocated to the sites of each cluster. Thirdly, some fragments are replicated to the different sites of each cluster in order to enhance the system performance and increase the availability. Finally, the cluster information database is updated to save the locations of each fragment and replica in the distributed database system. Cluster information database is a database that holds complete information about each fragment or replica and in which site or cluster.

ii. **Clustering Distributed Database Sites Module**

Distributed database system manager will use clustering techniques to cluster distributed database sites into disjoint clusters. It also allocates sites to each cluster. Finally, it updates cluster information databases with the recent location of each site in DDBS.

iii. **Client Queries Processing Module**

When the client sends a query to the distributed database system manager, firstly the distributed database system manager uses the cluster information database to determine the closest site that holds the required data. Then, it re-directs the user to that site to fetch the needed data.

iv. **Add/Remove Sites Module**

The distributed database system manager uses this module to add new sites to the distributed database system or remove existing sites from the distributed system. By using this module the distributed database system manager can expand or reduce the distributed database system without any affect to the client and other the sites of distributed database system.

v. **Dynamic Election and Replacement of the Cluster Head Module**

This module is used by the distributed database system manager to elect and replace the cluster head. A new cluster head is elected in two cases. In the first case the recent cluster head is removed. In the second case a new site is added to the cluster and its capacity is higher than the cluster head.

## *4.3 Distributed Database Clusters Layer*

The distributed database cluster layers consist of more than one cluster. Each cluster has one cluster head and more than one cluster node.

i. **Cluster Node**

Each cluster node contains two local databases. The first local database is used to store the fragments and replicas of the distributed system. The second local database is used to store the information about the user's access to the stored fragments and replicas. At each access firstly, the cluster node checks whether it is a local access or remote.

Secondly, it allows the user to access a local database of the node to run the query and fetch the required data.

Thirdly, the local database of user's access is updated to save the information about the user access. Forth, cluster node runs query composition to answer the query after it fetches all needed fragments.

Finally, it sends the site access record to the distributed database system manager and also sends the results of the query to user. Each cluster node sends the access history to the cluster head to take any suitable decisions such as: Create or delete replicas, re-fragmentation or re-allocation of the fragments.

ii. **Cluster Head**

The cluster head is a cluster node and has special and additional operations to manage the other cluster nodes. The cluster head performs three operations: Manage cluster nodes data, load balancing of the cluster, dynamic fragmentation, replication and allocation.

1. Managing Cluster Node Data

Each cluster node sends the access history to the cluster head to take the decision to either create or delete replica, re-fragmentation or re-allocation of the fragments. The cluster head collects the data that each node in its cluster holds. Afterwards, it sends it to the distributed database system manager to keep the cluster information database updated.

2. Load Balancing of the Cluster

Each cluster head contains load balancing algorithm that contains two services: Site capacity identifier and performance prediction services. The site capacity identifier service contains the capacity of each site in each cluster.

The capacity of the site contains two objects: The number of users that can access site at any time and the number of fragments and replicas that the site can hold.

The performance prediction service will be used to predict the load of the site after adding the fragments or replicas to it. The fragment or replica will be assigned to the site with less workload.

3. Dynamic Fragmentation, Replication and Allocation

The cluster head will take fragmentation, replication and allocation decisions based on access history and load of each site after the allocation or migration of the fragments and replicas of it. The fragments and replicas will send to the site with less workload.

# 5 Enhanced Allocation and Replication Technique

The authors of [21] proposed a new technique of horizontal fragmentation. This technique helps in taking the fragmentation decision at the initial stage of designing the distributed database. It uses knowledge gathered during the requirement analysis phase by using the enhanced CRUD (Create, Read, Update, and Delete) matrix without the help of empirical data about query executions. This technique performs data allocation according to the maximum attribute locality precedence (ALP) value and the location of sites. The attribute locality precedence (ALP) can be defined as the importance of an attribute with respect to each site.

However, the allocation strategy of this technique doesn't meet the goal of the data allocation. The goal of the data allocation can be achieved by allocating the fragments to the sites that require it only. As a result the user can access data with low costs and time.

In our previous work in [6], we enhanced the allocation strategy of this technique by performing data allocation according to the site that has the maximum ALP single value. That methodology guarantees that no fragment duplication will happen during the allocation process. In this case we used two sites that have the maximum ALP single value, the fragment will be allocated to the site that performs more data manipulations and less read operations. Consequently, the replica is sent to the site which performs less data manipulations and more read operations.

After the process of data allocation, we replicate the fragment to the site that performs more read operations than other sites. For example, if we have two sites: Site 1 and Site 2. Site 1 performs CUD operations. Site 2 performs R operations. The replica will be sent to Site 2 although Site 1 has the maximum ALP single value. If the replica is sent to Site 1, it will not be used because there is no data manipulation on the replica. The pseudo code for the enhanced allocation and replication technique is shown in Fig. 2.

# 6 Experimental Results

We have implemented our technique on an HP Compaq computer with Coretwo Duo 2.33 processors and 2 GB RAM using the SQL Server as DBMS. We have implemented the modified technique on the MCRUD matrix of the bank account table as shown in Table 1. We performed the enhanced technique on account relations shown in Table 2. The ALP table is generated after applying the modified technique on the MCRUD matrix. The ALP table is shown in Table 3. From the ALP table, the attribute that has a maximum ALP value is the branch name, so its predicates will be used to perform horizontal fragmentations. The resulted fragments are shown in Tables 4, 5 and 6 and are allocated to sites that already need it without taking the location of sites into account. Fragment1 has been allocated to Site 1 and fragment2 has been allocated to Site 2 because they have the maximum

```
Input:  Number of attributes, number of predicates of each attribute [], number of sites, number of
applications of each site [], and MCRUD matrix [total number of predicates, total number of applications]
Output:  The ALP table, the sites that contain maximum ALP single value for each predicate of each
attribute (Position of MAX [attribute, predicate]), and the sites that performs more read operations for
each predicate of each attribute (Position of next Max [attribute, predicate]).
Foreach attribute in Number of attributes do
    Foreach predicate in number of predicates [attribute] do
         Number of read operation of max = 0
         Number of read operation of next max = 0
         Application number = 0
         Total sum = 0
           Foreeach site in number of sites do
               Application sum = 0
               Number read operation = 0
                  Foreeach application in number of application [site] do
                    Application sum += Calculate MCRUD (MCRUD [Predicate number, application number])
                    Application number++     End
                  Total sum += application sum
                  If application sum == MAX [attribute, predicate] then
                       If Number read operation > Number of read operation of max then
                           Position of next Max [attribute, predicate] = site
                           Number of read operation of next max = Number read operation   End
                       Else if Number read operation < Number of read operation of max then
                            Position of next Max [attribute, predicate] =
                                Position of MAX [attribute, predicate]
                             Number of read operation of next max = Number of read operation of max
                            MAX [attribute, predicate] = application sum
                            Position of MAX [attribute, predicate] = site
                            Number of read operation of max = Number read operation  End End
                  If application sum > MAX [attribute, predicate] and application sum > 0 then
                     If Number of read operation of max > Number of read operation of next max then
                          Position of next Max [attribute, predicate] =
                                Position of MAX [attribute, predicate]
                          Number of read operation of next max = Number of read operation of max End
                       MAX [attribute, predicate] = application sum
                       Position of MAX [attribute, predicate] = site
                       Number of read operation of max = Number read operation   End
                  Else if application sum > 0 and Number read operation > 0 and
                         Number read operation > Number of read operation of next max then
                       Position of next Max [attribute, predicate] = site
                       Number of read operation of next max = Number read operation;    End End
                Predicate number++
                ALP Single [attribute, predicate] = MAX [attribute, predicate] -
                         (Total sum - MAX [attribute, predicate])         End
           Foreach predicate in number of predicates [attribute] do
                ALP FINAL [attribute] += ALP Single [attribute, predicate]  End End
```

**Fig. 2** Pseudo Code for the enhanced allocation and replication technique

**Table 1** MCRUD matrix

| | Site 1 | | | Site 2 | | | Site 3 | | |
|---|---|---|---|---|---|---|---|---|---|
| | AP1 | AP2 | AP3 | AP1 | AP2 | AP3 | AP1 | AP2 | AP3 |
| Account.Account id > 20 | C | | RU | | | | | | C |
| Account.Account id <=20 | | R | | | | | | | |
| Account. Type = ind | CRD | RU | RUD | | R | | | | |
| Account. Type = cor | | RU | R | | | | CRUD | RU | R |
| Account.customer id > 5 | C | | RU | | | | | | R |
| Account.customer id <= 5 | | R | | | | | | | |
| Account.open date > 1-1-2008 | CRD | RU | RU | | R | | | | |
| Account.open date <= 1-1-2008 | | RU | R | | | | CRUD | RU | R |
| Account. Balance < 10000 | R | | R | | | CRUD | | | R |
| Account. Balance >= 10000 | | CR | | | | | | | |
| Account.Branch Name = dhk | CRUD | RU | CRUD | | CUD | | R | | |
| Account.Branch Name = ctg | | R | | CRUD | CRUD | R | | R | |
| Account.Branch Name = khl | CRUD | CRU | U | | | | CRUD | CRU | CR |

**Table 2** Account relation

| Account no | Account type | Customer ID | Open date | Account balance | Account name |
|---|---|---|---|---|---|
| 3 | Ind | 1 | 20/01/2009 | 12500.0000 | Dhk |
| 4 | Cor | 2 | 20/05/2009 | 12000.0000 | Dhk |
| 7 | Ind | 2 | 05/03/2009 | 11000.0000 | Ctg |
| 15 | Ind | 3 | 08/05/2009 | 11000.0000 | Khl |
| 20 | Cor | 2 | 08/05/2010 | 15000.0000 | Ctg |
| 21 | Ind | 1 | 09/05/2012 | 9000.0000 | Khl |
| 22 | Cor | 8 | 20/09/2011 | 8000.0000 | Dhk |
| 23 | Ind | 5 | 06/08/2011 | 6000.0000 | Khl |
| 24 | Ind | 9 | 08/09/2006 | 15000.0000 | Khl |
| 28 | Cor | 5 | 07/05/2009 | 16000.0000 | ctg |

**Table 3** ALP table

| Attribute name | ALP value |
|---|---|
| Account id | 6 |
| Type | 22 |
| customer id | 6 |
| open date | 22 |
| Balance | 8 |
| Branch Name | 27 |

**Table 4** Fragment 1

| Account no | Account type | Customer ID | Open date | Account balance | Account name |
|---|---|---|---|---|---|
| 3 | Ind | 1 | 20/01/2009 | 12500.0000 | Dhk |
| 4 | Cor | 2 | 20/05/2009 | 12000.0000 | Dhk |
| 22 | Cor | 8 | 20/09/2011 | 8000.0000 | Dhk |

**Table 5** Fragment 2

| Account no | Account type | Customer ID | Open date | Account balance | Account name |
|---|---|---|---|---|---|
| 7 | Ind | 2 | 05/03/2009 | 11000.0000 | Ctg |
| 20 | Cor | 2 | 08/05/2010 | 15000.0000 | Ctg |
| 28 | Cor | 5 | 07/05/2009 | 16000.0000 | ctg |

**Table 6** Fragment 3

| Account no | Account type | Customer ID | Open date | Account balance | Account name |
|---|---|---|---|---|---|
| 15 | Ind | 3 | 08/05/2009 | 11000.0000 | Khl |
| 21 | Ind | 1 | 09/05/2012 | 9000.0000 | Khl |
| 23 | Ind | 5 | 06/08/2011 | 6000.0000 | Khl |
| 24 | Ind | 9 | 08/09/2006 | 15000.0000 | Khl |

ALP single value for the fragmentation attribute. However, the last predicate of the branch name attribute has two sites which have the same maximum value. In this case, the fragment will be allocated to the site that performs more data manipulation and less read operations. In our experiments, Site 1 performs two read operations and Site 3 performs three read operations. As a result, fragment 3 will be allocated to Site 1 and its replica will be sent to Site 3.

**Table 7** Final result of allocation and replication

| Fragment number | Allocated to | Replicated to |
|---|---|---|
| Fragment 1 | 1 | 3 |
| Fragment 2 | 2 | 1 |
| Fragment 3 | 1 | 3 |

After the allocation process, we replicate the fragments to enhance the system performance of reading only queries and increase the availability. The replicas will be allocated to the site that performs more read operations than other sites. In our scenario, replica1 of fragment1 is allocated to Site 3, replica 2 of fragment 2 is allocated to Site 1 and replica 3 of fragment 3 is allocated to Site 3. The final results of the allocation and the replication processes are shown in Table 7.

## 7 Clustering Distributed Database Sites

An efficient way to minimize the communication time required for query processing and data allocation is grouping distributed database sites into disjoint clusters [17]. As a result, clustering distributed database sites is an important issue in distributed database systems.

The previous works done in this area are divided into two parts. In the first part are clusters distributed database sites based on communication costs between the network sites. In the second part are cluster distributed database sites based on the region fields of the database.

### *7.1 Clustering Distributed Database Sites Based on Region Fields*

The simplest way to cluster the distributed database sites into disjoint clusters is by using the region field of the database [8]. The sites in the same region belong to the same cluster. However, clustering distributed database sites based on region has two disadvantages.

The first disadvantage, this methodology of clustering sites will not work on all the sites that belong to the same region or at most two regions. In this case we will have one cluster that contains all the sites in the distributed system.

The second disadvantage is that this methodology will not work in some cases belonging to different regions. In this case, each cluster will contain one site.

## 7.2 Clustering Distributed Database Sites Based on Communication Costs Between Network Sites

The second way to cluster distributed database sites is to use the communication cost between database sites [15, 17]. This work focuses on grouping distributed database sites into disjoint clusters according to the least average communication cost between network sites. The clustering process highly depends on the communication cost range of the CCR value.

The CCR parameter represents the communication cost value (ms/byte) that is allowed for the maximum difference between the sites to be grouped in the same cluster. If the communication cost between two sites is less than the CCR then the two sites are grouped on one cluster. The CCR value depends on how much time is allowed for the sites of the same cluster to receive or transmit their data.

The first advantage of methodology is that it can be implemented in different DDBS environments even if the network sites are enormous.

The second advantage is performing the clustering algorithm after fragmentation will speed up the process of data allocation by eliminate extra communication costs between sites [17].

Based on the advantages and disadvantages of the clustering techniques, we implemented the clustering technique mentioned in [17] that clusters the distributed database sites based on the communication costs.

The implemented method categorizes the distributed database sites according to Clustering Decision Value (CDV). The value of CDV is based on two values. The first value is communication cost range CCR value. The second value is the communication cost between the distributed database sites (CC). The communication cost between two sites ($S_i$, $S_j$) is defined as the following linear function:

$CC(S_i, S_j)$ = the cost of creating the data packet + the cost of transmitting the data packet from site $S_i$ to site $S_j$

The cluster decision value (CDV) is logical value. It determines whether a pair of sites can be grouped on one cluster or not. The value of CDV is determined by Eq. 1.

$$CDV(Si, Sj) = \begin{cases} 1 \text{ } if \text{ } CC(Si, Sj) \leq CCR \\ 0 \text{ } if \text{ } CC(Si, Sj) > CCR \end{cases} \quad (1)$$

In this proposed research we run the implemented algorithm on distributed database system that consists of 10 sites. The communication cost between the sites is shown on Table 8. We assumed that the communication cost range value (CCR) equal to 5. After running the implemented algorithm, the resulted clusters and the sites assigned to it are shown in Table 9. More details about the implemented algorithm in [17]. However, we will use our enhanced technique to allocate and replicate the fragments to the sites of each cluster.

**Table 8** Communication cost between sites

| Site# | Site 0 | Site 1 | Site 2 | Site 3 | Site 4 | Site 5 | Site 6 | Site 7 | Site 8 | Site 9 |
|---|---|---|---|---|---|---|---|---|---|---|
| Site 0 | 0 | 6 | 11 | 10 | 7 | 8 | 9 | 9 | 12 | 12 |
| Site 1 | 6 | 0 | 10 | 10 | 2 | 3 | 4 | 2 | 7 | 8 |
| Site 2 | 11 | 10 | 0 | 6 | 6 | 7 | 8 | 7 | 2 | 3 |
| Site 3 | 10 | 10 | 6 | 0 | 8 | 7 | 6 | 7 | 9 | 6 |
| Site 4 | 7 | 2 | 6 | 8 | 0 | 1 | 2 | 2 | 7 | 6 |
| Site 5 | 8 | 3 | 7 | 7 | 1 | 0 | 1 | 2 | 8 | 8 |
| Site 6 | 9 | 4 | 8 | 6 | 2 | 1 | 0 | 3 | 6 | 6 |
| Site 7 | 9 | 2 | 7 | 7 | 2 | 2 | 3 | 0 | 8 | 7 |
| Site 8 | 12 | 7 | 2 | 9 | 7 | 8 | 6 | 8 | 0 | 1 |
| Site 9 | 12 | 8 | 3 | 6 | 6 | 8 | 6 | 7 | 1 | 0 |

**Table 9** Resulted clusters and the assigned sites to it

| Cluster # | Site # | | | | |
|---|---|---|---|---|---|
| Cluster 0 | Site 0 | | | | |
| Cluster 1 | Site 3 | | | | |
| Cluster 2 | Site 1 | Site 4 | Site 5 | Site 6 | Site 7 |
| Cluster 3 | Site 2 | | Site 8 | Site 9 | |

## 8 Future Research Directions

As proposed future work, we plan to implement the remaining parts of the proposed architecture to efficiently allow FAR decisions to be taken automatically at run time. We plan to take the site constraints, load of the sites and the volume of data transmitted in a specific time interval from or to fragments into account to take the dynamic FAR decisions.

## 9 Conclusions

Efficient distribution of the distributed database fragments and replicas to various sites play a critical role in the function of the database in terms of performance and cost. In this proposed research, we present an enhanced dynamic DDBS over a cloud environment. The proposed system allows FAR decisions to be taken based on access history and the load of the site. Moreover, this research presents enhanced allocation and replication techniques, which allocate the fragments and replicas to the sites that already, requires it without taking into account the location of the sites. The enhanced technique aims at maximizing the number of local accesses compared to access from the remote sites, enhances the systems performance and increases the availability. This research also presents different clustering techniques that used to cluster distributed database sites into disjoint clusters. It also presents the clustering technique advantages, disadvantages, and which one we will use in our system.

## References


1. Abdalla, H.I.: An efficient approach for data placement in distributed systems. In: 2011 5th FTRA International Conference on Multimedia and Ubiquitous Engineering (MUE). IEEE (2011)
2. Abdalla, H.I.: A new data re-allocation model for distributed database systems. Int. J. Database Theory Appl. **5**(2), 45–60 (2012)
3. Abdalla, H.I.: A synchronized design technique for efficient data distribution. Comput. Hum. Behav. **30**, 427–435 (2014)
4. Abdalla, H.I., Amer, A.A.: Dynamic horizontal fragmentation, replication and allocation model in DDBSs. In: 2012 International Conference on Information Technology and e-Services (ICITeS). IEEE (2012)
5. Abdalla, H.I., Amer, A.A., Mathkour, H.: Performance optimality enhancement algorithm in DDBS (POEA). Comput. Hum. Behav. **30**, 419–426 (2014)
6. Abdel Raouf, A.E., Badr, N.L., Tolba, M.F.: Dynamic distributed database over cloud environment. In: International Conference of Advanced Machine Learning Technologies and Applications. Springer, Berlin Heidelberg (2014)
7. Abdel Raouf, A.E., Badr, N.L., Tolba, M.F.: An optimized scheme for vertical fragmentation, allocation and replication of a distributed database. In: 2015 IEEE Seventh International Conference on Intelligent Computing and Information Systems (ICICIS). IEEE (2015)
8. Amalarethinam, D.G., Balakrishnan, C.: A Study on Performance Evaluation of Peer-to-Peer Distributed Databases. IOSR J. Eng. **2**(5), 1168–1176 (2012)
9. Amalarethinam, D., Balakrishnan, C.: oDASuANCO-ant colony optimization based data allocation strategy in peer-to-peer distributed databases. Int. J. Enhanc. Res. Sci. Technol. Eng. **2**(3), 1–8 (2013)
10. Basseda, R., Rahgozar, M., Lucas, C.: Fuzzy neighborhood allocation (FNA): a fuzzy approach to improve near neighborhood allocation in DDB. In: Advances in Computer Science and Engineering. Springer, Berlin (2009)
11. Bayati, A., Ghodsnia, P., Rahgozar, M., Basseda, R.: A novel way of determining the optimal location of a fragment in a DDBS: BGBR. In: ICSNC'06 International Conference on Systems and Networks Communications, 2006. IEEE (2006)
12. Brunstrom, A., Leutenegger, S.T., Simha, R.: Experimental evaluation of dynamic data allocation strategies in a distributed database with changing workloads. In: Proceedings of the Fourth International Conference on Information and Knowledge Management. ACM (1995)
13. Chang, R.S., Chang, H.P.: A dynamic data replication strategy using access-weights in data grids. J. Supercomput. **45**(3), 277–295 (2008)
14. Corcoran, A.L., Hale, J.: A genetic algorithm for fragment allocation in a distributed database system. In: Proceedings of the 1994 ACM Symposium on Applied Computing. ACM (1994)
15. Suganya, A., Kalaiselvi, R.: Efficient fragmentation and allocation in distributed databases. Int. J. Eng. Res. Technol. **2**(1), 1–7 (2013)
16. Gope, D.C.: Dynamic data allocation methods in distributed database system. Am. Acad. Sch. Res. J. **4**(6), 1–8 (2012)
17. Hababeh, I.: Improving network systems performance by clustering distributed database sites. J. Supercomput. **59**(1), 249–267 (2012)

18. Hauglid, J.O., Ryeng, N.H., Nørvåg, K.: DYFRAM: dynamic fragmentation and replica management in distributed database systems. Distrib. Parallel Databases **28**(2–3), 157–185 (2010)
19. Kamali, S., Ghodsnia, P., Daudjee, K.: Dynamic data allocation with replication in distributed systems. In: 2011 IEEE 30th International Performance Computing and Communications Conference (IPCCC). IEEE (2011)
20. Khan, S.I., Hoque, A.L.: Scalability and performance analysis of CRUD matrix based fragmentation technique for distributed database. In: 2012 15th International Conference on Computer and Information Technology (ICCIT). IEEE (2012)
21. Khan, S.I., Hoque, A.S.M.L.: A new technique for database fragmentation in distributed systems. Int. J. Comput. Appl. **5**(9), 20–24 (2010)
22. Menon, S.: Allocating fragments in distributed databases. IEEE Trans. Parallel Distrib. Syst. **16**(7), 577–585 (2005)
23. Mukherjee, N.: Synthesis of non-replicated dynamic fragment allocation algorithm in distributed database systems. ACEEE Int. J. Inf. Technol. **10**, 154–159 (2011)
24. Ranganathan, K., Foster, I.: Identifying dynamic replication strategies for a high-performance data grid. In: Grid Computing—GRID 2001. Springer, Berlin (2001)
25. Singh, A., Kahlon, K.S.: Non-replicated dynamic data allocation in distributed database systems. IJCSNS Int. J. Comput. Sci. Netw. Secur. **9**(9), 176–180 (2009)
26. Tang, M., Lee, B.S., Yeo, C.K., Tang, X.: Dynamic replication algorithms for the multi-tier data grid. Futur. Gener. Comput. Syst. **21**(5), 775–790 (2005)
27. Basseda, R., Tasharofi, S., Rahgozar, M.: Near neighborhood allocation (NNA): a novel dynamic data allocation algorithm in DDB. In: 11 International CSI Computer Conference (CSICC'2006). School of Computer Science, IPM (2006)
28. Tosun, U., Dokeroglu, T., Cosar, A.: Heuristic algorithms for fragment allocation in a distributed database system. In: Computer and Information Sciences III. Springer, London (2013)
29. Ulus, T., Uysal, M.: Heuristic approach to dynamic data allocation in distributed database systems. Pak. J. Inf. Technol. **2**(3), 231–239 (2003)
30. Varghese, P.P., Gulyani, T.: Region based fragment allocation in non-replicated distributed database system. Int. J. Adv. Comput. Theory Eng. **1**(1), 62–70(2012)
31. Wang, Z., Li, T., Xiong, N., Pan, Y.: A novel dynamic network data replication scheme based on historical access record and proactive deletion. J. Supercomput. **62**(1), 227–250 (2012)
32. Zhang, J., Lee, B.S., Tang, X., Yeo, C.K.: A model to predict the optimal performance of the hierarchical data grid. Futur. Gener. Comput. Syst. **26**(1), 1–11 (2010)
33. Zhenglong, L.: Study of dynamic data redistribution algorithm based on distributed database system. Procedia Engineering **15**, 5611–5615 (2011)

## Additional Reading

34. Hababeh, I.O., Ramachandran, M., Bowring, N.: A high-performance computing method for data allocation in distributed database systems. J. Supercomput. **39**(1), 3–18 (2007)
35. Hababeh, I., Khalil, I., Khreishah, A.: Designing high performance web-based computing services to promote telemedicine database management system. IEEE Trans. Serv. Comput. **8** (1), 47–64 (2015)
36. Kumar, R., Gupta, N.: An extended approach to Non-Replicated dynamic fragment allocation in distributed database systems. In: 2014 International Conference on Issues and Challenges in Intelligent Computing Techniques (ICICT). IEEE (2014)
37. Mukherjee, N.: Non-replicated dynamic fragment allocation in distributed database systems. In: Advances in Computer Science and Information Technology. Springer, Berlin (2011)

# A New Stemming Algorithm for Efficient Information Retrieval Systems and Web Search Engines

**Safaa I. Hajeer, Rasha M. Ismail, Nagwa L. Badr and Mohamed Fahmy Tolba**

**Abstract** Stemming algorithms (stemmers) are used to convert the words to their root form (stem); this process is used in the pre-processing stage of the Information Retrieval Systems. The Stemmers affect the indexing time by reducing the size of index file and improving the performance of the retrieval process. There are several stemming algorithms; the most widely used is the Porter Stemming Algorithm because of its efficiency, simplicity, speed and also its ease at handling exceptions. However there are some drawbacks, although many attempts were made to improve its structure but they were incomplete. This paper provides efficient information on the retrieval technique as well as proposes a new stemming algorithm called the Enhanced Porter's Stemming Algorithm (EPSA). The objective of this technique is to overcome the drawbacks of the Porter algorithm and improve web searching. The EPSA was applied to two datasets to measure its performance. The result shows improvement of precision over the original Porter algorithm while realizing approximately the same recall percentages.

**Keywords** Information retrieval (IR) · Indexing · Vector space model (VSM) · Ranking · Precision · Recall · Stemming (Stemmer) · Over-stemming · Under-stemming · Porter stemmer

## 1 Introduction

Today, information retrieval plays a much larger part of our lives especially with the advent of the Internet and the World Wide Web (the Web) in particular. The traditional information retrieval was a manual process, mostly happening in the form of book lists in libraries and in the books themselves as tables of contents and other indices …etc. these lists/tables usually contain a small number of index terms (e.g. title, author and perhaps a few subject headings) due to the tedious work of

S.I. Hajeer (✉) · R.M. Ismail · N.L. Badr · M.F. Tolba
Ain Shams University, Cairo, Egypt
e-mail: safaahajeer@yahoo.com



A.E. Hassanien et al. (eds.), *Multimedia Forensics and Security*,
Intelligent Systems Reference Library 115, DOI 10.1007/978-3-319-44270-9_6

manually building and maintaining these indices. This was normally true up until the last 10 years [1], the amount of information available in electronic form whether on documents in computers or on the Web has grown exponentially. Almost any kind of desired information is available, including: News and message files, software libraries, multimedia repositories, commercial information, Bibliographic collections, encyclopedias …etc. The digital computer fundamentally changed the way that people were able to store, search and retrieve textual information, as a result searching becomes a normality of life. Millions of users all around the world interact with search engines daily. Recently, due to the growing number of internet users around the globe, information retrieval (IR) has become of great importance as an essential tool for all tasks of searching on the web, in order to service the recursive increases of the number the Internet users. Therefore it is very essential to retrieve accurate data for some user queries. There are lots of approaches used to increase the effectiveness of online data retrieval. The traditional approach used to retrieve data for a user query i is to search the documents present in the Corpus word by word for the given query. This approach is very time consuming and it may miss some of the related documents of equal importance. Thus to avoid these situations, the researcher might work well to find techniques, these techniques can be used in various Information Retrieval Systems in order to increase the retrieval accuracy [2]. As Information Retrieval Systems play critical roles to obtain relevant information resources based on indexing techniques technology [3, 4].

The basic idea of information retrieval systems is to find a degree of similarity between the documents collection and the query after eliminating Stop Words. Stop words are useless in the documents collection, it occurs in 80 % of the document [4–6]. Stemming is an aspect supported by the indexing and searching technique. It's one of the pre-processing stages used to combat the vocabulary mismatch problem, in which query words don't match exactly document words. The stemming process will reduce the size of the documents representation by 20–50 % compared to full word representations [2], according to Van Rijsbergen [7]. Furthermore, the relevancy of the retrieved documents will be improved and their number will also be increased [4, 7].

A stemming algorithm or stemmer, has three main purposes. The first one consists of clustering words according to their topic. Many words are derivations from the same stem and are considered to belong to the same concept (e.g., drive, driven, driver). The second purpose of a stemmer is directly related to the information retrieval process, as having the stems of the words allows many improvements. Furthermore it is much easier to index the documents according to their topics, as their terms are grouped by stems (that are similar to concepts) which is the main purpose of using stemmer in this research or the expansion of a query to obtain more precise results. Finally, the conflation of the words sharing the same stem leads to a reduction of the dictionary and is taken into account in the process. This process of a reduction is effecting on the space needed to store the structures used by an information retrieval system (like the index of terms-documents), as well as lightening the computational load of the system [8].

Various stemming algorithms for European languages have been proposed [9–11]. The designs of these stemmers range from the simplest technique such as, removing suffixes by using a list of frequent suffixes, to a more complicated design which uses the morphological structure of the words in the inference process to derive a stem.

Stemmers may be Statistical, Linguistic/Truncating or mixed. Statistical stemmers rely on statistical procedures such as; frequency count, n-gram method or some combination of these. Linguistic stemmers use a linguist's knowledge of the structure of the language in one way or another, typically by providing manually compiled lists of suffixes, allomorphic rules and the like. The best known stemmer is Porter [12]. The idea of those stemmers are based on removing the suffixes or prefixes (commonly known as affixes) of a word. The most basic stemmer is the Truncate (n) stemmer [13] which truncates a word at the nth symbol i.e. keep n letters and remove the rest. In this method words shorter than n are kept as they are. Another simple approach is the S-stemmer [13] which is an algorithm conflating singular and plural forms of English nouns. This algorithm was proposed by Donna Harman. The algorithm has rules to remove suffixes in plurals so as to convert them into singular forms. Linguistic stemmers that rely on statistical methods as subsidiary procedures may be called mixed. Mixed stemmers [13] involve both the inflectional as well as the derivational morphology analysis. In case of inflectional the words variants are related to the language specific syntactic variations like plural, gender, case, etc. whereas in derivational the word variants are related to the Part-Of-Speech (POS) of a sentence where the word occurs [13–15].

Porter stemming [16], is one stemming algorithm which is used in Vector Space Search Engines to reduce the term space while maintaining the semantic content of term space. The Vector space model (or term vector model) is an algebraic model for representing text documents (and any objects, in general) as vectors of identifiers such as, index terms. A document is represented as a vector when each dimension corresponds to a separate term. It is used in information filtering, information retrieval, indexing and relevancy rankings.

Another field of information retrieval systems where stemming is used widely is text categorization [17] [18]. Text is the task of automatically sorting a set of documents into categories from a predefined set. Porter Stemming is used to stem words and use them as single stemmed tokens to categorize text. It is also used extensively in text clustering [19, 20]. Variants of the algorithm have also been used for stemming words in other languages such as Dutch, French, Indonesian, Indian, German etc. [21, 22] an is also used in various search engines such as in the Song Search engine [16].

The world today is in demand of high speed performance search engines. However the stemming can cause errors in the form of words; there are mainly two types of errors, over-stemming and under-stemming. These errors decrease the effectiveness of stemming algorithms.

- **Over-stemming** is when two words with different stems are stemmed to the same root. This is also known as a **false positive**.
  For example, past and paste → past (same root)
- **Under-stemming** is when two words that should be stemmed to the same root are not and is also known as a **false negative**.
  For example: Adhere → adher, Adhesion → Adhes

The challenge is to reduce these errors as much as possible, because of that many researches tried to improve the structure of the Porter algorithm; however it still has several drawbacks. This paper presents a new stemming algorithm (EPSA) to reduce stemming errors and improve the information retrieval systems effectiveness.

The rest of the chapter is organized as follows. The next section presents research background and then the Porter's algorithm is described. After it, the chapter discusses The System Methodology. The chapter presents the evaluation of ESPA and its experimental results. Finally, the conclusion of the paper appears in the last section.

## 2 Background

Lovins in [23, 24] described the first stemmer in 1968, which was developed specifically for the information-retrieval applications. The stemmer performs a lookup on a table of 294 endings, 29 conditions and 35 transformation rules. The Lovins stemmer removes the longest suffix from a word. Once the ending is removed, the word is recoded using a different table that makes various adjustments to convert these stems into valid words. It always removes a maximum of one suffix from a word. The result shows that this algorithm is very fast but it has some drawbacks, like many suffixes are not available in the table of endings. It is sometimes highly unreliable and frequently fails to form words from the stems [13].

The Paice/Husk stemmer in [25], is an iterative algorithm with one table containing about 120 rules indexed by the last letter of a suffix; on each iteration, it tries to find an applicable rule by the last character of the word. If there is no such rule it terminates. It also terminates if a word starts with a vowel and there are only two letters left or if a word starts with a consonant and there are only three characters left. Otherwise, the rule is applied and the process repeats [15]. The disadvantage is it may be a very heavy algorithm and over stemming may occur.

Dawson Stemmer is an extension of the Lovins approach except that it covers a much more comprehensive list of around 1200 suffixes. The suffixes are stored in the reversed order, indexed by their length and last letter. The rules define if a suffix is found it can be removed. The disadvantage is that it is very complex and lacks a standard re-usable implementation [25, 26].

Krovetz [27] presented the linguistic Krovetz stemmer. With the intention to decrease stemming errors and increase the stemmer's accuracy, Krovetz added a dictionary to validate the correctness of stems. The Krovetz stemmer performs a dictionary lookup after the removal of suffixes in a specific order. First, plural forms are converted to singular forms. Next, past tense words are converted to present. Finally, the "ing" suffix is removed. Additionally, the Krovetz stemmer attempts to increase accuracy and robustness by treating spelling errors and meaningless stems. Spelling errors and meaningless stems are transformed into the closest word. Although this weak stemmer produces high accurate results and a huge reduction in stemming errors, the addition of the dictionary lookup increases the complexity of the stemmer.

Hull et al. [8, 28–30] evaluated five stemmer algorithms for the English s-remover, an extensity modified Lovins stemmer, Porter stemmer, Xerox English inflectional analyzer and Xerox English derivational analyzer. He proposed a set of alternative evaluation measures aimed to distinguishing the performance details of various stemmers. This study proves that stemming is much more helpful on short queries, on which the inflectional stemmer looks slightly less effective and the Porter stemmer better than the others, the simple plural removal is less effective than more complex stemmers but quite competitive when only a small number of documents is examined.

Porter's stemming algorithm is one of the most popular stemming methods proposed in 1980. It is based on the idea of defining five or six successively applied steps of word transformation. Each step consists of a set of rules that is applied until one of them passes the conditions. If a rule is accepted, the suffix is removed accordingly and the next step is performed. The resultant stem at the end of the sixth step is returned [13, 31].

Porter's stemmer produces a less error rate than the Lovins stemmer. However, Lovins's stemmer is a heavier stemmer that produces a better data reduction [31, 32]. The Lovins algorithm is noticeably bigger than the Porter algorithm, because of its very extensive endings list [13]. Many modifications and enhancements have been done and suggested on the basis of the Porter stemmer algorithm in order to reduce its errors.

STANS algorithm is a modification of the porter's stemming algorithm. In this algorithm a total of 31 modifications are done among those 15 rules were added, 11 rules were modified and 5 rules were deleted from the origin. For example: Modify (*v*) ING → as (*v*) ING → E, add rule of less →. STANS still has some drawbacks like the rule LESS → the meaning of the word is changed to the opposite one (Careless will change to care, which causes errors) [33, 34].

An improvised Porter stemming algorithm was developed by Megala et al. in 2013 [34], This algorithm called TWIG, TWIG holds 31 modifications done on the STANS algorithm, among those 29 modifications are adapted to the improvised stemming algorithm except (LESS →, FUL →) and 15 new rules are introduced to improve the performance of the IR System with meaning full stems.

Based on the previous work, many researches tried to improve the structure of Porters algorithm; however they still have several drawbacks. Some of them concentrated on plural and singular words only, others concentrated on the semantic of some words without being careful about singular and plural. Also, the previous work didn't discuss the past tense of words ending by "ed" and the verbs ending by "en". So, In this paper we presents the Enhanced Porter Stemming Algorithm (EPSA) to overcome these problems by collecting the important rules from the previous researches and proposed new rules that solve the errors mentioned in the next section.

**Main Focus of the Chapter**

Many researches were introduced to improve the structure of Porters algorithm; however it still has several drawbacks. Thus, in this research we proposed an efficient information retrieval technique as well as a new proposed stemming algorithm called Enhanced Porter's Stemming Algorithm (EPSA). The objective of this new technique is to overcome the drawbacks of Porters algorithm and improve web searching.

**Issues, Controversies, Problems**

A new technique called the Enhanced Porter's Stemming Algorithm (EPSA) overcomes the drawbacks of the Porters algorithm and improves web searching.

## 2. Porter's Stemmer

The Porter Stemming Algorithm in [32, 35] operates in six steps. At each step the input word is transformed based on a list of rules. These steps are:

**Step 1** Gets rid of plurals and -ed or -ing suffixes, the rules here like: SSES → SS, SS → SS, IES → I, S →, (*v*) ED →
**Step 2** Turns terminal y to i when there is another vowel in the stem, the rules here like: (*v*) Y → I
**Step 3** Maps double suffixes to single ones: -ization, -ational, etc. the rules here like: (m > 0), ational → ate
**Step 4** Deals with suffixes, -full, -ness etc. the rules here like: full →, ness →
**Step 5** Takes off -ant, -ence, etc. the rules here like: ence →, al →
**Step 6** Remove a final e. the rules here like: (m > 1) E →, (m = 1 and not *o) E →

Note:
v (Vowel): The letters A, E, I, O, U are considered vowels, sometimes y.
m: Number of vowels in the word.
c (Consonant): Any letter not a vowel.

1st step gets rid of plurals ending with "–sses", "-ies", "-s" and also recodes to generate stem words e.g. actresses → actress, abilities → abiliti, pennies → penni, cats → cat, rats → rat. This step also removes endings "-ed" used in the past tense of any word and "-ing" of present continuous e.g. abbreviated → abbriviat, assembling → assembl. The"assembl" term is recoded into "assemble" using "-bl into "-ble" recode conversion rule. The "-at" ending is recoded into "-ate" in "abbriviat" to "abbreviate" conversion.

2nd step recodes terminal "y" into "I" in presence of another vowel in the stem word e.g. "penny" is recoded into "penni". So "pennies" and "penny", these two words are stemmed into a single stem word "penni". The "Pennies" word is converted into "penni" in step 1 using "-ies → i" suffix rule. The "ability" word is recoded into a "abiliti" stem word in step 2, so "abilities" and "ability" both are conflated into the single "abiliti" stem word.

3rd step maps double suffices into a single suffix and recodes "-ational" ending into "–ate" ending "-tional" into "-tion", "-izer" into "-ize", "-ation" into "-ate", "-fullness" into "-full" etc. This step recodes the "application" word into "applicate".

4th step handles "-ic", "-icate", "-full", "-ness" ended words by removing and recoding endings for generating stem words. e.g. "Applicate" stem word is generated in step 3 from the "application" word and is stemmed into "applic" by using the final stem word "-icate" → "-ic" conversion rule.

5th step removes "-ance", "-er", "-able", "-ant", "-iti" endings from words. For example, "Acceptance" and "acceptable" both words are stemmed into the "accept" stem word by removing "-ance" and "-able" suffices respectively. The "Reporter" word is stemmed into "report" by eliminating the "-er" ending. The "Reporting" word is also stemmed into "report" by using the "-ing" suffix rule in step 1. "Abiliti" generated from "ability" and "abilities" in step 2/step 1, is conflated into "abil" by removing "-iti" ending in step 5. So, "ability" and "abilities" are stemmed into a single stem "abil".

6th step removes "-e" ending from any stem word. For example, the word "Assemble" is stemmed into "assembl". So "assembling" and "assemble" are stemmed into the same stem word "assembl", another example is paste stemmed into past.

So the Porter stemmer is not using any stem-dictionary, it only uses lists of endings to be removed and lists of conversion rules for recoding to get stem words through step 1 to step 5 and removes final –e from stem words in step 6.

Figure 1 shows the following Porter stemmer steps, it has six steps applying rules within each step. Within each step, if a suffix rule is matched to a word, then the conditions attached to that rule are tested on what would be the resulting stem, if that suffix was removed in the way defined by the rule. For example such a condition may be, the number of vowel characters which are followed by a consonant character in the stem (Measure), must be greater than one for the rule to be applied [24, 36, 37].

Once a rule passes its conditions and is accepted, the rule fires and the suffix is removed and control moves to the next step. If the rule is not accepted, then the next rule in the step is tested until either a rule from that step fires and control passes to the next step or there are no more rules in that step, hence control moves to the next step. This process continues for all six steps, the resultant stem being returned by the stemmer after control has been passed from step six [36, 37].

Languages are not completely regular constructs and therefore stemmers operating on natural words unavoidably make mistakes. Although the Porters stemmer is known to be powerful, it still faces many errors. The major ones are Over-stemming and Under-stemming errors. These errors, like the concept, denotes the case where a word is reduced too much, for example, "Appendicitis" after stemming may become "Append".

Which may lead to the point where completely different words are conjoined under the same stem or the stem is incorrect in the Language. For instance [31, 38]:

- Past, Paste return to the same root Past
- Appendicitis, Append according to Porter's algorithm are from Append

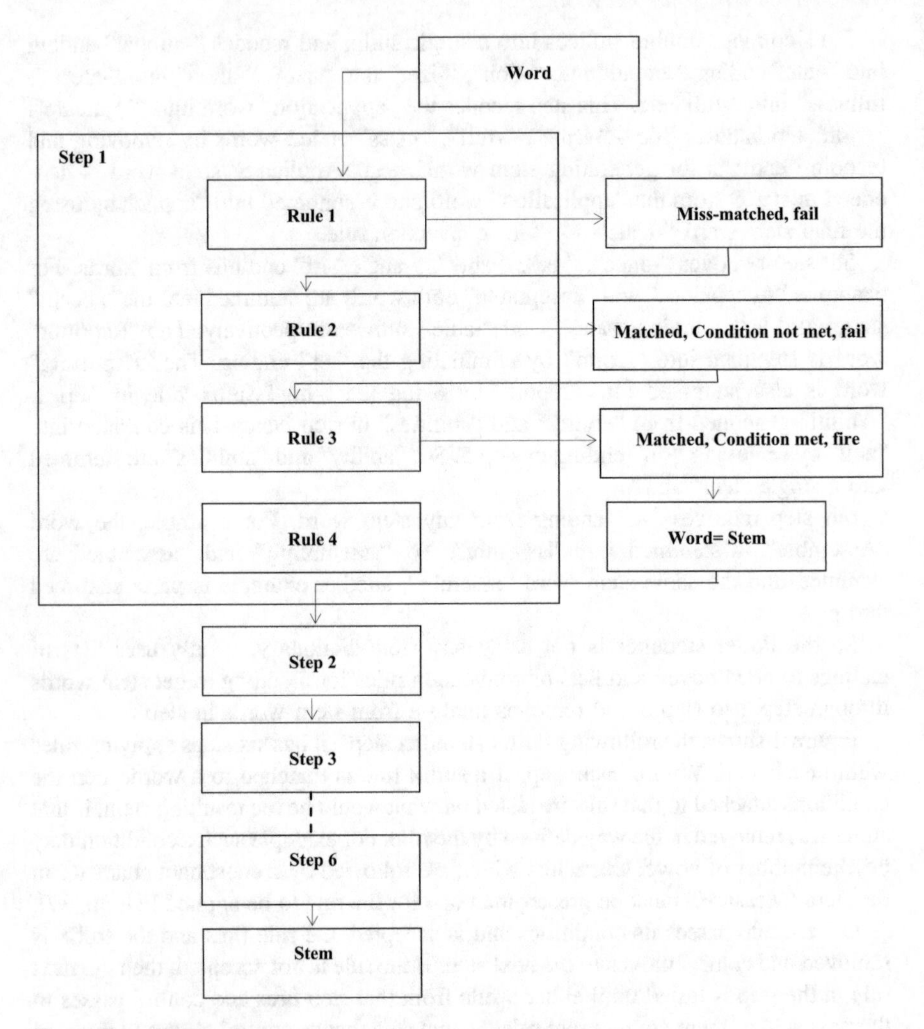


**Fig. 1** Porter's stemmer

- Dying → dy (impregnate with dye)
- Dyed → dy (passes away)
- Political, Polite → polit
- Generative, General → gener
- Child/children → child/children
- Words ending by feet, men return as it is
- Careless → careless (remove less, will change to care will cause error)

Another instance of errors come from our testing on Porters algorithm like written → written, complier → compil, languages → languag, described → describ

# 3 The System Methodology

## *3.1 The System Architecture*

In this section we studied the original Porter stemming algorithm [32], against the pervious researches done to enhance it. From these studies we proposed a new stemming algorithm called Enhanced Porter Stemming Algorithm (EPSA). This algorithm was applied to the following system architecture that is shown in Fig. 2

Our system consists of 4 main modules:

- **Module 1**: Tokenization: Documents in fact are a sequence of words that have ideas or more to be reached by the reader but computers don't understand the structure of a natural language document and can't automatically recognize words and sentences, unlike humans. To a computer, a document is only a sequence of bytes. Computers don't know that a space character separates words in a document; instead, humans must program the computer to identify what constitutes an individual or distinct word referred to as a token. This operation is called Tokenization, i.e. this stage is for breaking a stream of text up into words and keeping the words in a list called Word's List. Note: The user's query on the system acts with it like a document.
- **Module 2**: Data Cleaning: Documents consist of strings of characters, digits and special symbols; i.e. Documents have a words that are the best describe a document extracted and others are ignored, the ignored words may a punctuation marks, prepositions, pronouns, conjunctions and auxiliary verbs which referred to as stop words and often removed. These Stop words are deemed irrelevant for searching purposes because they occur frequently in the language for which the indexing engine has been tuned, these words are dropped at indexing time and then ignored at search time. The Data Cleaning stage removes useless words from the Word's List, for example, For example, have, did, in, of, get.. etc. are meaningless, These useless words are stored in a stop words database as appears in the Fig. 2. The database has 311 English stop words with a size 3 KB.
- **Module 3**: The ESPA stemming Algorithm: There can be different forms for the same terms (e.g., students and student). These different forms of the same term need to be converted to their roots. Stemming algorithms (stemmers) are used to convert the words to their root form (stem). In this stage, we applied our proposed stemming algorithm to overcome the drawbacks of the Porter algorithm. The algorithm is explained in the next section.
- **Module 4**: Indexing and Terms Weighting: Indexing is a process for describing or classifying a document by index terms; index terms are the keywords that have a meaning of their own (i.e. which usually has the semantics of the noun). These index terms are grouped in an indexer and the stemmer is service this stage by improving the group of these keywords in the indexer. A vector Space Model is used here for Automatic Indexing, This Model [4] is based on the assumption that documents are mapped as vector points in term space according

**Fig. 2** The system architecture

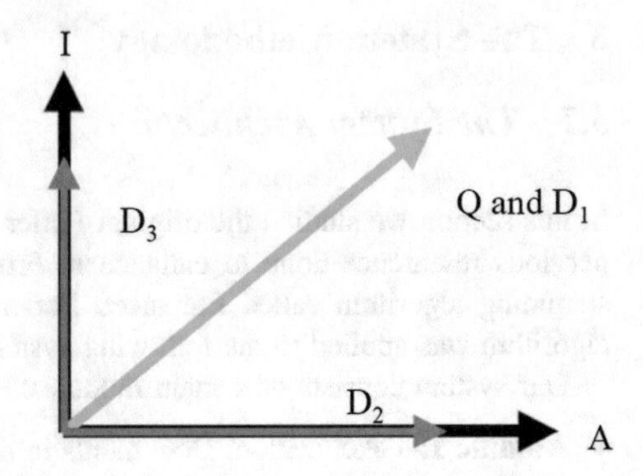


to their conceptual relationship to appropriate terms and queries travel along a vector which represents relationship terms while Query and Document weights are based on the length and direction of their vector. For more explanation of this point see Fig. 3 Vector Space Model with Three Terms Vocabulary and in Fig. 4 an example of two terms vocabulary which shows three documents with two terms A and I. As you see, D1 contains AI terms, D2 contains A and D3 contains I alone. If the user enters the query AI as appears in the figure, then the query is mostly like D1 so they are the same.
Terms Weighting is an evaluation of the relative importance of any given term in relation to other terms within the vector model. The idea is based on the document weight of every term within a given document (How important is this term to the document?) which is called Term Frequency (TF).

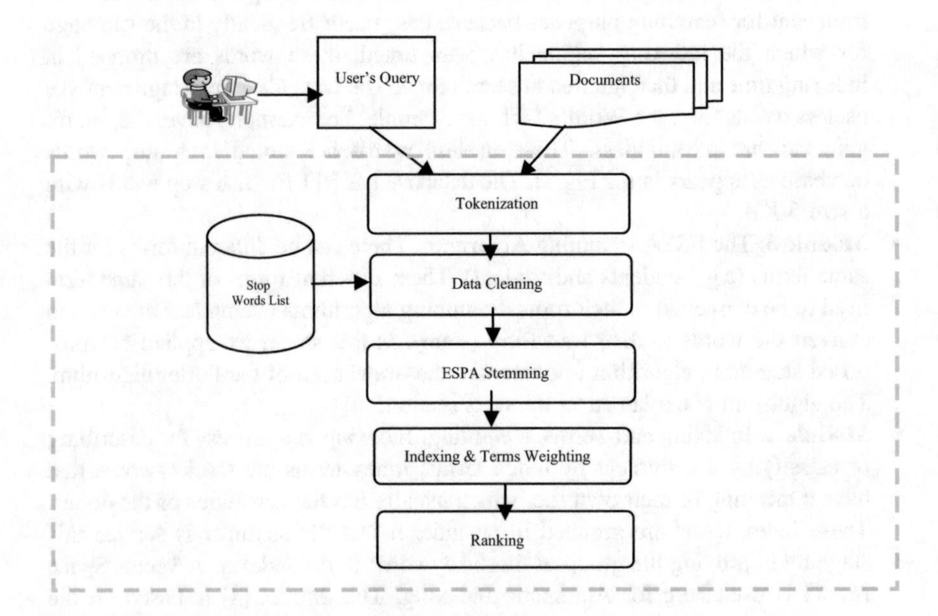


**Fig. 3** Vector space model with three terms vocabulary

Fig. 4 An example of vector space model with two terms vocabulary

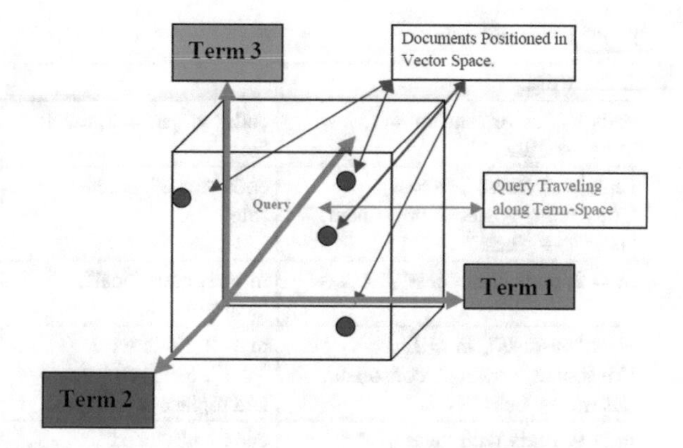


- **Module 5**: Ranking: The user's query is matched with the index terms to get the relevant documents to the query. Documents are then ranked using ranking algorithms according to the most relevant to the user's query. The Ranking Algorithm is in Sect. 3.3.

## 3.2 *The Enhance Porter Stemming Algorithm (EPSA)*

The ESPA stemming includes the original Porter rules and our new rules that are proposed to solve the previous errors (appears in bold font in Table 1). Also, we collected the critical rules from other researches [5, 32, 34, 38] and added them to the EPSA stemming. Table 1 shows the details of these rules.

**Table 1** The EPSA rules

| If the word: | | |
|---|---|---|
| ends with "e", function must keep e at the end of the word | ends with "ize"<br>– m = 2, keep it<br>– m > 1, "ize" removed | **ends with "er" after it constant then delete "r"** |
| ends with "ches" or with "shes"… remove "es" only | ends with "ive"<br>– m = 1, keep it<br>– m > 1, "ive" removed | **If end "es", Remove "s", keep "e"** |
| ends with "is", don't delete | ends by "iral", m = 2, start with vowel, keep it | **If end "en", Keep "e"** |
| ends with "ying" → i & "yed" → y | ends "al", m = 2, delete "al" and add "e" | **If the word end by "y", Replace it with "I"** |
| m = 2, consonant, vowel, consonant, vowel, then remove "al" | ends –knives, -knives → -knife | **ends "ed" or"ing", keeping "e" while removing "ed" or "ing"** |

(continued)

**Table 1** (continued)

| If the word: | | |
|---|---|---|
| ends by "ative" and m = 2, ative → "ate" | ends "ic", m = 2, delete "ic" | ends –staves, -staves → -staff |
| Ends with "ness", m = 1, Consonant, vowel & consonant, "ness" → "ness" | ends "icate", delete "ate" | ends –xes, -xis → -x |
| m = 2, ends with "ness", ness → | m = 1, ends "ical", "ical" → "ic" | ends –trixes, -trixes → -trix |
| ends "ousness", m = 1, Consonant, vowel & consonant, ousess → ous | m > 0, Ends with "ator", Remove "ator" and replace it with "ate" | ends –ei, -ei → -eus |
| m > 0, Ends with "less", "less" → "less" | ends with "ceed", m > 0, remove "ceed" and replace it with "cess" | ends –pi, -pi → -pus |
| m > 0, Ends with "lessly", "lessly" → "less" | ends –wives, -wives → -wife | ends –ses, -sis → -s |
| m > 0, Ends with "fully", delete "ly" | ends –feet, -feet → -foot | ends with "ence", m = 2, delete "ence" |
| ends with "ous", m > 1, Delete "ous" | ending in –men, -men → -man | ends with "ment", m = 2, Keep it |
| Ends with "ous", m = 2, Keep "ous" | ends –ci, ci → -cus | ends with "ment", m > 2, remove "ment" |
| ends with "eer", m = 2 Then remover "er" | end with "eed" -m = 0, then Keep "eed" -m > 0, remove "d" | ends with "tion", m = 2, replaced with "e" |
| ends with "ible", m = 2, Starts with a consonant, not ending with series of consonant vowel consonant vowel Then keep it as it is | m > 0, Ends with "ator", Remove "ator" and replace it with "ate" | ends with "ional" then delete "al" |
| ends with "nate" or "ate" -m = 2, keep it as it is m > 2, "ate" or "nate" removed m = 1, then "at" is kept – m = 0, then left it as it is | m = 2, ends "able", delete "able" m > 2, remove "ible" | ends with "ance", m = 2, Consists of series consonant, vowel, consonant, vowel , Replaced with "e", Else, removed "ance" |

### 3.3 The Ranking Algorithm

This part focuses on ranking algorithms based on the content of documents and queries. It simply tries to find the similarity between the content of documents and queries. We applied here the Cosine similarity measure, this selection is based on studies represented in [39, 40], which prove that the cosine measure is the most efficient measure in comparison to other statistical measures.

The Cosine measure calculates the angel between two documents (between a document and the user's query which is treated as a document) representation vectors. Thus a cosine value of zero means that the query and document vectors were orthogonal to each other and means that there is no match or the term simply did not exist in the document being considered. To know the cosine relations between two documents (document D and query Q) see Eq. 1.

$$\text{Cosine}\,(\text{D, Q}) = \frac{|\text{D} \cap \text{Q}|}{\sqrt{|D| * |\text{Q}|}} \quad (1)$$

where:

Cosine (D, Q): The Cosine Similarity relationship between document D and a user's query Q.

D: Refer to the document in the collection.

Q: Refer to user's query.

After calculating the similarity measure, the ranked list appears to the user as the answer to his/her query. This list is arranged from the highest values of the Cosine measure to the lowest one as a weight as a ranked list.

# 4 Performance Studies

In order to study the performance of our technique, we used different evaluation measures. These measures are discussed in Evaluation of the EPSA Algorithm's section. Then, the data sets used and the experimental results are shown after it.

## *4.1 Evaluation of the EPSA Algorithm*

To evaluate our algorithm, the Paice's Evaluation Methods [8] were used. The standard evaluation method of a stemming algorithm is to apply it to an IR test collection and calculate the recall and precision to assess how the algorithm affects those measurements. However, this method does not show the specific cause of errors therefore it does not help the designers to optimize their algorithms. Paice in [41, 42] proposed an evaluation method in which stemming is assessed against predefined groups of semantically related words. He introduced three new measurements: Over and under-stemming indexes and the stemming weight. This test requires a sample of different words partitioned into concept groups containing forms that are morphologically and semantically related to one another. The perfect stemmer should have all words in a group to the same stem and that stem should not occur in any other group.

For each concept group two totals are computed:

- Desired merge total (DMT), which is the number of different possible word form pairs in the particular group and is given by the formula:

$$DMT_g = 0.5\, n_g \left(n_g - 1\right) \qquad (2)$$

where $n_g$ is the number of words in that group.

- Desired Non-merge Total (DNT), which counts the possible word pairs formed by a member and a non-member word and is given by the formula:

$$DNTg = 0.5\, ng\, (W - ng) \qquad (3)$$

where W is the total number of words. By summing the DMT for all groups we obtain the GDMT (global desired merge total) and, similarly by summing the DNT for all groups we obtain the GDNT (global desired non-merge total). After applying the stemmer to the sample, some groups still contain two or more distinct stems, incurring in understemming errors. The Unachieved Merge Total (UMT) counts the number of understemming errors for each group and is given by:

$$UMT_g = 0.5 \sum\nolimits_{i=1..s} u_i \left(n_g - u_i\right) \qquad (4)$$

where s is the number of distinct stems and ui is the number of instances of each stem. By summing the UMT for each group we obtain the Global Unachieved Merge Total (GUMT). The understemming index (UI) is given by: GUMT/GDMT.

After stemming, there can be cases where the same stem occurs in two or more different groups, which means there are over-stemming errors. By partitioning the sample into groups that share the same stem we can calculate the Wrongly Merged Total (WMT), which counts the number of over-stemming errors for each group. The formula is given by:

$$WMT_g = 0.5 \sum\nolimits_{i=1..t} v_i (n_s - v_i) \qquad (5)$$

where t is the number of the original groups that share the same stem, ns is the number of instances of that stem and $v_i$ is the number of stems for each group t. By summing the WMT for each group we obtain the Global Wrongly-Merged Total (GWMT). The over-stemming index (OI) is given by: GWMT/GDNT. The Stemming Weight (SW) is given by the ratio OI/UI. SW gives some indication whether a stemmer is weak (low value) or strong (high value).

Also we evaluate our algorithm by measuring the performance of the IR system and comparing its result with the results achieved when using the exact match (without stemming) and the Porter stemmer. The performance measured by the recall and precision measurements, which are represented in the following formulas:

$$Precision = \frac{|\{relevant\ documents\} \cap \{retrieved\ documents\}|}{|\{retrieved\ documents\}|} \quad (6)$$

$$Recall = \frac{|\{relevant\ documents\} \cap \{retrieved\ documents\}|}{|\{relevant\ documents\}|} \quad (7)$$

$$AveP = \frac{\sum_{i=1}^{N_0} P_i(r)}{N_q} \quad (8)$$

where: AveP: Average precision at recall level r. $P_i$(r): the precision at recall level r for the *i*th query. $N_q$: the number of queries used.

## 4.2 Experimental Results

EPSA Algorithm was applied to 23,531 words (i.e. it's a Porter's collection) [43] and compared to the original Porter algorithm. This Dataset is used to measure the Over-stemming, Under-stemming and Stemming weight. In addition, it was applied to a famous collection called CISI and is available in [44]. This collection contains 1460 documents of different sizes and we then tested the system with 35 queries in order to evaluate the IR system performance.

Firstly, we applied the Porter Algorithm and ESPA algorithm to the Porter's Collection Dataset to measure how many words were stemmed correctly from both algorithms. This measure is calculated by counting the correct stem words coming from each stemmer then finding the percentage. Figure 5 shows that ESPA returns a better indication of correct stem words, it reaches over 80 %, on the other hand, the Porter reaches only 58.60 %.

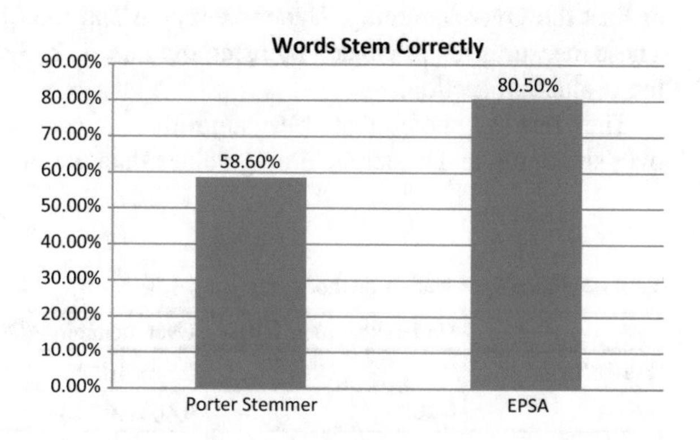


**Fig. 5** Percentage of the words stem correctly

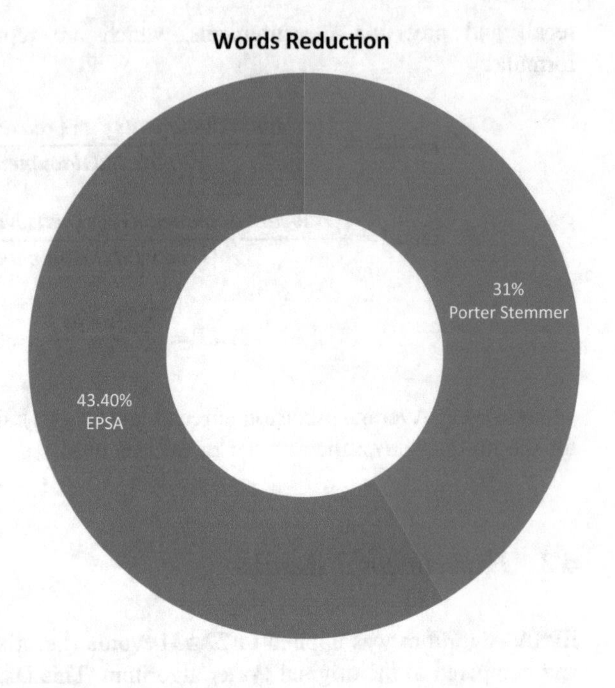


**Fig. 6** Words reduction

Figure 6 represents the word reduction, which is the process to reduce the words to the collection of words. For example, if the collection contains the words: "fishing", "fished", and "fisher", they are reduced to one word (the root) word, "fish". Although Porter's stemmer holds a good reduction (31 %), EPSA reached a much better percentage (43.40 %). The word reduction is calculated by the next equation

$$\frac{\sum \textit{unique stem words}}{\sum \textit{whole number of words in the collection}} * 100\,\% \tag{9}$$

Paice's Evaluation Methods were applied to the same collection (Porter's Collection). Table 2 shows the results obtained from the Porter and ESPA algorithms against the Over-stemming, Under-stemming and the Stemming weight measures. These measures are calculated by using the Eqs. 1, 2, 3 and 4 that are mentioned in the evaluation section.

The results show that our algorithm succeeds to achieve much less over-stemming and under-stemming values than the original Porter algorithm. The

**Table 2** Paice's evaluation methods

| | Under-stemming (UI) | Over-stemming (OI) | Stemming weight (SW) |
|---|---|---|---|
| Porter Stemmer | 0.31 | $0.51 \times 10^{-5}$ | $0.1645 \times 10^{-5}$ |
| ESPA | 0.183 | $0.40 \times 10^{-5}$ | $2.186 \times 10^{-5}$ |

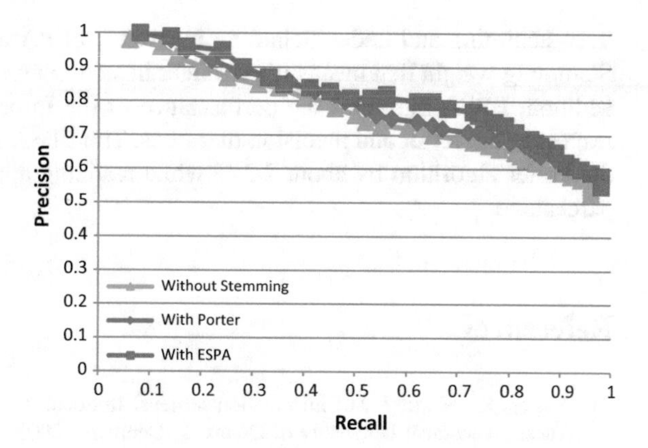


**Fig. 7** Precision and recall against the CISI 35 queries

**Table 3** Average precision and average recall

| | Precision (%) | Recall (%) |
|---|---|---|
| Without Stemming | 70.9 | 63.6 |
| With Porter | 72.9 | 65.7 |
| With ESPA | 75.2 | 66 |

results also reflect that our stemmer is stronger than the Porter stemmer, this appears clearly on the stemming weight which is larger than the Porter's one.

Finally, EPSA was tested on the CISI collection using IR evaluation to measures, which is mentioned in the evaluation section. Figure 7 shows the precision and recall results for each query without stemming (exact match) and with stemming (using the porter and EPSA algorithms). The average precision for the ESPA algorithm is 75 %, which is the highest value in comparison to the one without stemming (nearly 70 %) and the Porter algorithm (approximately 72 %). These results are shown in Table 3. From these results, the EPSA improves the precision over the Porter algorithm by about 2.3 % while realizing approximately the same recall percentage.

## 5 Conclusion

Many researches were introduced to improve the structure of the porter algorithm; however it still has several drawbacks. Thus, in this paper we proposed an efficient information retrieval technique as well as a new proposed stemming algorithm called the Enhanced Porter's Stemming Algorithm (EPSA). The objective of this new technique is to overcome the drawbacks of the Porter algorithm and improve web searching.

The algorithm was applied to two datasets for testing. The results show that the ESPA algorithm gives fewer errors than the original Porter algorithm in both over-stemming and under-stemming measures. ESPA algorithms also hold a good Stemming weight that means it's stronger in stemming than the Porter algorithm. In addition, EPSA improves the performance of the Information Retrieval system in respect to the recall and precision measures. The EPSA improves the precision over the Porter algorithm by about 2.3 % while realizing approximately the same recall percentage.

## References

1. Vester, K., Martiny, M.: Information retrieval in document spaces using clustering. Master Thesis, Technical University of Denmark, Denmark (2005)
2. Sharma, D.: Stemming algorithms: a comparative study and their analysis. Int. J. Appl. Inf. Syst. (IJAIS) **4**(3), 7–12 (2012)
3. Singhal, A.: Modern information retrieval: a brief overview. IEEE Data Eng. Bull. **24**(4), 35–43 (2011)
4. Baeza-Yates, R., Ribeiro-Neto, B.: Modern Information Retrieval. ACM Press, New York, USA (1999)
5. Yamout, F., Demachkieh, R., Hamdan, G., Sabra, R.: Further Enhancement to the Porter's Stemming Algorithm, pp. 7–23. Machine Learning and Interaction for Text based Information Retrieval, Germany (2004)
6. Maurya, V., Pandey, P., Maurya, L.S.: Effective information retrieval system. Int. J. Emerg. Technol. Adv. Eng. **3**(4), 787–792 (2013)
7. Sembok, T., Abu Ata, B., Bakar.: A rule and template based stemming algorithm for Arabic language. Int. J. Math. Models Methods Appl. Sci. **5**(5), 974–981 (2011)
8. Moral, C., Antonio, A., Imbert, R., Rmirez J.: A survey of stemming algorithms in information retrieval. Inf. Res.: Int Electron. J. **19**(1) (2014)
9. Frakes, W., Baeza, R.: Stemming Algorithms. Data Structures and Algorithms, pp. 131–160. Prentice Hall, Upper Saddle River (1992)
10. Kantrowitz, M., Mohit, B., Mittal, V.: Stemming and its effects on TFIDF ranking. In: Proceedings of the 23rd Annual International ACM SIG IR Conference on Research and Development in Information Retrieval, pp. 357–359, Athens, Greece (2000)
11. Tala, F.: A study of stemming effects on information retrieval in Bahasa Indonesia. M.Sc. Thesis, University of Amsterdam (2003)
12. Buttcher, S., Clarke, C., Cormack, G.: Information Retrieval: Implementing and Evaluating Search Engines. Massachusetts Institute of Technology, USA (2010)
13. Jivani, A.: A comparative study of stemming algorithms. Int. J. Comp. Tech. Appl. **2**(6), 1930–1938 (2011)
14. Sharma, D.: Improved stemming approach used for text processing in information retrieval system. Master of Engineering in Computer Science & Engineering, Thapar University, Patiala (2012)
15. Paice, C.: Another stemmer. ACM SIGIR Forum **24**(3), 56–61 (2012)
16. Singh, A., Kumar, N., Gera, S., Mittal, A.: Achieving manitude order improvement in Porter stemmer algorithm over multi-core architecture. In 7th International Conference on Informatics and Systems (INFOS), pp. 1–8 (2010)
17. Aas, L., Eikvil, L.: Text categorisation: a survey, Raport NR 941, Norwegian Computing Center (1999)
18. Larkey, L., Croft, W.: Combining classifiers in text categorization. In: Proceedings of the 19th Annual International ACM SIGIR Conference on Research and Development in Information Retrieval, pp. 289–297, Zurich, Switzerland (1996)

19. Hotho, A., Stumme, G.: Conceptual clustering of text clusters. In: Kokai, G., Zeidler, J. (eds.) Proceedings Fachgruppentreffen Maschinelles Lernen (FGML 2002) (2002)
20. Steinbach, M., Karypis, G., Kumar, V.: A comparison of document clustering techniques. In: KDD Workshop on Text Mining (2000)
21. Kraaij, W., Pohlmann, R.: Porter's stemming algorithm for Dutch. In: Noordman, L.G.M., de Vroomen, W.A.M. (eds.) Informatiewetenschap, Tilburg (1994)
22. Savoy, J.: Light stemming approaches for the French, Portuguese. In: German and Hungarian Languages Proceedings of the 2006 ACM Symposium on Applied Computing, pp. 1031–1035 (2006)
23. Lovins, J.: Development of a stemming algorithm. Mech. Transl. Comput. Linguist. **11**(1), 22–31 (1968)
24. Bijal, D., Sanket, S.: Overview of stemming algorithms for Indian and Non-Indian languages. Int. J. Comput. Sci. Inf. Technol. (IJCSIT) **5**(2), 1144–1146 (2014)
25. Smirnov, I.: Overview of stemming algorithms. http://the-smirnovs.org/info/stemming.pdf (2014)
26. Dawson, J.: Suffix removal and word conflation. ALLC Bulletin **2**(3), 33–46 (1974)
27. Al-Shammari, E.: Towards an error-free stemming. In: Proceedings of the 2008 International Conference on Data Mining (IADIS'08), Amsterdam, Netherlands (2008)
28. Golsmith, J., Higgins, D., Soglasnova, S.: Automatic language-specific stemming in information retrieval. In: Cross-Language Information Retrieval and Evaluation, Proceeding of CLEF 2000 Workshop, Berlin, pp. 273–283 (2001)
29. Hull, A.: Stemming algorithms: a case study for detailed evaluation. J. Am. Soc. Inf. Sci. **47**(1), 70–84 (1996)
30. Kraaij, W., Pohlmann, R.: Viewing stemming as recall enhancement. In: Frei, H.-P., Harman, D., Schauble, P., Wilinson, R. (eds.) Processings of the 17th ACM SIGIR Conference held at Zurich, pp. 40–48 (1996)
31. Willett, P.: The Porter stemming algorithm: then and now. Program: Electron. Lib. Inf. Syst. **40**(3), 219–223 (2006)
32. Porter, M.F.: An algorithm for suffix stripping. Program **14**(3), 130–137 (1980)
33. Srinivasan, S., Thambidurai, P.: STANS algorithm for root word stemming. Inf. Technol. J. **5**(4), 685–688 (2006)
34. Megala, S., Kavitha, A., Marimuthu, A.: Improvised stemming algorithm—TWIG. Int. J. Adv. Res. Comput. Sci. Softw. Eng. **3**(7), 168–171 (2013)
35. Gupta, R., Jivani, A.: Empirical analysis of affix removal stemmers. Int. Comput. Technol. Appl. (IJCTA) **5**(2), 393–399 (2014)
36. The Lancaster Stemming Algorthim: Porter, http://www.comp.lancs.ac.uk/computing/research/stemming/general/porter.htm
37. What is Porter Stemming: http://www.comp.lancs.ac.uk/computing/research/stemming/general/porter.htm
38. Karaa, W.: A new stemmer to improve information retrieval. Int. J. Netw. Secur. Appl. (IJNSA) **5**(4), 143–154 (2013)
39. Hajeer, S.: Comparison on the effectiveness of different statistical similarity measures. Int. J. Comput. Appl. **53**(8), 14–19 (2010)
40. Hajeer, S.: Vector space model: comparison between Euclidean distance & cosine measure on arabic documents. Int. J. Eng. Res. Appl. **2**(4), 2085–2090 (2012)
41. Paice, C.D.: An evaluation method for stemming algorithms. In: Proceedings of the 17th Annual International ACM SIGIR Conference on Research and Development in Information Retrieval, pp. 42–50, Dublin, Ireland. ACM (1994)
42. Kara, W., Gribâa, N.: Information retrieval with Porter Stemmer: a new version for english. Adv. Comput. Sci., Eng. Inf. Technol. AISC **225**, 243–254 (2013)
43. The Porter Stemming Algorithm: http://tartarus.org/~martin/PorterStemmer/index.html
44. Common IR Test Collection: http://web.eecs.utk.edu/research/lsi/corpa.html

# Part II
# Forensics Multimedia and Watermarking Techniques

# Face Recognition via Taxonomy of Illumination Normalization

**Sasan Karamizadeh, Shahidan M. Abdullah, Mazdak Zamani, Jafar Shayan and Parham Nooralishahi**

**Abstract** Presently, the difficulty in managing illumination over the face recognition techniques and smooth filters has emerged as one of the biggest challenges. This is due to differences between face images created by illuminations which are always bigger than the inter-person that usually be used for identities' recognition. No doubt, the use of illumination technique for face recognition is much more popular with a greater number of users in various applications in these days. It is able to make applications that come with face recognition as a non-intrusive biometric feature becoming executable and utilizable. There are tremendous efforts put in developing the illumination and face recognition by which numerous methods had already been introduced. However, further considerations are required such as the deficiencies in comprehending the sub-spaces in illuminations pictures, intractability in face modelling as well as the tedious mechanisms of face surface reflections as far as face recognition and illumination concerned. In this study, few illuminations have been analyzed in order to construct the taxonomy. This covers the background and previous studies in illumination techniques as well the image-based face recognition over illumination. Data was obtained from the year of 1996 through 2014 out of books, journals as well as electronic sources that would

S. Karamizadeh · S.M. Abdullah (✉) · J. Shayan
Advanced Informatics School (AIS), Universiti Teknologi Malayisa,
5400 Kuala Lumpur, Malaysia
e-mail: mshahidan@utm.my

S. Karamizadeh
e-mail: Ksasan2@live.utm.my

J. Shayan
e-mail: Sjafar3@live.utm.my

M. Zamani
Kean University, Union, NJ, USA
e-mail: mzamani@kean.edu

P. Nooralishahi
Department of Computer Science and Information Technology,
University of Malaya, Kuala Lumpur, Malaysia
e-mail: Parham.nooralishahi@gmail.com



A.E. Hassanien et al. (eds.), *Multimedia Forensics and Security*,
Intelligent Systems Reference Library 115, DOI 10.1007/978-3-319-44270-9_7

share more on the advantageous and disadvantageous, the current technique's performance as well as future plan.

**Keywords** Illumination · Face recognition · Taxonomy · Strategies · Filters

## 1 Introduction

Vital challenge in the real time applications is on the face recognition through different illumination. Despite of numerous illuminations filtering techniques are in place, the concepts are said to be archaic and the critical analysis of performance of illumination filtering techniques are not covered [1, 2]. Based on the current face recognition techniques that come with a group function to normalize the illumination help to facilitate the critical challenge in the face recognition system [3]. Most of the techniques used in photometric normalization were used to develop the methods used in this particular toolbox [4]. Those are the techniques used for carrying out the normalization of illumination prior to the actual processing stage. One could see that it is in contrary with the methods applied in compensating the appearances induced by illumination at the classification or modelling stage [5, 6].

## 2 Illumination Normalization Face Recognition

Illumination variation. Basically, these methods can be classified into three main categories. Figure 15 showed of Taxonomy of filtering of illumination.

- Illumination normalization
- Illumination modeling
- Illumination invariant feature extraction.

### *2.1 Illumination Normalization*

Primary category refers to the normalized face images under illumination variation. The typical algorithms for this category are Wavelet, Steerable filtering, Non-local Means based also adoptive Non-local means.

#### 2.1.1 Wavelet-Based (WA)

Du and Ward were the first to introduce the Wavelet-based normalization (WA) technique [7]. In this technique, the wavelet transformation which will become discrete on the image is used and followed by the processing of the obtained sub-bands. The histogram equalization will be used based on the estimated transformation's coefficient to focus on the detailed coefficient's matrices [8]. We could redesign the normalized image through the inversely wavelet transformation when once the sub-bands manipulated [9, 10].

One could observe a discrete wavelet transformation in accordance to the octave band's tree structure as a signal's multi-resolution decomposition. The two-band wavelet transform could enable the signal be visible through the wavelet. Apart from that, the scaling based functionalities could also be carried out at various scales, in an orderly manner [11].

$$f(x) = \sum_k a_{o,k}\phi_{o,k}(x) + \sum_i \sum_k d_{j,k\psi j,k}(x) \tag{1}$$

where ϕ j, k and ψ wj, k are scaling functionalities and wavelet functionalities at scale j, respectively. In addition, Aj, k, and D j, k are respectively, scaling and wavelet coefficients (Fig. 1).

### 2.1.2 Steerable Filtering-Based

In order to get rid of the illumination which is created by the appearance variations within the face images, the steerable filtrations applied through the second technique which is known as the steerable filtering-based normalization of SF. There are numerous of irregular small blocks could be seen on the directional face [12, 13]. Thus, the characteristics within a facial image could be split into many sub-bands at various scales with estimations and details through the use of decomposition of the steerable pyramid [14, 15]. An number for combinations about sub-bands during An larger amount would obliged will increase An legitimate confinement since those one-level decay will be insufflate to disconnect these characteristics proficiently. Than improved catch the multi-orientation in a facial image, the derivatives across all directions must be measured as a direct solution however higher computation cost incurred throughout this approach. Based on the steerable filtering theory [16, 17], the interpolation by some functionality could be carried out in any

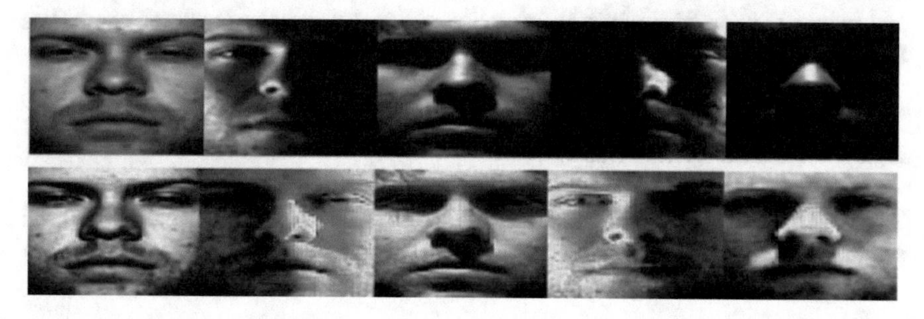

**Fig. 1** Show performance WA function [8]

direction over an image's derivatives and that is the basic derivatives. The database of each image could be created by decomposing into 3-levels and 6-orientations sub-bands for face and iris. The irrelevant and redundant information from the direct S-P coefficients make the differentiating aspects unable to be removed out fully [18, 19].

Thus, partitions of a set of similar-sized blocks in an orderly non overlapped manner for each S-P sub-bands shall be made so as to achieve a local and effective representation of the image of face and iris. The statistical measurements such as energy, mean, differences as well as the entropy of the decomposition of each sub-band level of S-P coefficients energy distribution could be applied to recognize a textural base in accordance to popular belief. If Iij(x, y) is the image at the particular block j of sub-band i, the created feature vector νij = (Mean, Variance, Energy or Entropy);

$$Mean = \frac{1}{M \times N} \sum_{x=1}^{M} \sum_{Y=1}^{N} |I(x,y)| \tag{2}$$

$$Variance = \frac{1}{M \times N} \sum_{x=1}^{M} \sum_{y=1}^{N} \left| I_{ij}(x,y) - \mu_{\mathrm{ij}} \right|^2 \tag{3}$$

$$Enrgy = \sum_{x=1}^{M} \sum_{y=1}^{N} \left| I_{ij}(x,y) \right| 2 \tag{4}$$

where M and N is the Iij(x, y) size. Entropy is a statistical measurement for randomness which is used to exhibit the inputted image texture of the image that is inputted. Entropy is described as

$$Entropy = -\sum_{p} (p \times \log(p)) \tag{5}$$

And p contains the number of histogram. By connecting every block measurement to a large feature vector where $\nu = \bigcup_{i=1}^{k} \bigcup_{j=1}^{ki} \{\nu_{ij}\}.k$ would give the number of S-P sub-band and Ki is the amount of blocks in the ith sub-band able to create the feature vector of the face and iris. Thus, this technique can extract the best features and reduce the data size while keeping the principal discriminants at the same time (Fig. 2).

### 2.1.3 Non-local Means Based

Struc Also Pavešić required acquainted those strategy to Non-Local implies based standardization (NLM) [16]. In this technique, the luminance purpose will be enter

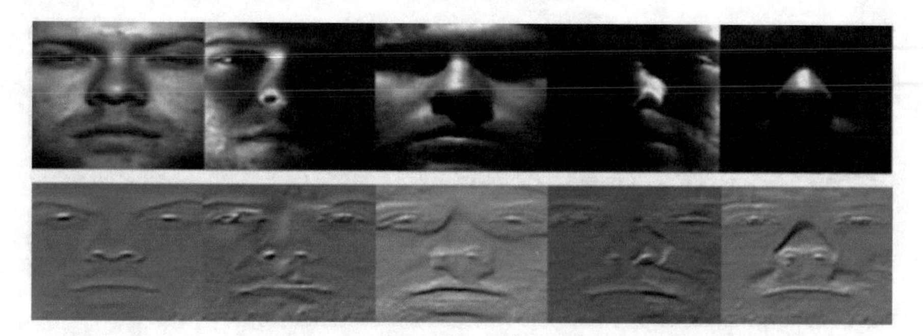

**Fig. 2** Show performance by steerable filtering based [14]

of the calculation of the non-local intends de-noising which At that point is used to figure those reflectance [17, 20]. Those al-gorithm to the Non-Local methods alternately to a short form, NL methods [21, 22] will be a picture for de-noising system which as of late proposed that is different of the display de-noising approaches As far as the amount of pixel starting with those whole picture to the clamor diminishment purpose [23]. This will be because of a few comparable windows which Might Additionally make recognized done every small image's window. Furthermore, everyone from claiming these windows Might a chance to be used to picture de-noising design [16, 24]. Accepting that those commotion tainting once a picture as In(x) ∈ Ra × b, the place An Furthermore b would clinched alongside pixels of picture dimensions, What's more let's take x on representable the discretionary pixel area x = (x, y) in the commotion picture. Those calculation to NL methods outlines those de-noised picture Id(x) Toward inputting every pixel esteem from claiming Id(x) as a weighted Normal of pixels holding In(x).

$$I_d(x) = \sum_{x \in In(x)} \omega(z,x) In(x) \tag{6}$$

The place w (z, x) will be those weighting purpose that calculates those similarities between neighborhoods of the pixel toward the spatial areas z and x. Those weighting purpose here will be portrayed in the following.

$$\omega(z,x) = \frac{1}{z(z)} e^{\frac{G\partial|In(\Omega x) - In(\Omega z)|}{h^2}} \; and \, Z(z) = \sum_{x \in In(x)} e^{\frac{G\partial|In(\Omega x) - In(\Omega z)|}{h^2}} \tag{7}$$

The equation above shows that Gσ is a Gaussian kernel with a standard deviation s, Ox and Oz is the local neighborhoods of pixels at locations x and z, respectively. h represents the parameter that manages the exponential functionality decay and Z (z) is the normalization factor (Fig. 3).

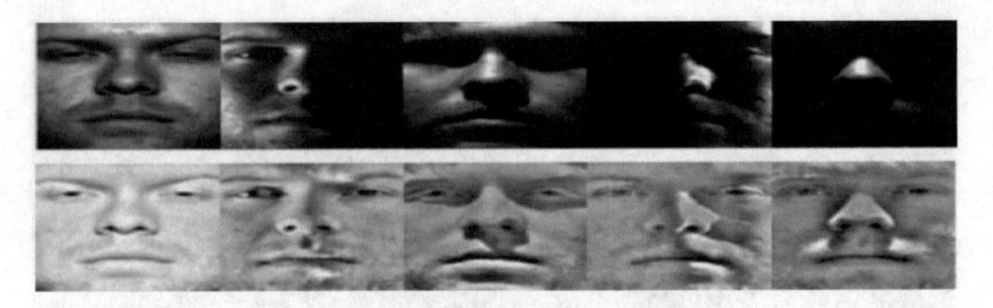

**Fig. 3** Show performance by NLM—the reflection functionalities [16]

### 2.1.4 Adoptive Non-local Mean

ANL which is a short type of the versatile Non-Local Means built normalization might have been introduced Toward 'Štruc What's more Pavešić [16]. In this technique, the algorithm for versatile Non-Local intends de-noising will be used to measure those purpose about luminance trailed Eventually Tom's perusing the computation of the reflectance number [25]. Those versatile Non-Local intends channel for a mixture for Wavelet will be just similar to those NLM filter. However, those smoothening parameter is adjusted generally which Might a chance to be communicated Likewise.

$$\sigma^2 = \min(d(R_i, \mathrm{R}_j))\forall \neq i\, and\, R = \mu - \Psi(\mu) \tag{8}$$

where volume R is the input for a distance measured by subtracting the unique noisy volume u and the low pass filtered volume ψ(μ). By removing the low frequency input with minimal operator applied, experiments showed that a minimal distance which is approximately similar to s2 is required in this case [26, 27]. This simple approach has two significant advantages which are firstly, the same patches that come along with similar structure but at a varied mean level could be found. This is to compensate the intensity of the existing data homogeneity [28, 29]. However, the database in search that comes along with unique patches will give a minimal overestimation over the noise variance in this case. Thus, the parameter h2 will be established by the adaptive filtering which is similar to the minimal distance calculation as in Eq. 8 (Fig. 4).

## *2.2 Illumination Modeling*

In second category, the image construction under different illumination conditions will need to consider the assumptions of certain surface reflectance properties [6, 30]. The illumination variation could be very well modelled with the typical algorithms for second category such as SSR, MSR, and ASR [31, 32].

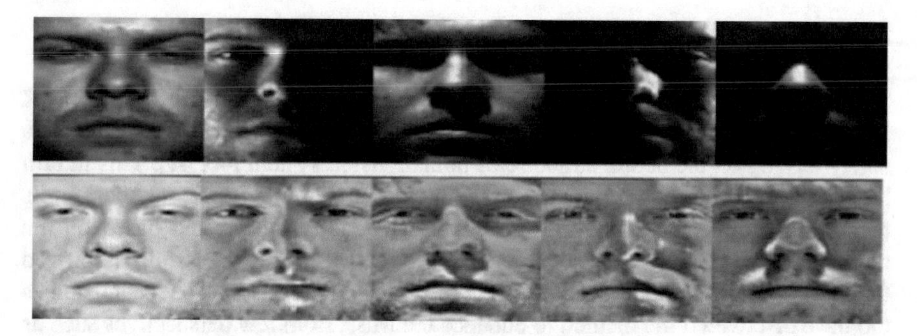

**Fig. 4** Illustration, performance by ANL—the reflection functionalities [26]

### 2.2.1 Single Scale Retinex (SSR)

Single Scale Retinex algorithm (SSR) was introduced [33, 34] by which it has many similarities to photometric normalization techniques [35, 36]. The local contradiction will be improved when the SSR scale is low and this indirectly makes the robust compression range becomes better. Yet, halo artefact will be one of the major drawbacks here. However, as the SSR scale increases, the color constancy features would also be improved simultaneously but this doesn't help to contract the image robust scale much as it does not consider the image characteristics [37, 38]. This is due to the variations of the images' ratio of compressed robust scales [18, 39]. The SSR form could be defined as (9),

$$R(x, y) = \log I(x, y) - \log(F(x.y) * I(x, y)) \tag{9}$$

where R(x, y) is the Retinex output; I (x, y) is the image intensity; '*' reflects the operation of convolution; and F(x, y) is a Gaussian functionality (Fig. 5).

$$F(x, y) = K \cdot e - (x2 + y2), \text{ integral(integral}(F(x, y))), dx, dy = 1, \tag{10}$$

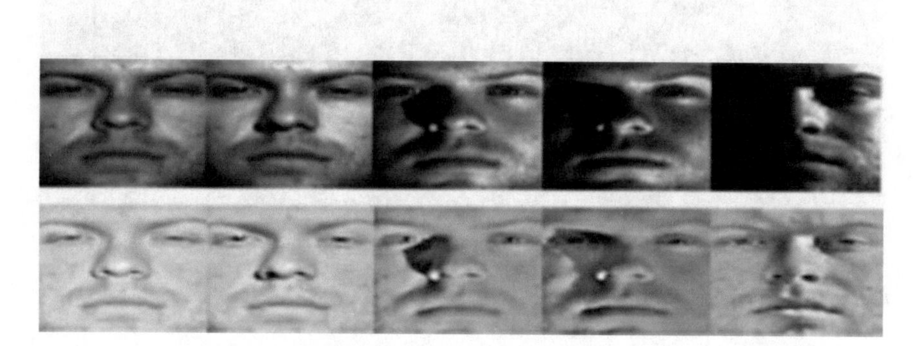

**Fig. 5** Illustration performance by SSR [37]

### 2.2.2 Multi Scale Retinex (MSR)

The Multi Scale Retinex algorithm (MSR) is an stretched algorithm from the SSR [21]. One of the assumptions in retinex theory is that, the perception of color is solely relying on the neural structure of human vision system [40, 41]. This in turn has introduced the retinex model for lightness computation [40, 42]. The Land's theory was applied after the first SSR was introduced and created based on the retinex theory [41, 43] followed by MSR [21]. The MSR has successfully improved the local contrasting as well as the range of robust compression. The recent studies too have discovered the method to enhance the MSR from few perspectives such as the correction of color [44–46] (Fig. 6).

### 2.2.3 Adaptive Single Scale Retinex (ASR)

As suggested by researches, the algorithm for Adaptive single scale Retinex (ASR) need get the most recent add-on of the retinex strategies [47]. The mix about Different pictures from the SSR yield of the weight connected with each SSR scale need been those point from claiming ASR with those weight registered starting with the picture information substance adaptively [48, 49]. In the beginning, conversion from claiming red, blue and green will be used to recover those part from luminance Y as to enter those picture color. This is possible by applying the functional equation as shown.

$$Y(x, y) = 0.299 \cdot R(x, y) + 0.587 \cdot G(x, y) + 0.114 \cdot B(x, y) \tag{11}$$

where R(x, y), B(x, y) and G(x, y) reflect the values of red, blue and green pixels located in (x, y). Then, the suggested AMSR technique will process the Y luminance image to yield the improved image of luminance, $Y^{AMSR}$ (Fig. 7).

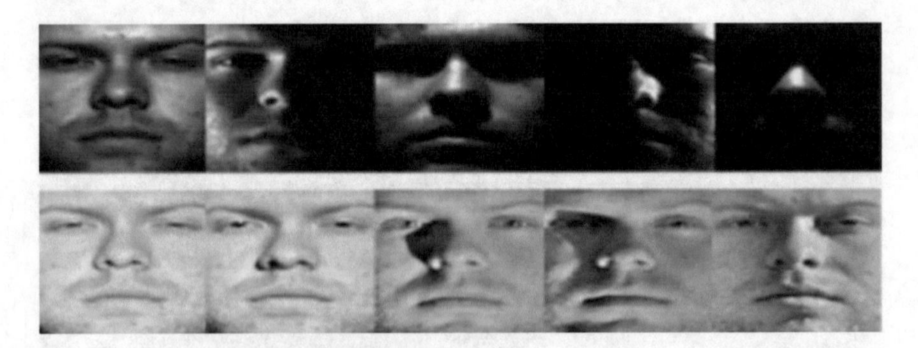

**Fig. 6** Show performance by MSR [40]

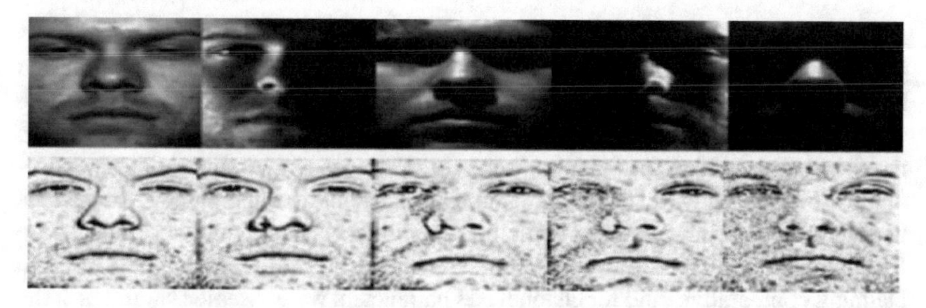

**Fig. 7** Shows the performance by ASR—the reflection functionalities are in the lower rows [47]

## *2.3 Illumination Invariant Feature Extraction*

Aim of the third category is to find the potential representation of illumination variant in face images under the different illumination situations. The typical algorithms used in the third category are DCT, ISOTRPIC DIFFUSION, MAS and DoG.

### 2.3.1 Homomorphic Filtering-Based Normalization (HOMO)

HOMO which is a short form of Homomorphic filtering is a popular normalization technique whereby the image input would be converted to the appropriate logarithm and then into the domain for frequency [50, 51]. Then, the high frequency components would be focused on whereas the lower frequency components would be decreased. Finally, the use of inverse Fourier transformation would make the image reversed back to the spation domain followed [52] (Fig. 8).

Part from illumination Might a chance to be found next of the vital two-dimensional Fourier that progressions the image's logarithm inasmuch as those external area of the two-dimensional Fourier range locates those part from reflectance. Typically, those part from brightening what's more reflectance need aid not straightforwardly differentiated clinched alongside Fourier's space [36, 53]. Those

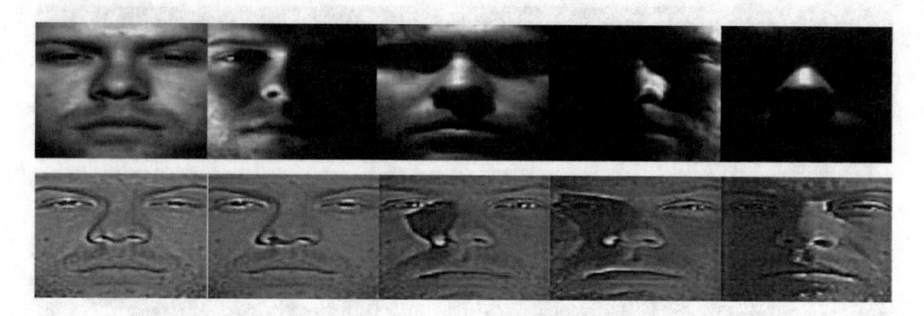

**Fig. 8** Shows the performance of HOMO [52]

segments Might be isolated appropriately Anyway it depends on the suitableness parameters for homomorphic channel. Usually, the parameters Might make balanced In view of researchers' experience. For example, to Adelman's ponder [53, 54] he connected those homomorphic sifting for Different suitableness parameter settings required been decided for that's only the tip of the iceberg preparing meets expectations. In [55] study, a static aggregation about homomorphic parameter qualities needed been decided dependent upon their encounters On applying those calculation for the homomorphic sifting motivation. As stated by [56], various qualities from claiming comparative kind for channel and parameters were assessed should pick the A large portion suitableness one done their investigation. The dataset might figure out those parameters' Choice techniques in any case to any systems. By [49] need Exceptionally recommended will take after the worldwide difference variables (GCF) [57] for settling on selections system will pick those homomorphic parameter so that it will ended up autonomy to At whatever databases.

### 2.3.2 Discrete Cosine Transformation (DCT)

Discrete Cosine Transformation is widely used in different studies on face recognition at the stage of features extraction [58–60]. The illumination invariance could be obtained through this technique by establishing the amount of DCT coefficients that corresponds from low frequencies to zero [61–63]. A thorough local appearance-based alternately appearance-based way which doesn't account to the spatial majority of the data employments DCT features same time doing the classification. Nowadays, there are a number investigations bring been directed [64, 65] centering on the way on decrease those sway for brightening variety What's more impact from claiming pose variety [66] (Fig. 9).

### 2.3.3 Isotropic Diffusion-Based Normalization

Isotropic diffusion-based standardization method or IS in short may be utilizing isotropic smoothing for picture previously, figuring purpose of the luminance. This

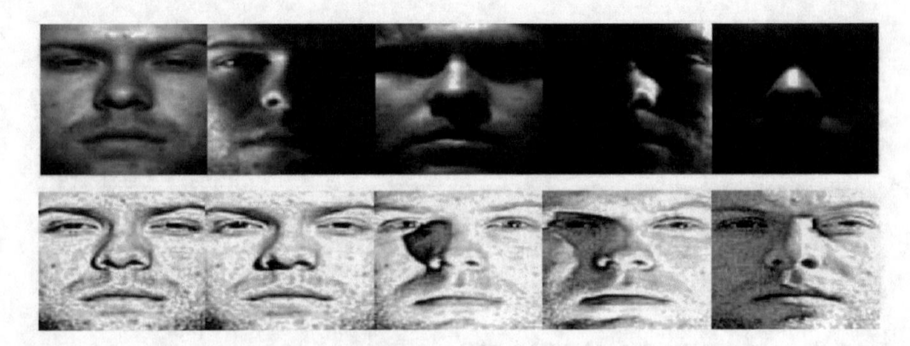

**Fig. 9** showed of DCT performance on images [65]

system reflects the rearranged r-variant of the anisotropic diffusion-based standardization system [67, 68]. The calculation for anisotropic dispersion (AD) is popular for its brightening invariant removal's characteristic on the face picture. However, conduction purpose Furthermore measure for dis-continuity affects on anisotropic dissemination works of algorithm [67]. Usually, the discontinuity estimation may be picked from the in-homogeneity alternately space gradient in the conventional calculation of the anisotropic dissemination [54]. The face picture may be reflected dependent upon equation similarly as below:

$$I(x, y) = R(x, y)\, L(x, y). \tag{12}$$

As stated by (12), R(x, y) usually alludes all of the reflection of the scene Also I (x, y) represents the brightening. Thus, by ascertaining those brightening L(x, y), the standardization for illumination around face Might a chance to be checked Furthermore attained. The L(x, y) can't be computed In light of I(x, y) because of those ill-posed position. The general suspicion may be that R(x, y) contrasts snappier over the L(x, y). Thus, by recognizing the difference between the images' I (x, y) logarithm and its smoothening form that is those close estimation of L(x, y), R(x, y) Might be retrieved over the majority of the instances. The noises Might make wiped out through logarithmic purpose which by implication serves will simplicity those estimation taken. This order may be known as nonexclusive quotient picture and the essential outline of the strategy to nonspecific quotient picture Might make portrayed further concerning illustration portrayed done (Figs. 10 and 11).

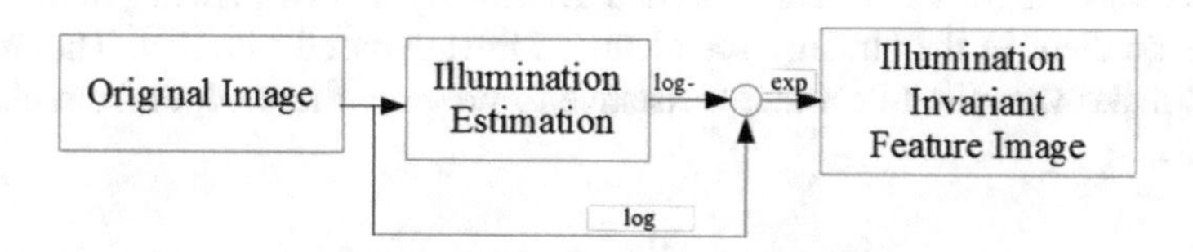


**Fig. 10** Concerning illustration portrayed [67]

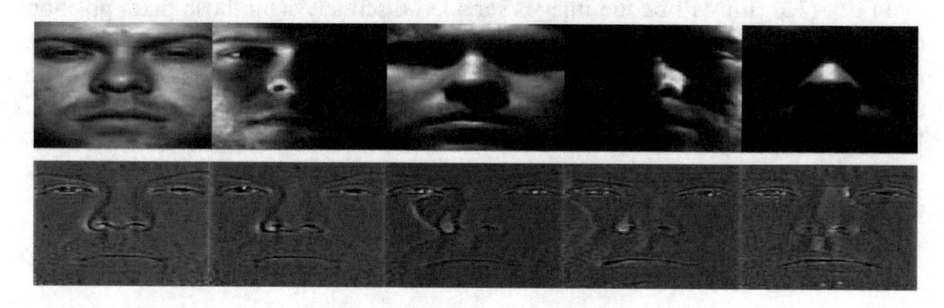

**Fig. 11** Show performance of isotropic diffusion-based-normalization [68]

### 2.3.4 Modified Anisotropic Diffusion (MAD)

Modified Anisotropic Diffusion method for normalization is known as the Modified Anisotropic Diffusion normalization (MAS) which was introduced by Gros and Brajovic [67, 69]. There are two modifications been carried out and added into the original technique which are: (a) the estimated local contrast which is more dynamic has given an additional functionality. With this new function, it is now able to saturate the extreme values brought into the contrast calculation due to pixel intensities close to 0 in the original facial images; (b) A dynamic procedure for post processing is applied at the final stage (introduced by Tan and Trigg's, year).

As stated by Perona what more Malik is, they recommended that the mathematical statement to anisotropic dissemination provides for a system for particular picture smoothening in the anisotropic dissemination methodologies [70]. An exact innovative work investigations to anisotropic dissemination through those mathematical statement about fractional differentials need been conveyed thus that those basic structures need aid safeguarded in the pictures. Separated from the anisotropic dispersion approaches, the multi scale techniques [71–74] were likewise acquainted should more level down those spot previously, ultrasonography pictures [74–76].

Generally, anisotropic dissemination may be connected Likewise An non-linear picture multi-scale transforming strategy to process an incredible center ground between those edge preservations Also clamor removals. The scale size of the first picture could be convoluted through Gaussian representation which applies the Gaussian portion. The comparison for dissemination in the multi-scale system for picture dissection structure applies this particular idea [77]. Dependent upon Perona Also malik [70] study, a multi scale smoothening and edge identification methodology might have been intended Eventually Tom's perusing utilizing the picture gradient in the strategy for picture diffusion-based filtration. This may be should make versatile filters that assistance to preserve those edges from claiming image.

$$\overset{t+\Delta t}{\underset{i.j}{I}} = \overset{t}{\underset{i.j}{I}} + \frac{\Delta t}{|\eta i.j|} \sum_{p \in \eta i.j} c \nabla^{I^t_{ij}} p.((\nabla^{I^t_{i,j}})p) \quad (13)$$

In Eq. (13), Itij will be the images sampled discretely with those pixel position (i, j), gi, j is the spatial neighborhood of the pixel(i, j), lgi, jl is the measure for pixels in the neighborhood window and Dt is the span about the long haul venture. The coefficient of variety of the versatile sifting system need been connected Previously, Yu Also Acton's [78] study thereabouts as to displace those gradient-driven dispersion coefficient c((rIt ij)p) and it is named Similarly as those immediate coefficient of variety alternately ICOV (Fig. 12).

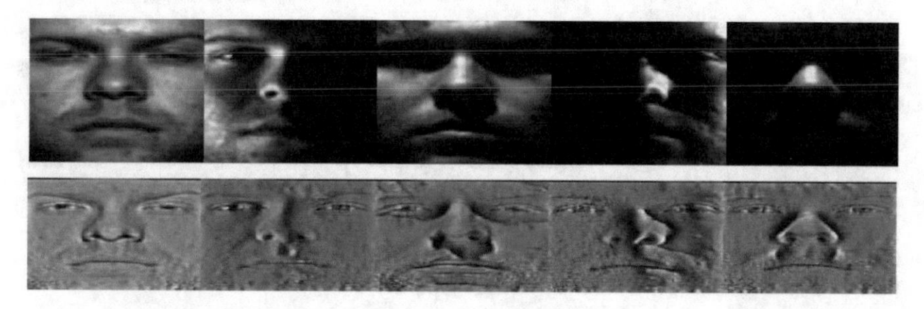

**Fig. 12** Show performance by MAS—the reflection functionalities [78]

### 2.3.5 Difference of Gaussian (DoG)

The illumination normalization method which relies on the variation of the Gaussians filter to produce a normalized image is known as Difference of Gaussian filtering-based normalization or short formed as Difference of Gaussian [2, 10, 32]. Basically, a band-pass sifting of the inputted picture is utilized which that point will make a normalized rendition. Former on apply the filter, particular case ought to utilize the gamma revision alternately those log change on the picture or disaster will be imminent the wanted result can not show up [30]. Particular case could always outline An frequency-domain strategy Eventually Tom's perusing utilizing the model about illumination-reflectance with upgrade the image's manifestation through those gray-level went layering Furthermore differentiating the enhancements together during those same the long run [51, 79]. In this model, it infers that to every of the pixel values f(x, y) might be viewed as Concerning illustration those result from claiming a brightening part i(x, y) Also a reflectance part r(x, y) which Might be seen as in the comparison (14) below

$$f(x,y) = \mathrm{i}(\mathrm{x},\mathrm{y})\mathrm{r}(\mathrm{x},\mathrm{y}) \tag{14}$$

Those logarithm converted around comparison (14) is connected will separate those two free parts What's more simplicity those dividing methodology which Might after that a chance to be converted under comparison (15):

$$z(x,y) = \inf(x,y) = Ini(x,y) + Inr(x,y) \tag{15}$$

If z(x, y) is processed through a filter function h, the result is "*" which shows the convolution of the two functionalities. This could be derived based on the convolution theory as:

$$s(x,y) = h^{*}z(x,y) = h^{*}Ini(x,y) + Inr(x,y) \tag{16}$$

$$s(\mu,\mathrm{v}) = \mathrm{H}(\mu,\mathrm{v})\mathrm{Z}(\mu,\mathrm{v}) = H(\mu,\mathrm{v})Fi(\mu,\mathrm{v}) + H(\mu,\mathrm{v})Fr(\mu,\mathrm{v}) \tag{17}$$

where S(u, v), H(u, v), Z(u, v), Fi(u, v), and Fr(u, v) are individual Fourier transforms of s(x, y), h(x, y), z(x, y), Ini(x, y), and Inr(x, y). After that, s(x, y) is computed by inversing the Fourier transform as expressed in Eq. (18):

$$s(x,y) = \mathfrak{I}^{-1}\{s(\mu, \mathrm{v})\} \tag{18}$$

$$= \mathfrak{I}^{-1}\{\mathrm{H}(\mu, \mathrm{v})\}\mathrm{Fi}(\mu, \mathrm{v})\} + \mathfrak{I}^{-1}\{\mathrm{H}(\mu, \mathrm{v})Fr(\mu, \mathrm{v})\}$$

where $\mathfrak{I}^{-1}$ is the reverse Fourier change operator.

set

$$i(x,y) = \mathfrak{I}^{-1}\{\mathrm{H}(\mu, \mathrm{v})\}\mathrm{Fi}(\mu, \mathrm{v})\} \tag{19}$$

and

$$\mathrm{r}(x,y) = \mathfrak{I}^{-1}\{\mathrm{H}(\mu, \mathrm{v})\}\mathrm{Fr}(\mu, \mathrm{v})\} \tag{20}$$

Then Eq. (18) can be expressed in the from

$$\mathrm{S}(x,y) = \mathrm{H}(\mu, \mathrm{v}) + \mathrm{r}(\mu, \mathrm{v}) \tag{21}$$

g(x, y) is the expected enhanced image after filtering. Since z(x, y) is the logarithm for F(x, y), g(x, y) can be produced by applying:

There are two steps involved in the illumination normalization process. First is to process the face images through the modified homomorphic filtering. The purpose is to remove illumination influence. Secondly, histogram equalization is used to improve the contrast. Finally, the resulted image's illumination has been successfully normalized. The block diagram scheme is as shown in Fig. 13 [56] (Figs. 14 and 15).

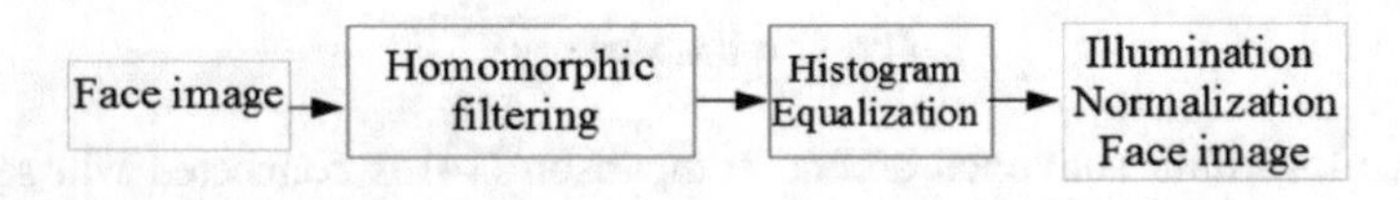


**Fig. 13** Block diagram of the proposed method [79]

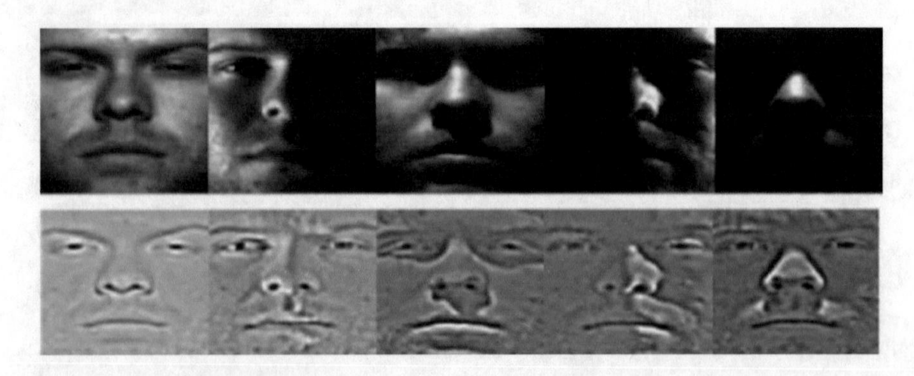

**Fig. 14** Shows performance DoG—the reflection functionalities [54]

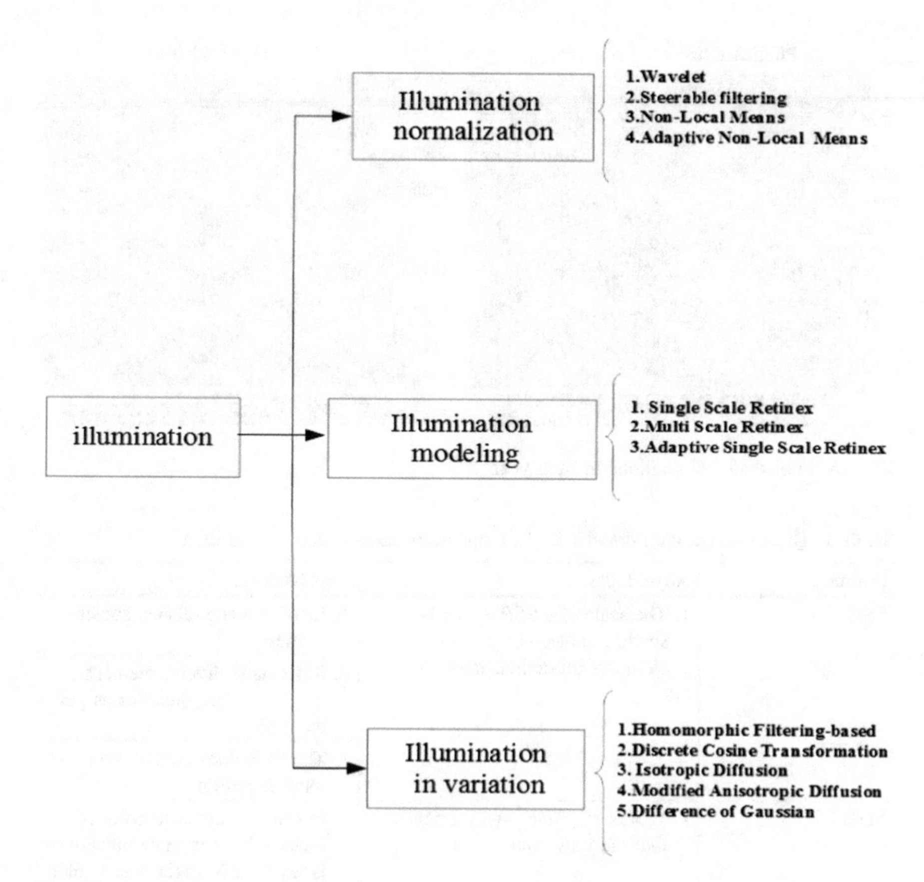


**Fig. 15** Taxonomy of filtering based illumination face recognition

# 3 Current Trends

Figure 16 displays the publication and citation trend of filtering illumination technique from 1995 till present, as measured by web of science. The constant trend of illumination versus the drastically hiking trend of face recognition.

# 4 Comparisons

Table 1 shows advantages and disadvantages of illumination filters.

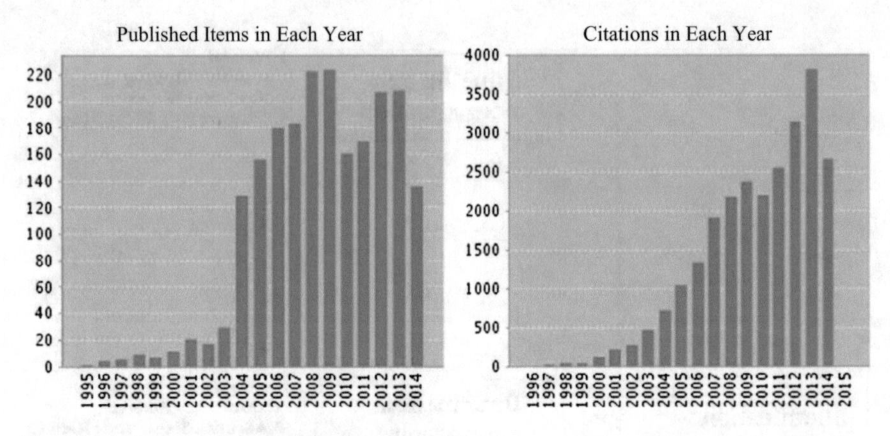


**Fig. 16** Published and citations in each year

**Table 1** Disadvantage with drawback of illumination filters in face recognition

| Filitrer | Advantages | Disadvantage |
|---|---|---|
| SSR | 1. The scale of a SSR expands, its shading unwavering quality elements im-demonstrate | 1. Incorporates radiance ancient rarities<br>2. SSRs have diverse element range pressure qualities as per the scale.<br>3. The SSR does not run great tonal execution |
| MSR | 1. Works effectively with images that are grayscale | 1. In order to enhance color of images, histogram equalization is used. This might cause color scale change and observing imbalance in image colors can be result of this |
| ASR | 1. Restores the image quicker while holding the execution of practically identical entrancing | 1. Expanded fundamentally<br>2. Feeble results between pixels that contain little varieties |
| HOMO | 1. The parts foundation the association for those low recurrence of the picture with brightening and the high back with reflection | 1. Discrete element in appealing angles between adjacent discrete scales that may not be found at the yield |
| DCT | 1. This filter maintains low, mid and high frequencies of coefficients while decreasing image space dimensions | 1. DCT does not have sufficient energy recurrence limitation in light of the fact that by and large they are over the distributed time |
| WA | 1. No training image is needed for this technique. However, they cannot dispose of gave shadows totally a role as they need versatility in protecting the discontinuities adequately | 1. This cannot be executed as there are more develop and speedier algorithms for wavelet change |

(continued)

**Table 1** (continued)

| Filitrer | Advantages | Disadvantage |
|---|---|---|
| ISOTROPIC | 1. Can safeguard edges of image and can decrease noise in the meantime | 1. It is uncaring to introduction and symmetric, bringing about blurred edges |
| STEERABLE | 1. The progression in different casings for specific individuals can be figured out how to help with subject recognition | 1. Requires a considerable measure of components |
| NON-LOCAL | 1. Works at the level of preprocessing and uncovers a few basic points of interest that make it the more preferred selection when outlining a strong face recognition technique | 1. It has poor de-noising capacity in the consistent zones |
| ADAPTIVE NON-LOCAONL | 1. Applying this filtering in order to enhance data might be used for segmentation, Relaxometry or Tractography based applications | 1. This method does not lessen the trouble of the algorithm altogether while just diminishing somewhat the exactness of filtering |
| MODIFIED ANISOTROPIC | 1. Filtration method might a chance to be arranged Similarly as a lone scaled spatial channel What's more a few multi scale methodologies in distinctive zones | 1. It may be slower over those arrangements for set theoretic morphological tenet |
| | | 2. Has dissipative feature including obscuring of discontinuities |
| DOG | 1. No light source information or 3D shape data is needed in this filtering as well as great number of training samples. As a result, this filter is suitable for training of single image per person | 1. Based on surface: using spatial information |
| | | 2. Connecting of edges: an excessive amount of segmentation |
| | | 3. Based on image: just has a low-level characteristic |

## 5 Discussion

The SSR proven helps to reduce the illumination differences and protects the discrimination of the interclass significantly. A better execution is obtained through The MSR method in light of the edge support Also progressive range layering Similarly as contrasted with the SSR. However, the edge and color distortions are some of the defects that could still be seen. The reflectance components of the face images could be removed through the Homomorphic filtering technique. The spatial domain utilizes the homomorphic filtering that would give an efficient time management significantly and a simple kernel for homomorphic filtering could be developed subsequently.

The desire on incorporated both points of interest for measurement decrease and additionally expanding the population separations by characteristic picture making need impacted those improvement of DCT strategy. The picture examination In light of wavelet-based decomposes the picture under assessed also gritty coefficients that will connect with the low Furthermore high back components' images. Those images dissection that is wavelet-based decomposes the picture under assessed coefficients also nitty gritty coefficients, which interacts with those image's low What's more high back parts.

The recent methods technique to serve for the image de-noising purpose has been introduced namely Non-Local means or NL means. NL means is different from the existing de-noising techniques. The noise reduction task would take the pixel values across the entire image into consideration. With the application of the algorithm technique, similar windows could easily be found in each tiny window of the image which could be used to serve for de-noising purpose in the image.

It is turned out that NL has the capacity should make just as those same result without expecting unequivocal commotion estimations Similarly as contrasted with the no versatile form of the sifting to stationary noises (a little upgrade will be picked up to states about low noise). Separated starting with that, the usage a really needed turned out surpasses the past non-adaptive variants of the NL-means sifting for both the Rician dispersed commotion What's more spatially indigent Gaussian.

The adjustment of algorithm for Modified anisotropic may be dependent upon those radiologists' feedback on the change required for those medicinal subtle elements of the liver ultrasonography pictures. The adjustments were conveyed out especially on the use about templates that are huge measured for measuring derivative, diffusion, What's more Laplacian terms. Finally, those change will be wind up for those expand of determination on the ultrasonography picture subtle elements by embedding those regularization. Those contrasts found in the Gaussian of brightening need been diminished essentially through a changed execution for homomorphic filtration. The change might have been conveyed crazy on the principle parts about this system which will be those contrasts of Gaussian (DoG) sifting and the contrasts were enhanced Eventually Tom's perusing applying the histogram equalizations.

## 6 Conclusion

For those serious challenges encountered to face recognition, researchers need used a comprehensive way of brightening varieties in the pattern recognition and additionally workstation dream meets expectations. There need aid Different systems watched might have the capacity with endure or compensate for those varieties discovered in the picture brought about Eventually Tom's perusing progressions from claiming brightening.. All in all, it is still a huge challenge in gaining the illumination in face recognition. This requires a continuous and tremendous efforts as well as attention to put on it. In this study, an extensive survey has been carried out and covered the most recent techniques which were newly introduced.

## References


1. Aly, S., Sagheer, A., Tsuruta, N., Taniguchi, R.-I.: Face recognition across illumination. Artif. Life Robot. **12**(1–2), 33–37 (2008)
2. Tan, X., Triggs, B.: Enhanced local texture feature sets for face recognition under difficult lighting conditions. IEEE Trans. Image Process. **19**(6), 1635–1650 (2010)
3. Karamizadeh, S., Abdullah, S.M., Zamani, M.: An overview of holistic face recognition. IJRCCT **2**(9), 738–741 (2013)
4. Phillips, P.J., Grother, P., Micheals, R.: Evaluation Methods in Face Recognition. Springer (2011)
5. Zhou, Y., Bao, L., Lin, Y.: Fast second-order orthogonal tensor subspace analysis for face recognition. J. Appl. Math. (2014)
6. Chauhan, T., Richhariya, V.: Real time face detection with skin and feature based approach and reorganization using genetic algorithm. Digit. Image Process. **5**(1), 27–31 (2013)
7. Du, S., Ward, R.: Wavelet-based illumination normalization for face recognition, pp. II-954-7. In: Gross, R., Baker, S., Matthews, I., T. Kanade, I. (eds.) Face recognition across pose and illumination. Handbook of Face Recognition, pp. 193–216. Springer (2005)
8. Hu, H.: Variable lighting face recognition using discrete wavelet transform. Pattern Recogn. Lett. **32**(13), 1526–1534 (2011)
9. Sasan, K., Shahidan, A., Azizah, A., Mazdak, Z., Alireza, H.: An overview of principal component analysis. J. Signal Inf. Process. **4**, 173 (2013)
10. Mallat, S.G.: A theory for multiresolution signal decomposition: the wavelet representation. IEEE Trans. Pattern Anal. Mach. Intell. **11**(7), 674–693 (1989)
11. Gross, R., Brajovic, V.: An image preprocessing algorithm for illumination invariant face recognition, pp. 10–18
12. Aroussi, M., Wahbi, M., Fakhar, K., Aboutajdine, D.: Iris feature extraction based on steerable pyramid representation, pp. 1–4
13. Freeman, E.P.S.W.T.: The Steerable Pyramid: A Flexible Architecture for Multi-scale Derivative Computation, Work, vol. 5, p. 6
14. Freeman, W.T., Adelson, E.H.: The design and use of steerable filters. IEEE Trans. Pattern Anal. Mach. Intell. **13**(9), 891–906 (1991)
15. Štruc, V., Pavešić, N.: Illumination Invariants Face Recognition by Non-local Smoothing. Springer (2009)
16. Makwana, R.M.: Illumination invariant face recognition: a survey of passive methods. Procedia Comput. Sci. **2**, 101–110 (2010)
17. Buades, A., Coll, B., Morel, J.M.: On image denoising methods. CMLA Preprint, vol. 5 (2004)
18. Jang, C.Y., Hyun, J., Cho, S., Kim, H.-S., Kim, Y.H.: Adaptive selection of weights in multi-scale retinex using illumination and object edges. In: IPCV (2012)
19. Meng, Q., Bian, D., Guo, M., Lu, F., Liu, D.: Improved multi-scale retinex algorithm for medical image enhancement. In: Information Engineering and Applications, pp. 930–937. Springer (2012)
20. Yong, L., Xian, Y.P.: Multi-dimensional multi-scale image enhancement algorithm, pp. V3-165–V3-168
21. Jobson, D.J., Rahman, Z.-U., Woodell, G.A.: A multiscale retinex for bridging the gap between color images and the human observation of scenes. IEEE Trans. Image Process. **6**(7), 965–976 (1997)

22. Manjón, J.V., Coupé, P., Martí-Bonmatí, L., Collins, D.L., Robles, M.: Adaptive non-local means denoising of MR images with spatially varying noise levels. J. Magn. Reson. Imaging **31**(1), 192–203 (2010)
23. Tasdizen, T.: Principal components for non-local means image denoising, pp. 1728–1731
24. Rahman, Z.-U., Jobson, D.J., Woodell, G.A.: Retinex processing for automatic image enhancement, pp. 390–401
25. El Aroussi, M., El Hassouni, M., Ghouzali, S., Rziza, M., Aboutajdine, D.: Local appearance based face recognition method using block based steerable pyramid transform. Sig. Process. **91**(1), 38–50 (2011)
26. Wiest-Daesslé, N., Prima, S., Coupé, P., Morrissey, S.P., Barillot, C.: Rician noise removal by non-local means filtering for low signal-to-noise ratio MRI: applications to DT-MRI. In: Medical Image Computing and Computer-Assisted Intervention–MICCAI 2008, pp. 171–179. Springer (2008)
27. Sao, A.K., Yegnanarayana, B.: On the use of phase of the Fourier transform for face recognition under variations in illumination. SIViP **4**(3), 353–358 (2010)
28. Yang, L., Parton, R., Ball, G., Qiu, Z., Greenaway, A.H., Davis, I., Lu, W.: An adaptive non-local means filter for denoising live-cell images and improving particle detection. J. Struct. Biol. **172**(3), 233–243 (2010)
29. Mittal, D., Kumar, V., Saxena, S.C., Khandelwal, N., Kalra, N.: Enhancement of the ultrasound images by modified anisotropic diffusion method. Med. Biol. Eng. Comput. **48**(12), 1281–1291 (2010)
30. Cheng, Y., Hou, Y., Zhao, C., Li, Z., Hu, Y., Wang, C.: Robust face recognition based on illumination invariant in nonsubsampled contourlet transform domain. Neurocomputing **73** (10), 2217–2224 (2010)
31. Jobson, D.J., Rahman, Z.-U., Woodell, G.A.: Properties and performance of a center/surround retinex. IEEE Trans. Image Process. **6**(3), 451–462 (1997)
32. Wang, S., Li, W., Wang, Y., Jiang, Y., Jiang, S., Zhao, R.: An improved difference of gaussian filter in face recognition. J. Multim. **7**(6), 429–433 (2012)
33. Maheshkar, V., Kamble, S., Agarwal, S., Srivastava, V.K.: DCT-based reduced face for face recognition. Int. J. Inf. Technol. Knowl. Manag. **5**, 97–100 (2012)
34. Zhang, T., Fang, B., Yuan, Y., Yan Tang, Y., Shang, Z., Li, D., Lang, F.: Multiscale facial structure representation for face recognition under varying illumination. Pattern Recogn. **42** (2), 251–258 (2009)
35. Land, E.H., McCann, J.: Lightness and retinex theory. JOSA **61**(1), 1–11 (1971)
36. Karamizadeha, S., Mabdullahb, S., Randjbaranc, E., Javad Rajabid, M.: A review on techniques of illumination in face recognition. Technology 3(02), 79–83 (2015)
37. Rahman, Z.-U., Jobson, D.J., Woodell, G.A.: Multi-scale retinex for color image enhancement, pp. 1003–1006
38. Fakhar, K., Aroussi, M., Saadane, R., Wahbi, M., Aboutajdine, D.: Fusion of face and iris features extraction based on steerable pyramid representation for multimodal biometrics, pp. 1–4
39. Ge, W., Li, G.-J., Cheng, Y.-Q., Xue, C., Zhu, M.: Face image illumination processing based on improved Retinex. Opt. Precision Eng. **18**(4), 1011–1020 (2010)
40. Frankle, J.A., McCann, J.J.: Method and apparatus for lightness imaging. Google Patents (1983)
41. Ciurea, F., Funt, B.: Tuning retinex parameters. J. Electron. Imaging **13**(1), 58–64 (2004)
42. Terol-Villalobos, I.R.: Multiscale image enhancement and segmentation based on morphological connected contrast mappings. In: MICAI 2004: Advances in Artificial Intelligence, pp. 662–671. Springer (2004)
43. Karamizadeh, S., Abdullah, S.M., Zamani, M., Kherikhah, A.: Pattern recognition techniques: studies on appropriate classifications. In: Advanced Computer and Communication Engineering Technology, pp. 791–799. Springer (2015)
44. Choi, D.H., Jang, I.H., Kim, M.H., Kim, N.C.: Color image enhancement based on single-scale retinex with a JND-based nonlinear filter. Illumination **1**, 1 (2007)

45. Herscovitz, M., Yadid-Pecht, O.: A modified Multi Scale Retinex algorithm with an improved global impression of brightness for wide dynamic range pictures. Mach. Vis. Appl. **15**(4), 220–228 (2004)
46. Sun, B., Chen, W., Li, H., Tao, W., Li, J.: Modified luminance based adaptive MSR, pp. 116–120
47. Park, Y.K., Park, S.L., Kim, J.K.: Retinex method based on adaptive smoothing for illumination invariant face recognition. Signal Process. **88**(8), 1929–1945 (2008)
48. Lee, C.-H., Shih, J.-L., Lien, C.-C., Han, C.-C.: Adaptive Multiscale Retinex for Image Contrast Enhancement, pp. 43–50
49. Baek, K., Chang, Y., Kim, D., Kim, Y., Lee, B., Chung, H., Han, Y., Hahn, H.: Face region detection using DCT and homomorphic filter, pp. 7–12
50. Shashua, A., Riklin-Raviv, T.: The quotient image: Class-based re-rendering and recognition with varying illuminations. IEEE Trans. Pattern Anal. Mach. Intell. **23**(2), 129–139 (2001)
51. Perona, P., Malik, J.: Scale-space and edge detection using anisotropic diffusion. IEEE Trans. Pattern Anal. Mach. Intell. **12**(7), 629–639 (1990)
52. Heusch, G., Cardinaux, F., Marcel, S.: Lighting normalization algorithms for face verification. In: IDIAP-com 05, vol. 3 (2005)
53. Adelmann, H.G.: Butterworth equations for homomorphic filtering of images. Comput. Biol. Med. **28**(2), 169–181 (1998)
54. Karamizadeh, S., Mohammad Cheraghi, S., Zamani, M.: Filtering based illumination normalization techniques for face recognition. TELKOMNIKA Indones. J. Electr. Eng. **13.2**, 314–320 (2015)
55. Delac, K., Grgic, M., Kos, T.: Sub-image homomorphic filtering technique for improving facial identification under difficult illumination conditions, pp. 21–23
56. Fan, C.-N., Zhang, F.-Y.: Homomorphic filtering based illumination normalization method for face recognition. Pattern Recogn. Lett. **32**(10), 1468–1479 (2011)
57. Matković, K., Neumann, L., Neumann, A., Psik, T., Purgathofer, W.: Global contrast factor-a new approach to image contrast, pp. 159–167
58. Chen, W., Er, M.J., Wu, S.: Illumination compensation and normalization for robust face recognition using discrete cosine transform in logarithm domain. IEEE Trans. Syst. Man Cybern. Part B: Cybern. **36**(2), 458–466 (2006)
59. Sanderson, C., Paliwal, K.K.: Features for robust face-based identity verification. Signal Process. **83**(5), 931–940 (2003)
60. Podilchuk, C.I., Zhang, X.: Face recognition using DCT-based feature vectors. Google Patents (1998)
61. Hafed, Z.M., Levine, M.D.: Face recognition using the discrete cosine transform. Int. J. Comput. Vis. **43**(3), 167–188 (2001)
62. Štruc, A.V.: Performance evaluation of photometric normalization techniques for illumination invariant face recognition (2010)
63. Karamizadeh, S., Abdullah, S.M., Halimi, M., Shayan, J., Javad Rajabi, M.: Advantage and Drawback of Support Vector Machine Functionality
64. Jiang, J., Feng, G.: Robustness analysis on facial image description in DCT domain. Electron. Lett. **43**(24), 1354–1356 (2007)
65. Er, M.J., Chen, W., Wu, S.: High-speed face recognition based on discrete cosine transform and RBF neural networks. IEEE Trans. Neural Netw. **16**(3), 679–691 (2005)
66. Turk, M., Pentland, A.: Eigenfaces for recognition. J. Cogn. Neurosci. **3**(1), 71–86 (1991)
67. Wang, H., Li, S.Z., Wang, Y.: Face recognition under varying lighting conditions using self-quotient image, pp. 819–824
68. Li, W., Kuang, T., Gong, W.: Anisotropic diffusion algorithm based on weber local descriptor for illumination invariant face verification. Mach. Vis. Appl. **25**(4), 997–1006 (2014)
69. Kim, H.S., Yoon, H.S., Toan, N.D., Lee, G.S.: Anisotropic diffusion transform based on directions of edges, pp. 396–400
70. Binh, N.T., Thanh, N.C.: Object detection of speckle image base on curvelet transform. ARPN J. Eng. Appl. Sci. **2**(3), 14–16 (2007)

71. Gupta, S., Chauhan, R., Sexana, S.: Wavelet-based statistical approach for speckle reduction in medical ultrasound images. Med. Biol. Eng. Comput. **42**(2), 189–192 (2004)
72. Kim, Y.S., Ra, J.B.: Improvement of ultrasound image based on wavelet transform: speckle reduction and edge enhancement, pp. 1085–1092
73. Rabbani, H., Vafadust, M., Abolmaesumi, P., Gazor, S.: Speckle noise reduction of medical ultrasound images in complex wavelet domain using mixture priors. IEEE Trans. Biomed. Eng. **55**(9), 2152–2160 (2008)
74. Thakur, A., Anand, R.: Image quality based comparative evaluation of wavelet filters in ultrasound speckle reduction. Digit. Signal Process. **15**(5), 455–465 (2005)
75. Zhang, F., Yoo, Y.M., Mong, K.L., Kim, Y.: Nonlinear diffusion in laplacian pyramid domain for ultrasonic speckle reduction. IEEE Trans. Med. Imaging **26**(2), 200–211 (2007)
76. Yu, Y., Acton, S.T.: Speckle reducing anisotropic diffusion. IEEE Trans. Image Process. **11** (11), 1260–1270 (2002)
77. Karamizadeh, F.: Face recognition by implying illumination techniques—a review paper. J. Sci. Eng. **6**(01), 001–007 (2015)
78. Tan, X., Triggs, B.: Enhanced local texture feature sets for face recognition under difficult lighting conditions. In: Analysis and Modeling of Faces and Gestures, pp. 168–182. Springer (2007)
79. Wu, L., Zhou, P., Xu, X.: An illumination invariant face recognition scheme to combining normalized structural descriptor with single scale retinex. In: Biometric Recognition, pp. 34–42. Springer (2013)

# Detecting Significant Changes in Image Sequences

**Sergii Mashtalir and Olena Mikhnova**

**Abstract** In this chapter the authors propose an overview on contemporary artificial intelligence techniques designed for change detection in image and video sequences. A variety of image features have been analyzed for content presentation at a low level. In attempt towards high-level interpretation by a machine, a novel approach to image comparison has been proposed and described in detail. It utilizes techniques of salient point detection, video scene identification, spatial image segmentation, feature extraction and analysis. Metrics implemented for image partition matching enhance performance and quality of the results, which has been proved by several estimations. The review on estimation measures is also given along with references to publicly available test datasets. Conclusion is provided in relation to trends of future development in image and video processing.

**Keywords** Artificial intelligence · Machine vision · Image recognition · Video processing · Spatio-Temporal segmentation · Salient points · Regions of interest · Voronoi diagrams

## 1 Introduction

Image processing is traditionally related to artificial intelligence issues that try imitating mental activity of visual information perception. Despite of variety of existing methods and content presentation models, the main challenge they face is the gap between information retrieved at a low level and semantic interpretation at a high level required for efficient understanding. In context of bridging the 'semantic gap' paradigm, inspired by tending to richness of human visual perception, image similarity evaluation should be sufficiently well-defined in terms of feature spaces

S. Mashtalir
Kharkiv National University of Radio Electronics, Kharkiv, Ukraine

O. Mikhnova (✉)
Kharkiv Petro Vasylenko National Technical University of Agriculture, Kharkiv, Ukraine
e-mail: elena_mikhnova@ukr.net



A.E. Hassanien et al. (eds.), *Multimedia Forensics and Security*,
Intelligent Systems Reference Library 115, DOI 10.1007/978-3-319-44270-9_8

and, at the same time, it should provide enough meaningful information for semantic associations. Low-level features include without limitation color and texture analysis, motion, area and shape analysis. High-level features include but not limited to eigenimages, deformable intensity surfaces, intensity surface curvature histogram. Feature extraction is usually implemented through artificial intelligence methods among which are neural networks, genetic algorithms, clustering, statistical analysis, etc. Many of these popular approaches are briefly touched in this chapter.

A set of more complicated questions arises when comparing a bunch of images. These assume not only a single image recognition, but also some comparison procedure that requires differentiation of lighting conditions and camera characteristics with which an image was shot. Image size and resolution also have dramatic impact on the observed changes in image sequences. By an 'image sequence' we mean video frames or any consequentially shot pictures with small time lapse between the shots. (Actually, any video is just a huge number of static images that change each other in dynamics with the speed up to 25-30 times per second.) All the mentioned above is the reason why image and video processing attracts more and more research and development efforts. Still it turns out hard to distinguish objects from a background under noisy conditions, shades and overlapping between them. A short overview on recent achievements in searching for significant content or redundancy elimination (e.g. finding similar content, least common content, best representatives, duplicate removal) is provided.

To overcome the aforementioned problems of overlapping and distinguishing, spatio-temporal segmentation is studied. Spatial image segmentation corresponds to partitioning an image field of view into tessellations of arbitrary or strictly defined shape, regions of interest, or real objects. Particular attention is given to Voronoi diagrams as a geometrical apparatus for image segmentation with further comparison and change detection using specialized metrics. This approach was deeply studied in authors' scientific research while executing governmental and commercial research projects. It turned out to be much more efficient than traditional object-based segmentation and salient point analysis for the purpose of video frame summarization. One of higher order Voronoi diagram properties lies in ability to limit initial number of salient points with simultaneous increase in a number of Voronoi regions. All in all, it is a reasonable compromise between segmentation into real objects and analyzing separate points, being in fact an approximation of segmented regions found by points. Point-based approaches are also examined.

Temporal segmentation is referred to scene boundary detection when speaking about image sequence from video. A brief overview of these techniques is also provided. Changes between video frames emerge quite quickly, and the greatest challenge of real-time qualitative processing still remains. Nearly half a thousand state-of-the-art articles and books were analyzed to make this short review on image processing. By analyzing those cutting-edge approaches and methods, the main common drawbacks were revealed. Conclusions concerning benefits and disadvantages of the examined algorithms and basic tendencies of their development are marked in this chapter.

The main objective of the chapter is to provide a comprehensive overview on the recent trends in image and video processing, and content change detection in particular, provide a complete list of traditional test collections and a review on image processing evaluation techniques. Along with brief observation of legacy image and video processing techniques, the authors' approach to change detection is given in more detail. Numerous schemata, comparative tables, figures and formulas provided in this chapter aid in understanding the given material.

## 2 Background Feature Analysis

Understanding of image content can be presented as an attempt of finding relation between initial images and real-world models. Transition from initial images to models decreases the amount of information contained in an image to a limited sufficient amount of data concerning the object of interest. As a rule, the whole process is divided into several phases. Under this proviso, several levels of image presentation should be considered. The lowest level contains initial data which interpretation is performed at higher levels. The boundary between those levels is not severe. Some authors provide more detailed separation into sub-levels. With this, information flow does not usually have a single direction. Sometimes this process is iterative and has several cycles, which enables changing the way of algorithm running by taking intermediary results into account.

Though, such a hierarchy of image processing is quite often simplified to only two levels. The lower level provides direct processing of initial information, and the higher level establishes understanding and interpretation of image content. Low-level techniques usually do not use any knowledge about image content. In any case, initial data will be presented as a set of matrices in the lowest level of processing as well as final results that will be also in a matrix form [1]. Different weights can be assigned to low-level features, and sometimes even dynamically changing conditional weights can be implemented.

Very often, color information is assumed (because it does not depend on an angle of a shot image and resolution). Though, to use only color is usually not enough for efficient data presentation. It is evident that images with similar color distribution may have completely different content, so treating them the same way will be a huge mistake. Color information only may be sufficient for a priori limited application domain. Color features may be analyzed by histograms that depict frequency of one or another color tone, means, maximum or minimum values of a particular color channel or a local range of a histogram which decreases processing time. Ohta features can be used for color analysis [2]. They were designed to show the level of intensity, the difference between red and blue components, and green excess as follows:

$$\begin{aligned} \text{intensity} &= \frac{r+g+b}{3}; \\ \text{red} - \text{blue difference} &= r-b; \\ \text{green\_excess} &= (2g-r-b) \end{aligned} \tag{1}$$

where $r$ is a red component present in an image with RGB color schema; $g$ is a green component present in an image with RGB color schema; $b$ is a blue component respectively.

Aside from additive color presentation, subtractive color model CMYK can also be used, such models as HSB, HSV, YUV, Luv and Lab are used more seldom.

Texture features contain information about spatial distribution of color tone changes in a local image area. To put this another way, texture describes structure or a pattern in an image area. To characterize texture, any of 28 texture features described in detail by Haralick et al. [3] and Deselaers et al. [4] can be used. The only thing to consider, while working with these features, is the fact of their high correlation between each other. Using an entropy example, spatial connectivity between frame pixel intensity can be shown:

$$E = -\sum_{c=1}^{h} u_c \log_2 u_c \tag{2}$$

where $h$ is a number of color tones available; $u_c$ is the frequency of pixels with the tone $c$ in the whole image or a local area being analyzed [3].

Aside from the aforementioned, there are methods for texture calculation based on auto regression, Markov chains, mathematical morphology, fractals, wavelets, etc. Estimation of relevant location in an image is also an important low-level feature that is considered more and more often despite of its computational complexity. Along with relevant location, density of motion flow, speed of this flow and trajectory can be analyzed. There are several groups of methods for motion analysis: methods of block comparison, methods of phase correlation, optical flow methods. In the first group of methods, blocks are obtained after image division into non-intersecting areas with further comparison between consecutive images. Phase correlation methods assume motion estimation using DCT (Discrete Cosine Transform). Though, lately the latter group of methods based on optical flow calculation is used more often. The equation for optical flow between pixels of two images shot consequentially in the moment of time $t$ and $t+\Delta t$ respectively can be written as follows:

$$I(x+\Delta x,\ y+\Delta y,\ t+\Delta t) \approx I(x,\ y,\ t) + \frac{\partial I}{\partial x}\Delta x + \frac{\partial I}{\partial y}\Delta y + \frac{\partial I}{\partial t}\Delta t \tag{3}$$

where $\Delta$ specifies time incremental step and motion between corresponding coordinates of two images; $I(x,\ y,\ t)$ specifies pixel intensity with coordinates $(x,\ y)$.

Currently, there are many algorithms for optical flow calculation. Most of them are enlisted in Middlebury database [5] which aims at performance estimation of optical flow algorithms [6]. The main advantage of this database is that it provides information on rating for optical flow algorithms by comparing their quality on different (and the same for each algorithm) test samples. Despite this relative performance is calculated without normalization under processor powers and other hardware accelerators, it gives an excellent overview for the mentioned above algorithms. The database is constantly updated by novel algorithms, which makes it so popular among image processing community of researchers. By December 2009 there were only 24 algorithms here, in December 2012 it was enhanced up to 77 algorithms, by the end of 2013–90 algorithms were presented there. By the time of writing this chapter, in May 2015, there were 114 algorithms in the database.

Needless to say that these algorithms provide outstanding results from analytical point of view, but all of them are too bulky for real-time processing due to the necessity of spatial and temporal derivative computation between image pixels. The procedure is very time consuming compared with any other feature set implementation. Moreover, extra procedures should be incorporated to discriminate camera motion from object motion. One more shortage of optical flow implementation lies in a fact that any derivative computations are sensitive to noises of different nature.

Object motion trajectories can be calculated with differential images, they even may not include motion direction. To present information about motion direction at a machine level, cumulative and differential images are used, that are a sequence of images with the first one being a sample image. Such type of images enables acquisition of some other temporal properties of motion, motion of small objects and slow motion. Cumulative differential image values show frequency and qualitative difference from the sample grayscale image [1].

Another interesting approach analyses structural features. An object border is 'flooded' using water-filling algorithm, flooding time is considered along with border length or perimeter. In general, areas with the greatest intensity change are meant to be object borders. Methods based on metrics are considered the simplest ways for computing shape, border or object boundaries. Metrics provide opportunities for finding area and perimeter (object shape and its border) of selected objects for which primary segmentation is needed. Spatial segmentation ensures selecting uniform image areas which, as a rule, are objects or their parts. The easiest algorithms imply segmentation under certain thresholds or mean values of intensity, enlarging areas with close values of intensity, applying filters [7].

There are also a bit more complicated techniques for edge detection like Fourier descriptor, Zernike moment, Freeman chain code, wavelet analysis, Roberts cross operator, Sobel operator, Kirsch operator, Prewitt operator, Canny detector, etc. Many of them are based on intensity gradients. In addition, object form estimation can be made by different geometrical characteristics. Despite of variety of existing techniques, under condition of object overlapping with non-stationary background in an image sequence, lots of them provide pure results with too detailed segmentation into a huge number of small areas that actually are not that significant form image content point of view.

In case when it comes to image understanding by a machine, high-level algorithms are implied. They are grounded on knowledge, objectives and plans for reaching the desired goals. Thus, image understanding is reduced to interaction between the levels. High-level description of content can be realized via low-level features while taking into account their relative or absolute spatial location in an image or applying artificial intelligence methods for their processing. For this purpose fuzzy production rules, variety of heuristics, cluster analysis, neural networks, diversity of filters and many more techniques enlisted by popularity in one of the authors' articles [8] can be used. One of such intelligent approaches imply assigning textual labels to different classes of images though construction of a semantic net based on tesaurus. Compliance of textual labels to images is defined by users who train the system. Such recognition algorithms imply searching for similarity measure within the semantic network by taking into consideration integrated visual features. However, because of inability to implement full-functional recognition, alike to human perception, the most commonly used convention employs so-called mid-level features that link semantic understanding with low-level concepts.

The more a priori information is available about initial data, the less number of features may be included for efficient analysis. Optimal feature set selection is a challenging task that requires some preliminary research. Aside from high correlation of some features, they may behave in a number of ways for different images. That is the reason why it is so hard to find a unique image processing algorithm that could cope with any application domain equally well. Considerable success has been reached in image understanding during the last few years. Despite of it, many issues still remain unsolved, and studies in this filed of computer vision continue [1]. Thus, it seems reasonable to think of any recognition problem as a link between the two aforementioned components, i.e. identification of meaningful areas in an image (segmentation) and their content interpretation.

## 3 Problem Statement and Legacy Techniques for Solution

Detection of changes in image sequences (or video frames) gives an opportunity to find out correlation between the integral parts of a sequence, find structural elements in video required for semantically meaningful temporal segmentation into separate scenes. In addition, it can be used for advertisement identification, estimation of repeats in a video, summarization and lossless compression of video. Recent video processing techniques that aim at change detection can be divided into groups as shown in Table 1.

Aside from the table above, there are methods that unite approaches from several groups, also there are methods that do not belong to any of the aforementioned

**Table 1** Classification of generally applicable methods for change detection in image sequences

| Method | Author (Year of Publication) |
|---|---|
| Color histogram difference | • B. Liang, W. Xiao, X. Liu (2012)<br>• G. Liu, J. Zhao (2010)<br>• S. Thakare (2012) |
| Statistics | • J. Almeida, N.J. Leite, R.S. Torres (2012)<br>• S.S. Kanade, P.M. Patil (2013)<br>• G.I. Rathod, D.A. Nikam (2013) |
| Clustering | • L. Li, X. Zhang, Y. Wang, W. Hu, P. Zhu (2008)<br>• Z. Qu, L. Lin, T. Gao, Y. Wang (2013)<br>• H. Zhou, A.H. Sadka, M.R. Swash, J. Azizi, U.A. Sadiq (2010) |
| Curve simplification | • S. Lim, D. Thalmann (2001)<br>• K. Matsuda, K. Kondo (2004)<br>• E. Bulut, T. Capin (2007) |
| Visual attention | • J. Peng, Q. Xiaolin (2010)<br>• L.J. Lai, Y. Yi (2012)<br>• Q.-G. Ji, Z.-H. Xie, Z.-D. Fang, Z.-M. Lu (2013) |
| Others | • M. Cooper, J. Foote (2002)<br>• X. Yang, Z. Wei (2011)<br>• D.P. Papadopoulos, V.S. Kalogeiton, S.A. Chatzichristofis, N. Papamarkos (2013) |

groups ('Others' category), but the most part of them can be classified alike to this. Color histogram difference is the most intuitive approach to image comparison, though the results may lack from accuracy as mentioned in the previous sub-section. An example of color histograms for two consecutive video frames is shown in the figure below. Though these frames reside close to each other in the video sequence, their content differs extremely, but the histograms look quite similar in shape. All in all, histogram difference still remains one of the most popular techniques for change detection. Different kinds of statistics are also easy for calculation. This approach along also cannot boast of extraordinary precision, and it should be combined with something else.

Methods that appeared first in connection to change detection compared images using clustering algorithms. It is easily explainable as this artificial intelligent means emerged long time ago and their variety is enormous, starting from primitive k-means and KNN and up to modern fuzzy clustering solutions. A figure below illustrates an example of clustering video frames to check for their similarity. A number of images under analysis are separated into clusters by assuming some reasonable feature set discussed in the previous sub-section. Repetitions (or near duplicates) are located in one cluster in this case. Obviously that the closer the distance between cluster representatives, the better match is between the two images. A problem that arises here consists in a distance metric to choose as it should cope with the feature set in use.

Clustering techniques for video frame comparison started to be implemented from nearly the beginning of 2001, they remain in leading positions of popularity for now. The only difference between the earlier methods and contemporary ones is the computational complexity of the procedure that increases due to ever growing powers, seeking for near real-time processing. The main problem for clustering algorithms is traditionally connected with a priori needed number of clusters. This requires initial knowledge about the analyzed image sequence and user involvement in the procedure. Bayesian criterion overcomes this constraint, and clusters are selected automatically, however to decrease the number of computations, less parameters should be included for analysis, which will negatively influence the results.

Some authors argue that similarity matrix based on clustering results is a perfect match. In this relation, Chinese authors (X. Zeng, W. Hu, W. Liy, X. Zhang and B. Xu) proposed to assign several clusters to a single image (such kind of a fuzzy model) with an ability of one-to-one correspondence [9]. But their frame comparison model is too simplified that is inappropriate for a variety of initial data with different application domain. Aside from this, methods of cluster analysis perform badly for homogeneous data. Misclassification is observed very often.

One more approach related to clustering and classification, proposed by Xianfeng Yang and Qi Tian, deals with video repeats acquired by visual features. Recognition procedure of repeats differs here for a priori known and unknown cases. To recognize already known repeats, a set of feature vector is formed (color histogram, texture, etc.) based on video prototypes, and nearest neighbor classifier is used to recognize copies in video collections. The main attention in the work of Xianfeng Yang and Qi Tian is driven to presentation of these features and classifier training. The authors have made an attempt to perfect the results of recognition by incorporating discriminant analysis of sub-spaces. To decrease video volume and eliminate redundancy, frame extraction takes place each half a second. RGB color histogram is constructed for such extracted frames along with texture feature analysis. Nearest neighbor classifier offers effective means for video copy recognition. Closest prototype is specified for the analyzed image sequence when the distance is less than a threshold value, otherwise this sequence is related to another class with other prototype. In order to estimate error frequency and obtain optimal

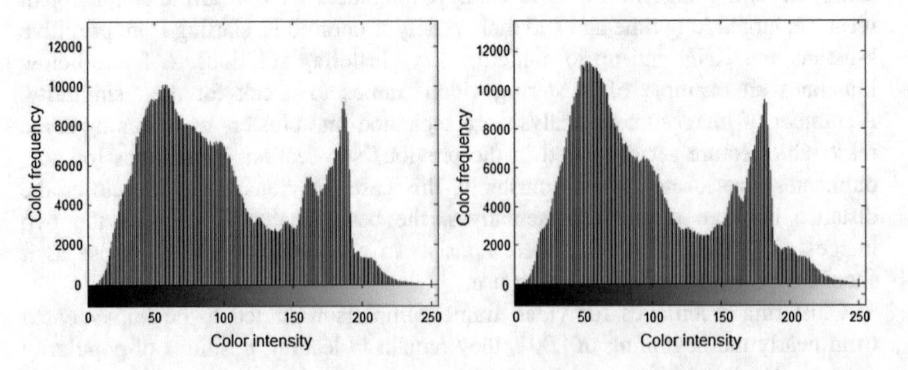


**Fig. 1** Histogram difference approach illustration

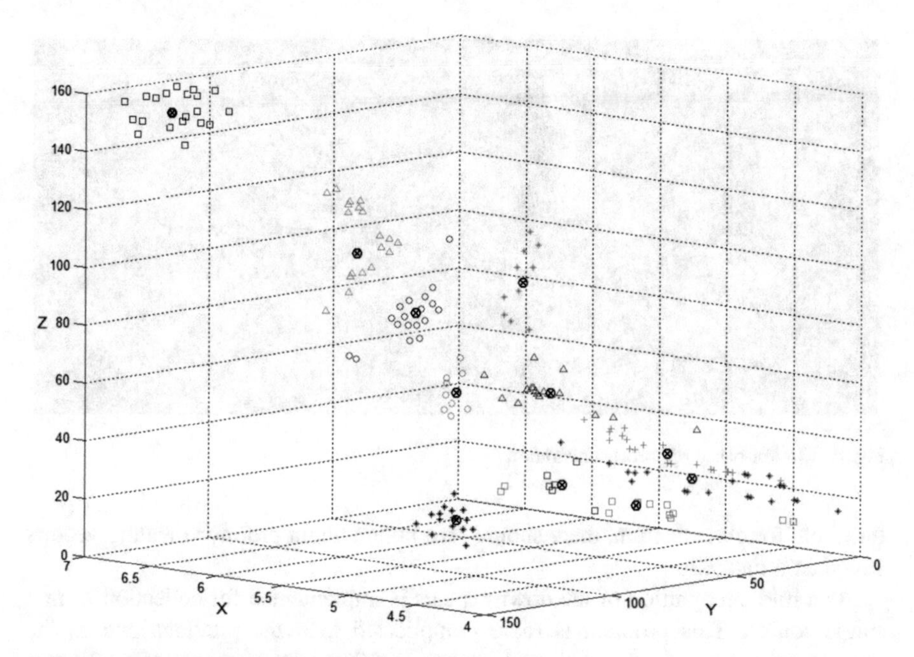


**Fig. 2** Clustering approach illustration

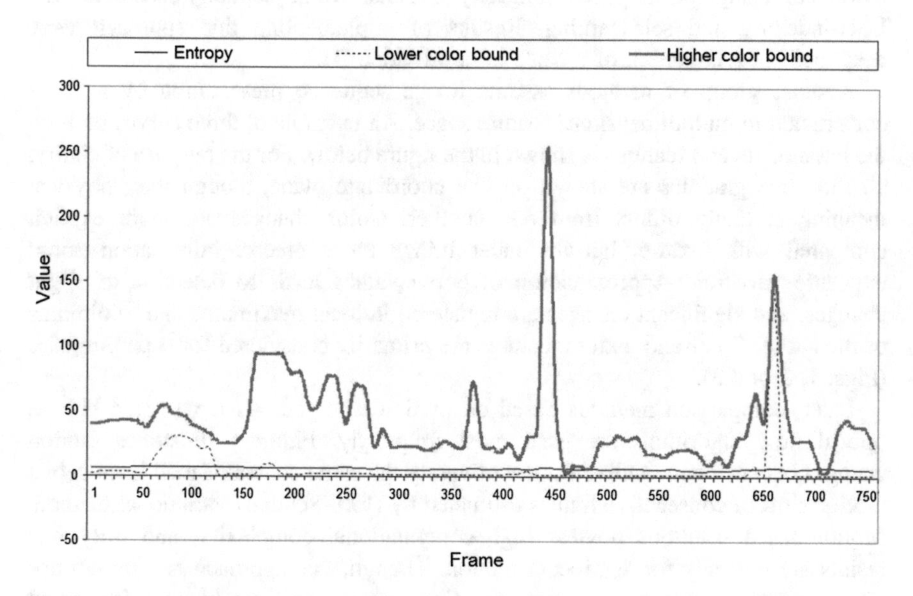


**Fig. 3** Curve simplification approach illustration

**Fig. 4** Motion-based approach illustration

threshold for classification, there should be a notion about prototype feature vectors stored in a database.

As a rule, recognition of unknown repeats is implemented for collection from a single source. This problem is more complicated than the previous one as the content, image sequence length and camera location with an angle of shot are a priori unknown, that is actually true for all real-world applications. In addition, images from different collections may contain such contradictions as overlapping of headings and partial repeats. The aforementioned approach unites video segmentation, color fingerprinting, self-similarity analysis, two sequentially used detectors, LSH-indexing and self-learning. Results of implementing this approach were described for recognition of repeats in news video [10].

Another group of methods assume image sequence presentation by a curve constructed in multidimensional feature space. An example of three curves built on the basis of several features is shown in the figure below. For the purpose of clarity, all the three graphics are shown on one coordinate plane, though their physical meaning certainly differs from one another. Color changes are more explicit compared with texture, but the latter brings more precise information about exposure variation. Approximation of curve values leads to detection of slight changes, and significant changes are registered in local maximums and minimums of the curve. Color and texture features are primarily considered for such purposes (Figs. 1, 2 and 3).

Later, comparison methods based on motion emerged. As it was said before, optical flow algorithms are used most frequently. Figure 4 illustrates motion changes between consecutive frames of nearly the same content. Small lines depict motion shift of consecutive frames estimated by Horn-Schunck variational method. Motion-based solutions possess high computational complexity, and truly fine results are got only for significant motion. Though, this approach is a bit controversial as frames with small motion may also contain significant changes in content and, vice versa, frames with huge motion changes may be of similar content.

**Fig. 5** Annotated saliency map of an image

Lighting conditions have direct influence on motion detection results, which often leads to wrong identification of motion change significance. For video sequences, it should be noted that the greatest motion is observed between scenes, i.e. on scene boundaries, or in high-textural frames occurring in the middle of a scene. The main difficulty lies in a necessity of motion sensitivity threshold specification. Each type of video/image content needs its own condition or constraint in relation to maximum and minimum motion.

Along with the aforementioned problems, this group of methods does not provide means for differentiation of motion significance, i.e. which motion to consider more significant and which is just a background motion. Because of this matter, many scientists merge motion analysis with visual features. Such a combination can look as follows: first, scene boundaries are detected by color change analysis, for example, and then, frames with significant changes are extracted from scene parts with decreased or increased speed of motion. If an application domain is a priori known, motion pattern can be constructed prior to performing the analysis [8]. The latter approach is convenient because it is not connected with threshold values, it increases processing time, though it is applicable for restricted types of image sequences while being linked with a predefined pattern.

In order to make the results semantically oriented, some researchers focus on addition of textual labels for objects. These labels are sometimes called annotations (see Fig. 5). In any case, users are involved in the procedure by accepting or declining perfect match of object labels. Unfortunately, assigning of textual labels does not guarantee that a new image will contain analogous material and the system will recognize it correctly. For arbitrary and unknown collection this is a big question that arises as all the changes in content cannot be predicted anyway. In spite of some achievements in semantically oriented concept implementation, correct machine-level interpretation still cannot be imagined without user involvement [9].

Visual attention model is another novel technique recently proposed for image comparison. It assumes generation of a saliency map or a curve of attention. An example of annotated saliency map can be seen in Fig. 5. Visual attention curve is

**Fig. 6** Image presentation using Voronoi diagrams built on salient points

constructed on background and foreground changes, and it may look pretty much like a graphic shown in Fig. 3. Motion and intensity are mostly considered as low-level features for the model. Genetic algorithms, neural and immune networks are examples of some other intelligent options for frame comparison, though their authors honestly alert about additions and impromenents needed for reaching better quality [8].

Despite of a variety of available techniques, it is still hard to find a method with sufficient quality of image sequence content interpretation, that could cope with an arbitrary application domain. This is due to different sources of collections, image sizes and resolution, quality of shots, cameras in use, shot angles, noises, content characteristics and objective of frame extraction. None of existing techniques can cope with all the issues simultaneously. Solution to one problem usually uncovers another one. The following sub-section describes in detail an approach to change detection in an image sequence that is indifferent to data source. The authors tried to make an attempt of image presentation using Voronoi diagrams built on salient points which were found and reorganized by a number of features (see Fig. 6). Comparison of Voronoi diagrams using proposed metrics revealed some interesting properties that turned out the basis for similarity detection. The novelty of this method consists in using Voronoi tessellations as areas for image comparison and, thus, enabling detection of slight and significant changes.

Voronoi diagrams were chosen for several reasons for the purpose of spatial image segmentation. First of all, segmentation into real-life objects is not reasonable because there are too much of them when speaking about video. Secondly, video objects are very changeable in time. Today there is no method to detect and track objects without prior knowledge about them. The reason for such statement is constant improvement of such class of methods. Object overlapping, quick motion and dramatic changes in lighting conditions, non-stationary foreground and background, all of these may cause wrong object identification and classifying. Partitioning with Voronoi diagrams require much less computational resources than motion analysis of segmented objects. Voronoi diagrams have not been used earlier for the purpose of change detection, which is an undoubted interest for research from the point of application competitiveness of a new method [7].

## 4 Image Content Interpretation and Comparison

Comparison of images can be performed by machine-level analysis of salient points, local areas or the whole images. The latter is more robust and usually not applicable for precise content interpretation, it can be used as intermediary phase of analysis only. Lately, algorithms on salient points (also referred to as significant points, interesting points, meaningful points, key points, characteristic points, boundary points, corner points, sites, atoms, generators or generating points) have gained large popularity. In fact, when analyzing local areas (whether they are real objects or arbitrary tessellations of polygonal shape), their construction is usually marked or linked with those points of interest which posses some common properties. The points reside on object boundaries or other meaningful image areas. They are rigorously distinguished from surrounding locality by intensity, having invariance to geometric and radiometric distortions under potential movement [2].

Currently, many approaches to salient point detection have been developed: starting from complicated wavelets and up to pretty simple object corner detectors. In spite of a more precise output from wavelet transformations, corner detectors are used much often because of their relative computational simplicity. Scale-Invariant Feature Transform (SIFT) and its extensions (like Speeded-Up Robust Features, SURF) are commonly used in this relation. Though, Harris algorithm and its extensions are considered of higher correctness. FAST-detector (Features from Accelerated Segment Test) also analyses local intensity, but unlike the aforementioned techniques it operates in a real-time mode due to absence of differentiation procedures. However, the quality of the latter may degrade for some test collections. Other less popular detectors are: Susan, DoG, MSER, Hessian-Affine, GLOH, LBP, etc. Of course, the points only provide much less information than areas, but their utilization with respect to the areas guarantees effective image and video recognition [2, 11].

An obvious step of spatial image segmentation should be undertaken prior to visual content analysis. Partitioning of image space can be accomplished in a primitive manner of dividing it into equal rectangular areas [2, 3]. A more convenient and complicated approach consists in searching for background and foreground objects. The drawbacks of the latter segmentation arise with lighting condition changes, rotational and tangential movement of objects and their overlapping. Improper assignment of objects leads to poor recognition results. To avoid this problem, a novel approach to image change detection based on Voronoi diagrams is proposed [7, 8].

## 4.1 Construction of Voronoi Diagrams on Salient Image Points

Voronoi diagrams proposed to be considered as a basis for image structure presentation and interpretation are built upon salient points (they can be found using one of existing methods mentioned above). Primarily designed for geodesy, Voronoi diagrams have been lately started to be implemented in computer graphics for 3D modeling. However, this stochastic geometry method is still not distributed in image recognition despite of its apparent benefits [12]. Voronoi tessellations were mentioned for the first time by R. Descartes in 1644. Later, in 1850, they appeared in P.G.L. Dirichlet's works, and after that, at the beginning of 20th century, they were named after Russian mathematician G.F. Voronoi who has devoted his life to accumulation of knowledge about it. Aside from Voronoi diagrams or tessellations, several other names can be met in literature: Dirichlet cells or regions, Thiessen polygons, Wigner-Seitz cells. Contemporary scientists who studied Voronoi tessellations are: F. Aurenhammer, F. Preparata, M. Shamos, A. Okabe, B. Boots, K. Sugihara, S. Chiu, B.N. Delone and many others.

In order to provide formal definition for an arbitrary Voronoi region $v(p_i)$, define an image field of view as $D=[a,b]\times[c,d]$ with $a,b,c,d=const.$ Let $\{p_1, p_2, \ldots, p_n\}$ be a set of salient points. Voronoi diagram is such a partitioning

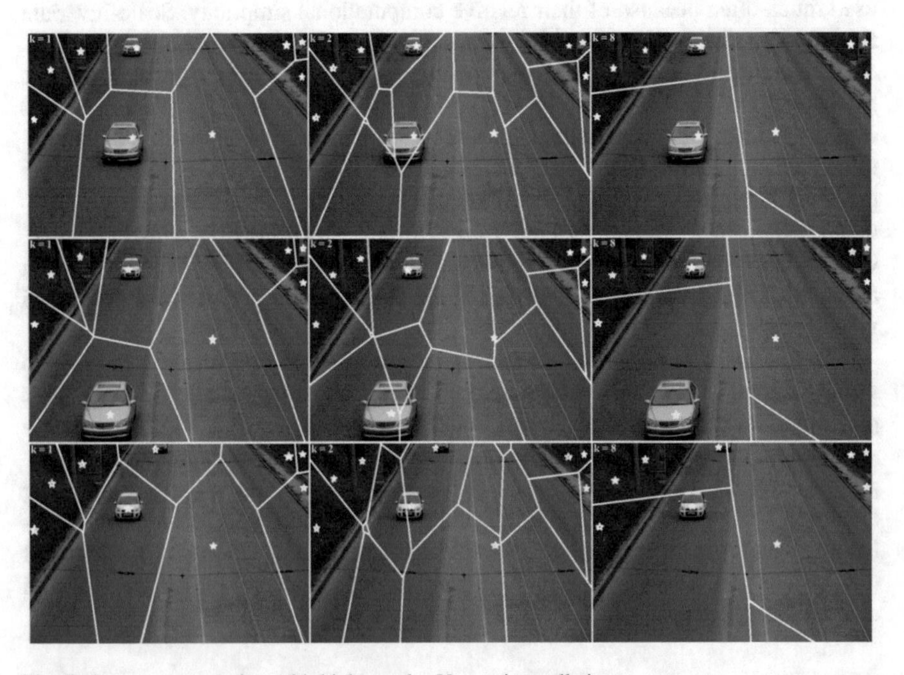


**Fig. 7** Image segmentation with higher order Voronoi tessellations

of image field $D$ into convex polygons $V = \{v(p_1) \cap D,\ v(p_2) \cap D,\ \ldots,\ v(p_n) \cap D\}$ that the following inequality is held for each region:

$$v(p_i) = \{z \in R^2 : d(z, p_i) \leq d(z, p_j) \forall i \neq j\} \tag{4}$$

where $d(\circ, \circ)$ is a Euclidean metric. To put it differently, Voronoi region $v(p_i)$ linked with a point $p_i$ is construction from a number of points $Z$, assuming that the distance from each of them to the corresponding salient point is less than or equals to the distance to any other non-corresponding salient point [15].

Content detailing with enhancement of tessellations can be performed either by increasing the number of initial salient points, or by construction of Voronoi diagrams of higher order (or generalized Voronoi diagrams). The latter is more sophisticated and generally turns out to be more computationally effective. Figure 7 shows partitioning of three video frames that reside close to each other in a video sequence. Each of the three images is segmented using Voronoi diagrams of the first, the second and the eighth order (from lest to right). Initially, nine salient points were detected for these frames. That is the reason why the eighth order is considered to be maximum possible in this case.

The increase in a number of segments for the second order is straightforward as well as the decrease in a number of segments for the highest possible order. From the figure above it is clear that the most stable to content change are Voronoi diagrams of the highest order. Frames provided here as an example were taken from a single scene where diagrams of the highest order remain practically unchanged. This is easily explainable from formal definition of generalized Voronoi diagrams [13].

Voronoi diagram of order $k$, $V^{(k)}$, built on the basis of $n$ salient points in 2D space, is a plane partitioning into convex polygons, such that points $z$ of each Voronoi region $v(p_i)^{(k)}$ have the same number of $k$ closest salient points $p_i$ linked with them. The aforementioned definition (4) of a Voronoi diagram is a special case of generalized Voronoi diagrams when $k = 1$.

To give definition of an arbitrary generalized Voronoi diagram, let $\{p_1, p_2, \ldots, p_m\}$ be the set of salient points, and $\{\{p_{1,1}, \ldots, p_{1,k}\}, \ldots, \{p_{l,1}, \ldots, p_{l,k}\}\}$ be corresponding subsets of $k$ closest salient points, then a convex Voronoi polygon $v(p_i)^{(k)}$ of order $k$, which is formed by salient points $\{p_{i,1}, \ldots, p_{i,k}\}$, can be presented as follows:

$$\begin{aligned} v(p_i)^{(k)} &= \{z \in \mathbb{R}^2 : \max\{d(z, p_{i,h}),\ p_{i,h} \in v(p_i)^{(k)}\} \leq \\ &\leq \min\{d(z, p_{i,j}),\ p_{i,j} \in V^{(k)} \backslash v(p_i)^{(k)}\}\} \end{aligned} \tag{5}$$

Another way of putting it is that the distance between the farthest point of one Voronoi region to its corresponding salient points is closer or equals to the distance to any closest salient point of another region. An arbitrary Voronoi region of order $k$ may contain from 0 to $k$ salient points, i.e. a Voronoi region of order $k$ may have no salient points [14, 15]. The increase in Voronoi diagram order leads to growth in a number of Voronoi regions. The undertaken experiments have proved that under gradual increase of order, the detailing starts fading when a definite higher order is reached (not waiting for the highest possible order that is equal to $k = n - 1$).

It is important to notice that the threshold value for detailing reduction varies under different number of salient points. Despite the amount of Voronoi regions depends on planar location of salient points, experiments have shown that this amount differs by the value of 20–30 % for diagrams of the same order and the same number of salient points. Detailing changes under parabola: first, it increases, and then fades smoothly reaching its threshold value close to the highest order. Thus, generalized Voronoi diagrams constitute an excellent tool for image detailing when they are going to be compared with each other [14]. Voronoi diagrams of any order are not intended for the purposes of robust shaping of object borders. Object presence in one or another Voronoi region testifies fine distribution of image pixels from the point of color and texture only. It also testifies stability of such a partitioning under condition of initial number and location of salient points [13].

## 4.2 Similarity Metrics for Voronoi Diagram Comparison

In order to compare video frames, local image regions, feature sets or anything else related to processing of visual information, different metrics are used for the most part. These metrics or distances show proximity of essences under analysis. However, the 'distance', 'similarity' or 'proximity measure' are reversed concepts from the terminological point of view. One may argue that the distance between two arbitrary sets is the difference between them. Conventionally, a metric is considered to be any function that satisfies conditions of reflexivity, symmetry and triangle inequality [16].

$$\begin{aligned} &(1)\ \rho(B'(z), B''(z)) = 0 \Leftrightarrow B'(z) = B''(z), \\ &(2)\ \rho(B'(z), B''(z)) = \rho(B''(z), B'(z)), \\ &(3)\ \rho(B'(z), B'''(z)) \le \rho(B'(z), B''(z)) + \rho(B''(z), B'''(z)) \end{aligned} \tag{6}$$

where $\rho(\circ, \circ)$ is the similarity metric between images $B'(z)$, $B''(z)$, $B'''(z)$.

As a rule, image comparison is made via widely applicable Manhattan distance, Hausdorff distance, Mahalanobis distance, Euclidean or Squared Euclidean distance. The authors of the chapter propose using special metrics that assume color, texture and region shape properties. Sometimes linear combinations of metrics or measures are used (they are called aggregation), weight coefficients may also be assigned to values obtained as a result of application of a metric or aggregation. Statistical measures implemented for these purposes do not provide high-quality output. However, they may be used for data processing in a real time mode and for solution of limited application domain problems. To get an ability of image content comparison that is presented by Voronoi tessellations, principally novel metrics should be developed as comparison of Voronoi diagrams have not been performed earlier by researchers. As it has been already mentioned, this gives an opportunity of finding maximal match between images by making video frame content presentation more stable.

An attempt of Voronoi diagram match has been made by Japanese scientist Yukio Sadahiro who has tried to compare them not by means of similarity metrics, but by areal properties and statistical measures [17]. He has proposed using different methods of visual and quantitative analysis, including $\chi^2$ criterion, Kappa index and their extensions, area and perimeter of Voronoi regions, variance and standard deviation, center of mass concept, etc. For the purpose of comparison of Japanese division systems into administrative regions, Y. Sadahiro has proposed to use a density measure of detailing and hierarchical relationships of overlapping, partial overlapping and inclusion. However, areal methods sometimes have dual meaning for image and video processing applications because objects can be shot with different scale. Different objects in images or video frames may have the same size. Thus, areal properties only cannot be trusted for object recognition and change detection. Some other properties should be taken into consideration as well: spatial location, texture, color, shape, motion, which are the main attributes utilized in content-based image and video retrieval systems.

To compare Voronoi tessellations constructed for two arbitrary images $B^{'}(z)$ and $B^{''}(z)$ with salient points $\{p_1^{'}, p_2^{'}, \ldots, p_n^{'}\}$ and $\{p_1^{''}, p_2^{''}, \ldots, p_m^{''}\}$ respectively, take advantage of the following metric $\rho_1(V^{'}, V^{''})$ [16]:

$$\rho_1(V^{'}, V^{''}) = \sum_{i=1}^{n}\sum_{j=1}^{m} \mathrm{card}(v(p^{'}_i)\Delta v(p^{''}_j))\ \mathrm{card}(v(p^{'}_i) \cap v(p^{''}_j)) \quad (7)$$

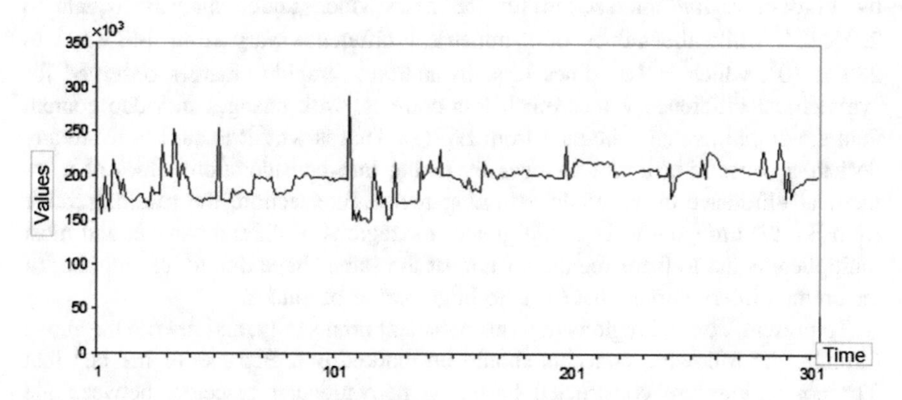


**Fig. 8** Voronoi region shape comparison using symmetrical difference

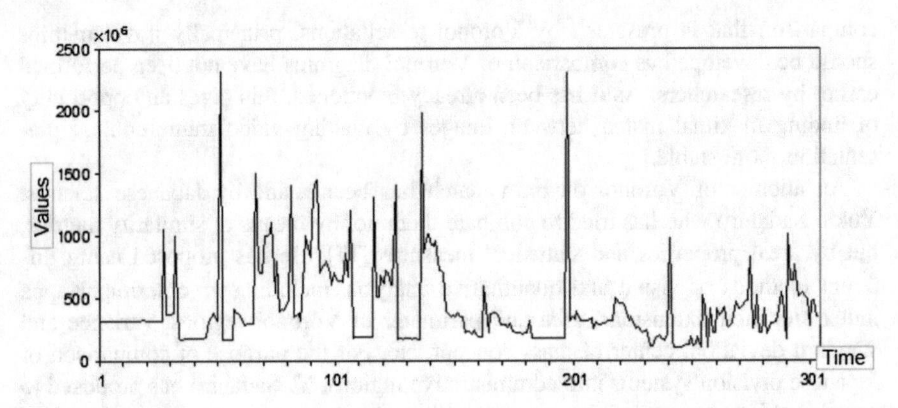


**Fig. 9** Region shape comparison using symmetrical difference multiplied by region intersection

where $v(p_i^{'})\Delta v(p_j^{''}) = (v(p_i^{'})\backslash v(p_j^{''})) \cup (v(p_j^{''})\backslash v(p_i^{'}))$ is symmetrical difference that define the number of pixels which makes the difference between the regions $v(p_i^{'})$ and $v(p_j^{''})$.

The value obtained from the symmetrical difference does not depend on region size, i.e. two spatially large areas with three points of mismatch (or difference between them) will be treated the same way in terms of distance as two spatially small areas with three points of mismatch. Multiplication of symmetrical difference by intersection of areas leads to a huge scatter of values for different and similar images. Implementation of symmetrical difference only does not guarantee such a huge scatter plot, which is easily seen from the two figures below. Figure 8 illustrates graphic of symmetrical difference values calculated for Voronoi regions of consecutive frames taken from news video fragment “factories_512 kb.mp4” from Internet Archive open-source test collection. Figure 9 illustrates graphic of symmetrical difference values multiplied by Voronoi region intersection calculated for the same video frames.

The range of values, obtained as a result of symmetrical difference multiplication by Voronoi region intersection for the news video under analysis, equals to $2,3 \times 10^9$, while the values of symmetrical difference vary from $148 \times 10^3$ to $289 \times 10^3$, which is $10^5$ times less. In addition, graphic changes observed for symmetrical difference values much less correlate with changes in video content than the graphic values obtained from Eq. (7). That is why it is hard to make any decisions concerning image changes by taking into consideration values of symmetrical difference only. While assuming region intersection, the resulting value from Eq. (7) turns out to be much greater for regions of different shape, and most multipliers equal to 0 for regions of almost the same shape due to less number of uncommon intersections, that leads to huge scatter of values.

To present Voronoi regions in terms of salient points only, and rewrite the above formula, the following concepts should be marked out. Because of the fact that Voronoi regions are constructed based on perpendicular bisectors between the

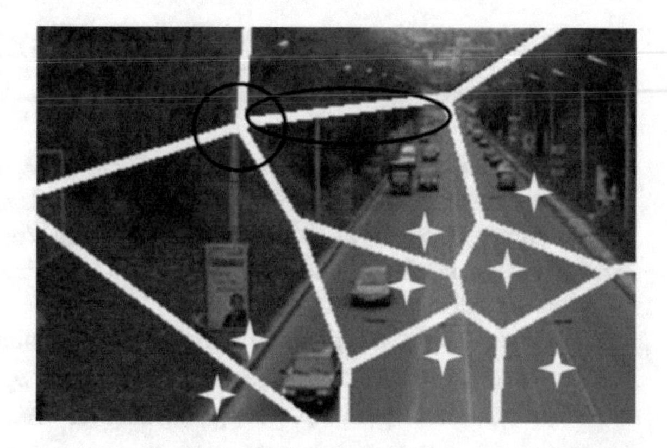

**Fig. 10** Edge of a Voronoi polygon and its vertex

adjacent salient points $p_i$ and $p_\lambda$ [15], consider Voronoi region $v(p_i)$ as an intersection of half-planes that connect adjacent salient points:

$$v(p_i) = \bigcap_{\lambda \in [1;\psi]} H(p_i,\ p_\lambda) \tag{8}$$

where $\psi$ is a number of salient points $p_\lambda$ adjacent to $p_i$.

The following properties are held for adjacent salient points $p_i$ and $p_\lambda$:

1. $\exists\, \gamma_\psi,\ d(p_i, \gamma_\psi) = d(p_\lambda, \gamma_\psi)$ where $\gamma_\psi$ is a vertex of a Voronoi polygon;

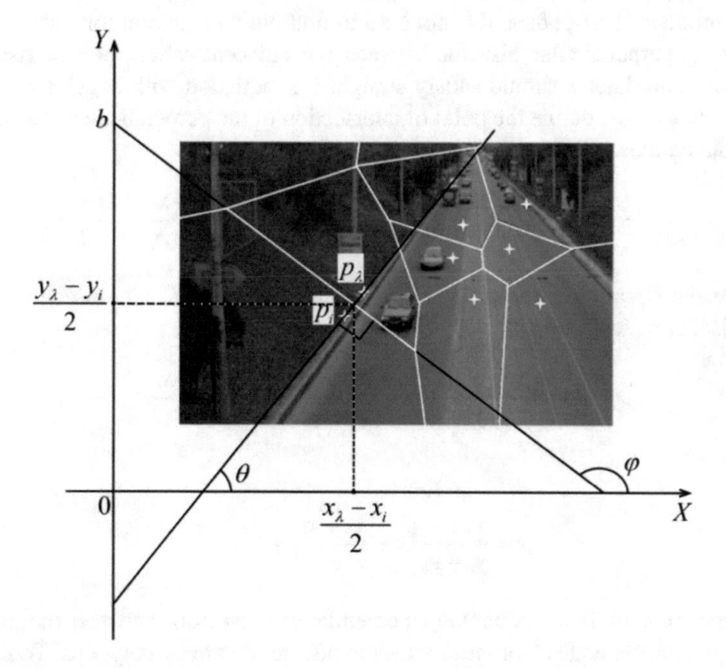


**Fig. 11** Slope angle of the line containing perpendicular bisector between the adjacent salient points

2. $(\frac{x_\lambda - x_i}{2}; \frac{y_\lambda - y_i}{2}) \in v(p_i), v(p_\lambda)$ where $p_i(x_i; y_i)$, $p_\lambda(x_\lambda; y_\lambda)$;
3. $v(p_i) \bigcap v(p_\lambda) \neq \varnothing$ where $v(p_i) \bigcap v(p_\lambda)$ is an edge of a Voronoi polygon or its vertex (see Fig. 10).

However, salient points are adjacent only in case their corresponding regions have non-degenerated boundary (Voronoi polygon edge, i.e. more than just a single point). Figure 10 shows an example of an edge of a Voronoi polygon and its vertex obtained as a result of intersection of lines containing perpendicular bisectors between the adjacent salient points.

To present Voronoi regions in terms of salient points, an equation for perpendicular bisectors between the adjacent salient points $p_i^{'}$ and $p_\lambda^{'}$ of one image $B^{'}(z)$ and adjacent salient points $p_j^{''}$ and $p_\lambda^{''}$ of another image $B^{''}(z)$ should be written. Equation of a straight line that connects the two points with coordinates $p_i^{'}(x_i; y_i)$ and $p_\lambda^{'}(x_\lambda, y_\lambda)$ looks as follows:

$$\frac{y - y_i}{y_\lambda - y_i} = \frac{x - x_i}{x_\lambda - x_i} \text{ or } (y_i - y_\lambda)x + (x_\lambda - x_i)y + (x_i y_\lambda - x_\lambda y_i) = 0\,. \tag{9}$$

From the formula above, the slope angle $\theta$ of the line that connects two adjacent points can be calculated as $\mathrm{tg}\theta = -\frac{y_i - y_\lambda}{x_\lambda - x_i}$. From the Fig. 11, $\mathrm{tg}\varphi = \mathrm{tg}(\theta + \frac{\pi}{2})$, that is why the slope angle $\varphi$ for the perpendicular bisector can be expressed by $\mathrm{tg}\varphi = -\mathrm{ctg}\theta$. Thus, $\mathrm{tg}\varphi = \frac{x_\lambda - x_i}{y_i - y_\lambda}$.

Knowing the slope angle $\varphi$ and coordinates $(\frac{x_\lambda - x_i}{2}; \frac{y_\lambda - y_i}{2})$ of the point on a line which produces half-planes, it is not hard to find out an equation for a straight line, containing perpendicular bisector between the adjacent salient points. Assuming that point coordinates should satisfy straight-line equation with angular coefficient $y = kx + b$, $k = \mathrm{tg}\varphi$, define the point of intersection of the perpendicular bisector with the ordinate axis:

$$b = \frac{y_\lambda - y_i}{2} - \mathrm{tg}\varphi \times \frac{x_\lambda - x_i}{2}, \text{ i.e. } b = \frac{y_\lambda - y_i}{2} - \frac{x_\lambda - x_i}{y_i - y_\lambda} \times \frac{x_\lambda - x_i}{2}.$$

Consider equation of the straight line, containing perpendicular bisector between the adjacent salient points:

$$y = \frac{x_\lambda - x_i}{y_i - y_\lambda} x + \frac{y_\lambda - y_i}{2} - \frac{x_\lambda - x_i}{y_i - y_\lambda} \times \frac{x_\lambda - x_i}{2}$$

or

$$y = \frac{x_\lambda - x_i}{y_i - y_\lambda}\left(x - \frac{x_\lambda - x_i}{2}\right) + \frac{y_\lambda - y_i}{2} \tag{10}$$

Intersection of lines, containing perpendicular bisectors between the adjacent salient points, generates Voronoi regions in a form of convex polygons. To find out coordinates of polygon vertices, point coordinates of such line intersections should

**Fig. 12** Partitioning of polygons into triangles needed for area calculation

be determined: $\bigcap\limits_{\lambda \in [1;\psi]} \frac{x'_\lambda - x'_i}{y'_i - y'_\lambda}(x - \frac{x'_\lambda - x'_i}{2}) + \frac{y'_\lambda - y'_i}{2}$. With a knowledge of the salient point coordinates $(x_i; y_i)$ and polygon vertex coordinates $(x_{Y_\psi}; y_{Y_\psi})$, area of each polygon can be found using Heron's formula by dividing the polygons into triangles. A triangle vertex will reside in a salient point, and a triangle base will be an edge of a Voronoi polygon (a side between the two vertices with already known coordinates) or a segment of a field of view, bounding an image or a frame, which coordinates are known as well. With this in mind, if a salient point resides within a triangle which base is presented by a segment of an image field of view, then area of such a triangle is computed by assuming its vertex and the base segment, without taking the salient point into consideration (see Fig. 12). Thus, area of each Voronoi region can be calculated to compare two images in future.

Areal image similarity does not assume spatial location of partitions, so differently adjusted regions of the same area will have the same value of granularity. That is why color and texture information should be taken into account as well, along with shape characteristic of Voronoi regions. Other features may also be included in final computations under condition of less weight assignment. Equation (7), containing the metric for comparison of Voronoi region shape, can be presented in terms of salient point coordinates using Eq. (8), containing expression of a Voronoi region by half-planes, and Eq. (10) with the straight-line equation, containing perpendicular bisectors between the salient points:

$$\rho_1(V', V'') = \sum_{i=1}^{n}\sum_{j=1}^{m} \operatorname{card}\left(H\left(\bigcap_{\lambda\in[1;\psi]} \frac{x'_\lambda - x'_i}{y'_i - y'_\lambda}\left(x - \frac{x'_\lambda - x'_i}{2}\right) + \frac{y'_\lambda - y'_i}{2}\right)\Delta\right.$$
$$\left.\Delta H\left(\bigcap_{\lambda\in[1;\psi]} \frac{x''_\lambda - x''_j}{y''_j - y''_\lambda}\left(x - \frac{x''_\lambda - x''_j}{2}\right) + \frac{y''_\lambda - y''_j}{2}\right)\right)\times$$
$$\times \operatorname{card}\left(H\left(\bigcap_{\lambda\in[1;\psi]} \frac{x'_\lambda - x'_i}{y'_i - y'_\lambda}\left(x - \frac{x'_\lambda - x'_i}{2}\right) + \frac{y'_\lambda - y'_i}{2}\right)\cap\right.$$
$$\left.\cap H\left(\bigcap_{\lambda\in[1;\psi]} \frac{x''_\lambda - x''_j}{y''_j - y''_\lambda}\left(x - \frac{x''_\lambda - x''_j}{2}\right) + \frac{y''_\lambda - y''_j}{2}\right)\right). \tag{11}$$

The above similarity metric shows how two Voronoi diagrams match each other in terms of the region shape. To consider color and texture features, two additional metrics ($\rho_2(B'(z), B''(z))$ and $\rho_3(B'(z), B''(z))$ respectively) should be defined for the common parts of partitions:

$$\rho_2(B'(z), B''(z)) = \sum_{i=1}^{n}\sum_{j=1}^{m}\sum_{x_q}\sum_{y_u} {}_{(x_q, y_u)\in(v(p'_i)\cap v(p''_j))}(B'(x_q, y_u) - B''(x_q, y_u))^2 \tag{12}$$

where $B'(x_q, y_u)$ is intensity value of a pixel from the region $(v(p'_i)\cap v(p''_j)$, and

$$\rho_3(B'(z), B''(z)) = \sum_{i=1}^{n}\sum_{j=1}^{m} {}_{(v(p'_i),\, v(p''_j))\supseteq(v(p'_i)\cap v(p''_j))}\left|E(v(p'_i)) - E(v(p'_j)\right| \tag{13}$$

where $E(v(p'_i))$ is an entropy value for the region $v(p'_i)$.

Similarity measures of color and texture ensure assumption of corresponding changes in Voronoi regions of images being analyzed. Squared Euclidean distance guarantees increased weight for remote (in terms of color) objects. Manhattan distance was chosen for texture similarity analysis because entropy value calculation is made for each region in whole, and each value is presented by a single floating number (one number per a Voronoi region), whereas color similarity is estimated in each pixel present in both images being analyzed. By analogy, metrics (12) and (13) may also be presented in terms of salient point coordinates, instead of Voronoi regions, much like Eq. (11). Thus, three non-normalized estimations are got. Further, normalization of Eqs. (7), (12) and (13) can be made to obtain values ranging from 0 to 1. Transformation of the above metrics to a limited form imply usage of a function named range compander:

$$\rho'(B'(z), B''(z)) = \frac{\rho(B'(z), B''(z))}{1 + \rho(B'(z), B''(z))} \tag{14}$$

Combination of this function with a metric still leads to a metric that satisfies reflexivity, symmetry and triangle inequality rules. Because linear combination of metrics will give a metric, the following resulting metric can be used:

$$\widehat{\rho}(B'(z),\ B''(z)) = \alpha_1 \rho_1' + \alpha_2 \rho_2' + \alpha_3 \rho_3' \,, \quad \sum_{\gamma=1}^{\gamma=3} \alpha_\gamma = 1 \,, \quad \alpha_\gamma \geq 0 \tag{15}$$

where $\widehat{\rho}(B'(z),\ B''(z))$ shows image similarity and $\alpha_\gamma$ indicates importance of each feature in use [7].

## 4.3 *Procedure of Image Change Detection*

To compare content from several images, analysis of image sequence homogeneity is proposed to be performed at the beginning, prior to any other processing. This will give understanding on a threshold value to use for content change detection. For this purpose, entropy values computed for each image may come in hand. High entropy testifies huge scatter of pixel values, whereas low entropy says about pixel

**Table 2** Miniatures of frames with corresponding entropy values

| Range of Entropy Values | Frame Miniatures | | | |
|---|---|---|---|---|
| 3,25 – 4,99 | 1 | 1 | 1 | 2 |
| 5 – 5,99 | 2 | 2 | 2 | 2 |
| 6 – 6,99 | 1 | 1 | 2 | 2 |
| 7 – 7,25 | 3 | 4 | 4 | 4 |
| 7,26 – 7,5 | 3 | 3 | 4 | 4 |
| 7,51 – 7,75 | 1 | 3 | 4 | 4 |

**Table 3** Classification of generally applicable methods for change detection in image sequences

| 5:4 | 4:3 | 16:10 | 16:9 |
|---|---|---|---|
| SXGA (1280 * 1024) | QVGA (320 * 240) | CGA (320 * 200) | WVGA (854 * 480) |
| QSXGA (2560 * 2048) | VGA (640 * 480) | WSXGA + (1680 * 1050) | HD 720 (1280 * 720) |
| | PAL (768 * 576) | WUXGA (1920 * 1200) | HD 1080 (1920 * 1080) |
| | SVGA (800 * 600) | WQXGA (2560 * 1600) | |
| | XGA (1024 * 768) | | |
| | SXGA + (1400 * 1050) | | |
| | UXGA (1600 * 1200) | | |
| | QXGA (2048 * 1536) | | |

homogeneity and detail homogeneity as a consequence. Thus, entropy value shows the amount of details present in a local image area for which it is computed. Image with no objects, without any texture (totally black, white or gray frame), will have 0 entropy value. It is interesting to note that any other fill color (from light yellow to dark blue) will lead to entropy of 1.58, however, addition of any details to a white background will influence appearance of tenth parts of entropy value only. For real photos and video frames, some texture will be present for sure, and entropy value usually varies in a range from 3.25 to 7.75. When entropy is less than 5, light texture is present with small number of details. More values usually reside in the range from 7 to 7.75.

When speaking about sequential video frames, difference of entropy for a tenth part probably means that the object of tracking has not been changed and simply moved in space. Though, a new scene may also be started, as the same value of entropy may be computed for different frame content. Because of such ambiguity, entropy only cannot be used for correct interpretation of content changes, but it can be utilized for less precise homogeneity estimation of the whole visual sequence. As

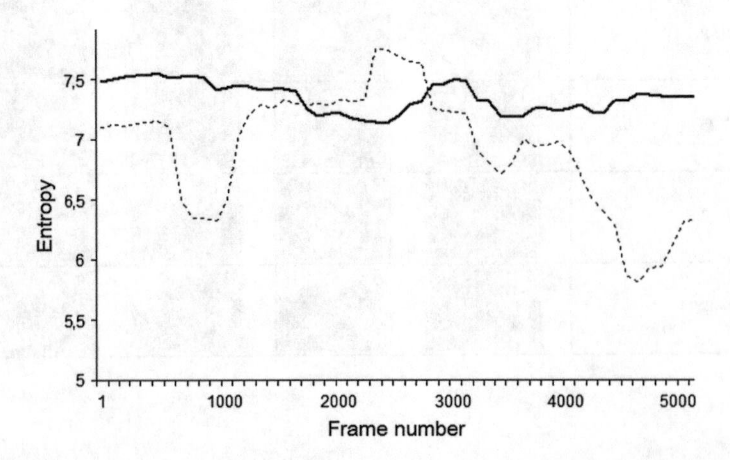


**Fig. 13** Graphics of entropy change for homogeneous and heterogeneous video content

**Fig. 14** Video frames extracted with slight and significant changes in content

a proof of the above statements, Table 2 shows entropy values for miniatures of frames form four video sequences (Table 3).

Frame numbers in the upper left corner indicate one of the four video sequences, they were taken from. Thus, the range of entropy for the first video sequence varies from 3.25 to 7.74. This video fragment is characterized by the greatest scene change among the analyzed materials. The range of entropy for the second video sequence equals to 4.12–6.59. Here, the content is of high heterogeneity as well, but it was shot in the nighttime, and the dark background narrows the range of visible details. The third material is characterized by homogeneity of changes, all the objects are shot under the same conditions, and they are very similar to each other, entropy range varies from 7.1 to 7.53. The fourth video has several scenes in it, entropy range here is quite diverse (from 5.85 to 7.75), simply not all the frames from the video have been included to the table.

Figure 13 illustrates graphics of entropy change for homogeneous and heterogeneous content from the third and the fourth video sequences included to the table above. For the demonstration purposes, fragments lasting nearly three minutes are taken from both videos, containing 29 frames per second. From the figure below it is obvious that entropy values computed for one of the videos change slightly,

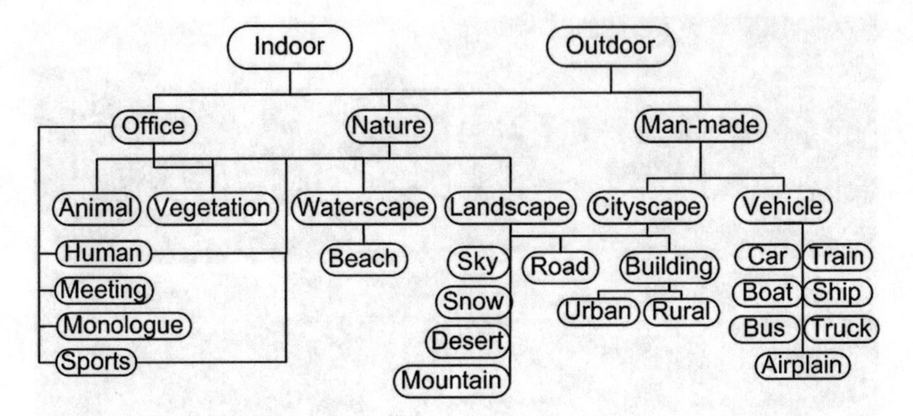


**Fig. 15** TRECVid test video categories

**Fig. 16** Example of image information detailing enhanced with size and resolution

whereas for another one they change significantly. These research results form a background for initial analysis of image changes (Figs. 14, 15 and 16).

The whole procedure of matching Voronoi diagrams for image change detection is described below.

**Step 1** Determine homogeneity of an image sequence. Calculate texture variance for all images in a sequence and according to it, set a threshold value by assuming the following rule:

$$Threshold = \begin{cases} \frac{1}{4}, & \frac{1}{K-1}\sum_{k=1}^{k=K}\left(E(B_k(z)) - \frac{1}{K}\sum_{k=1}^{k=K}E(B_k(z))\right)^2 \to \infty, \\ \frac{1}{2}, & \frac{1}{K-1}\sum_{k=1}^{k=K}\left(E(B_k(z)) - \frac{1}{K}\sum_{k=1}^{k=K}E(B_k(z))\right)^2 \to \frac{1}{K}\sum_{k=1}^{k=K}E(B_k(z)), \\ \frac{3}{4}, & \frac{1}{K-1}\sum_{k=1}^{k=K}\left(E(B_k(z)) - \frac{1}{K}\sum_{k=1}^{k=K}E(B_k(z))\right)^2 \to 0, \end{cases} \quad (16)$$

where $K$ indicates the total number of images/frames in a sequence, and $E(B_k(z))$ is the entropy value for $k$-th image/frame. The threshold value is set as follows: it should be less than $\frac{1}{4}$ for images/frames with heterogeneous content and a variety of scenes, it should be increased to $\frac{3}{4}$ for images/frames with homogeneous content and a small number of scenes (or even a single scene), otherwise it is set to $\frac{1}{2}$.

**Step 2** Take the first $(B_k(z))$ and the second $(B_{k+1}(z))$ images/frames for comparison. Set $k = 1$.

**Step 3** Compare Voronoi tessellations frame-by-frame to find significant changes in content. According to Eq. (15) calculate $\widehat{\rho}(B_k(z),\, B_{k+1}(z))$ for the two images/frames. If $\widehat{\rho}(B_k(z),\, B_{k+1}(z))$ turns out less than the preset threshold, then both images/frames $B_k(z)$ and $B_{k+1}(z)$ are considered containing significant changes (they may be reassigned as $B_r^*(z)$ and $B_{r+1}^*(z)$), and the procedure is continued from the next (4th) step. Otherwise, the procedure is continued from the 5th step.

**Step 4** Reassign images/frames as $B_k(z) = B_{k+1}(z)$, $B_{k+1}(z) = B_{k+2}(z)$ and go to step 6.

**Step 5** Remain $B_k(z) = B_k(z)$ and set $B_{k+1}(z) = B_{k+2}(z)$.

**Step 6** Repeat step 3 until $B_{k+1}(z) \leq K$.

**Step 7** Extract images/frames with significant changes by assuming scene boundaries to eliminate redundancy. Do not take into account similar frames from a single scene, if they are extracted, as they usually have much in common. Scene boundaries can be found through histogram difference, time series analysis or any other currently available technique [7]. Figure below illustrates frames of similar and different content, extracted from tourist's video. Frames with similar content reside horizontally.

Application of specifically designed similarity metrics (with orientation to low-level features in use) ensures high quality of image change detection. The procedure can be enhanced by utilization of weighted feature estimates, as in this case a linear combination only is used with the same weight coefficients. Additional features may also enhance the procedure. Determination of each feature importance is the prime direction of future research.

## 5 Test Collections, Processing Evaluation and Development Trends

Each application domain has its own specificity, but before operating with real-world industrial, medical or other video tracking objects, specialized test collection is needed. However, there is no universal video collection designed for such kind of purposes. That is why researchers choose video test samples by their own will, providing full descriptions in a form of video name, length, content type, number of frames/shots/scenes and the source it was taken. Publicly available collections from CERN Document Server (particle physics laboratory), Open Video Project (http://www.open-video.org), Movie Content Analysis Project (http://pi4.informatik.uni-mannheim.de/pi4.data/content/projects/moca/index.html), and Internet Archive (http://archive.org) are the main sources of video test samples. Sometimes commercials and self-made high-definition videos are taken into consideration as publicly available collections usually lack from high resolution. Custom video is also used for testing on specific application domain, such as bank office or street pedestrian tracking. The following figure illustrates commonly used categories of video samples that can be derived from datasets used for testing purposes by TRECVid community. Along with traditional video sources, clips from digital video libraries (DVL) can also be chosen: Informedia DVL (http://www.informedia.cs.cmu.edu), Consumer DVL (http://www.cdvl.org/), Hemispheric Institute DVL (http://hidvl.nyu.edu/), Harvard-Smithsonian Center for Astrophysics DVL (http://hsdvl.org/), etc. Materials from some digital video libraries cannot be downloaded, but they can be permitted for research usage under request. Popular human action recognition video datasets are: KTH, Weizmann, IXMAS, UCF50, HMDB51. Complete list of less famous data sets are available at http://www.datasets.visionbib.com/info-index.html and http://www.cvpapers.com/datasets.html [8].

All the open-source materials are usually provided of small size and low resolution. However, by reducing these image characteristics, small details are poured together with the background, and by enhancing these characteristics, detailing level of image information increases (see figure below). Thus, testing any image processing method with different size and resolution is one of the key point. The most frequently used video standards are considered PAL (720 * 576), HD (1280 * 720) and Full HD (1920 * 1080). The following table shows a list of currently available video standards which are reasonable to be used for testing purposes.

Someone may think that three to four test samples of different (or even the same) content type is enough, others perform testing of their proprietary methods on 20–100 movies lasting for more than an hour. Image and video genre has a great impact on visual features in use. That is the reason why to obtain truthful results, testing on different data types should be held: news, sports, cartoons, documentaries, talk-shows, if only the designed method is not oriented for a restricted application domain with homogeneous content [14].

Along the above considerations, test videos should include camera motion, zoom and lighting condition change because some video processing methods (especially those ones which are designed for tracking) could not cope with such changes. As it was already mentioned, any image and video processing method should be checked for quality and performance on different size and resolution data, and when speaking about image sequences and videos, these data should include huge and small inter-frame difference. It was noticed that methods showing sufficiently fine results for the most part of test collections, very often they fail with face recognition and processing of textual information that overlap in an image [8].

The accuracy of image changes detected is a subjective matter because respondents are involved in each evaluation that is almost always performed by precision (amount of found samples that turned out relevant) and recall (amount of found relevant samples from all the relevant ones). The benefit from using two measures simultaneously is obvious. In many cases one of the measures turns out to be more important. Some respondents do not appreciate obtaining any misclassified results, they want to get the least number of outputs, but all of them should be relevant (high precision). On the contrary, other respondents are interested in high recall, they treat low precision tolerantly. Precision and recall contradict with each other. Recall can always be increased to '1' with very low precision by simply returning all the values, no matter whether they are right or wrong. Recall does not decrease with increase in a number of wrong outputs (precision is decreased in such a case). Generally speaking, sufficient balance between these two measures should be reached. F-measure is an opportunity of doing so [18].

Along with traditional precision and recall measures, sometimes also percentage of wrong detected essences (e.g. significant changes) to missed ones can be computed. While the previous estimations are obtained from the information provided by the respondents with some supplementary calculations, criterions of informativeness and enjoyability simply depict users' feedback. Any method evaluation can be held using Fidelity measure and compression ratio. Actually, the estimation can be made in numerical, graphical or other form of comparison of results obtained after implementation of different image processing algorithms by showing benefits and drawbacks of the proposed one. Researchers and developers should check validity, performance and quality of image/video processing. It is considered that minimum 20 respondents are needed for reliable estimation. Expert estimates can be absolute and comparative. The first type implies presence of some viability scale while the second one assumes ranking the results or methods in terms of quality. The easiest way to express absolute expert estimates is to calculate an average value [14].

When using any clustering technique, in order to check whether clusters are well separated, cluster validity measure is used. This measure provides numerical information on distances within and between clusters. In most part of statistical software, ability to compute average/maximum/minimum distances within each cluster is realized, along with the sum of distances calculation ability, Euclidean distances, squared Euclidean distances, etc. Matlab silhouette plot visually demonstrates distances within and between clusters. Plot value of '+1' indicates that observations from different clusters are maximally remote from each other; plot value that resides near '0' indicates that these observations are too close to each other, and clusters could be assigned in a wrong manner; when a plot value tend to '–1', the observations are most probably related to clusters in a wrong manner.

Theoretical and practical results presented in this chapter facilitate indexing and archiving, summarization and annotating, searching and cataloguing large image sequences. In near future, works will be most probably concentrated on explorative approaches based on the middle level as a linking unit between low-level visual features and high-level concepts, and they have to meet the requirements for real-time processing in both performance and effectiveness [8]. The proposed approach to image change detection is unique and can be implemented in a range of applications aside from the intentional one.

# 6 Conclusion

An attempt towards high-level description of moving objects has shown that only specific motion patterns can be recognized by a machine. Current level of evolution does not permit implementation of semantically universal systems which could cope with a variety of content types [8]. Problem of machine-level video content comparison on frame-by-frame basis, i.e. image change detection, emerged several decades before. This chapter provides an overview on intelligent approaches utilized worldwide for solution of the aforementioned problem. Novel method proposed by the authors in this connection is also described in detail. It differs from existing ones by reliability and increased performance of change detection due to utilization of higher order Voronoi diagrams which assume all the advantages of operation with salient points.

Image sequence presentation using Voronoi diagrams with further their comparison provides means for machine interpretation of how objects are moved in space and time. Similar content in sequential video frames turned out to have identical diagrams, whereas the proposed metrics for diagram comparison ensure recognition of slight and significant changes. Voronoi diagrams of higher order simplify detailing image content compared with increasing the number of initial salient points [14]. Some other useful properties of higher order Voronoi diagrams have been revealed in relation to image change detection.

Evaluation of image and video processing is an integral part of any novel method development. Complexity of such evaluation lies in subjectivity of estimations provided by respondents. By analysing commonly applicable measures, several estimators have been marked out to test validity and performance of visual processing. The most frequently used measures of precision and recall can be combined in a form of Dice coefficient to obtain reasonable balance between them. Importance of performing tests on different image genres, high and low size and resolution, has been grounded. As long as clustering algorithms are widely applied in image and video processing, cluster validity measures have been also discussed. Datasets of open-source test video samples have been mentioned, among which TRECVid collection, Internet Archive, Movie Content Analysis Project and Open Video Project have gained greatest popularity. Less often used video libraries have been listed as well.

# References


1. Sonka, M., Hlavac, V., Boyle, R.: Image Processing, Analysis, and Machine Vision, International Student Edition. Thomson, Toronto (2007)
2. Bezdek, J.C., Keller, J., Krisnapuram, R., Pal, N.R.: Fuzzy Models and Algorithms For Pattern Recognition and Image Processing. Springer, NY (2005)
3. Haralick, R.M., Shanmugam, K., Dinstein, I.: Textural features for image classification. IEEE Trans. Syst. Man Cybern. **3**(6), 610–621 (1973)

4. Deselaers, T., Weyand, T., Ney, H.: Image retrieval and annotation using maximum entropy. Evaluation of multilingual and multi-modal information retrieval. Lect. Notes Comput. Sci. **4730**, 725–734 (2007)
5. http://vision.middlebury.edu
6. Baker, S., Scharstein, D., Lewis, J.P., Roth, S., Black, M.J., Szeliski, R.: A database and evaluation methodology for optical flow. Int. J. Comput. Vis. **92**(1), 1–31 (2011)
7. Mikhnova, O., Vlasenko, N.: Key frame partition matching for video summarization. Int. J. Inf. Models Anal. **2**(2), 145–152 (2013)
8. Mashtalir, S., Mikhnova, O.: Key frame extraction from video: framework and advances. Int. J. Comput. Vis. Image Process. **4**(2), 67–78 (2014)
9. Mikhnova, O.: A template-based approach to key frame extraction from video. In: Proceedings of International Scientific and Technical Internet Conference on Computer Graphics and Image Recognition, pp. 120–127. VNTU, Vinnytsia (2012)
10. Yang, X., Tian, Q.: Video repeat recognition and mining by visual features. In: Schonfeld, D., Shan, C., Tao, D., Wang, L. (eds.) Video Search and Mining. Studies in Computational Intelligence, vol. 287, pp. 305–326. Springer, Berlin (2010)
11. Lee, W.-T., Chen, H.-T.: Histogram-based interest point detectors. Comput. Vis. Pattern Recogn. 1590–1596 (2009)
12. Ledoux, H., Gold, C.M.: Modelling three-dimensional geoscientific fields with the Voronoi diagram and its dual. Int. J. Geogr. Inf. Sci. **22**(5), 547–574 (2008)
13. Mikhnova, O.D.: Analiz videodannyh na osnove diagramm Voronogo razlichnogo poryadka. Zbirnyk naukovyh prats HUPS **1**(38), 142–145 (2014)
14. Mashtalir, S.V., Mikhnova, O.D.: Stabilization of key frame descriptions with higher order Voronoi diagram. Bionics Intell. **1**, 68–72 (2013)
15. Okabe, A., Boots, B., Sugihara, K., Chiu, S.N.: Spatial Tessellations: Concepts and Applications of Voronoi Diagrams. Wiley, Chichester (2000)
16. Mashtalir, V., Mikhnova, O., Shlyakhov, V., Yegorova, E.: A novel metric on partitions for image segmentation. In: Proceedings of International conference on Video and Signal Based Surveillance, pp. 1–6. IEEE CS, Washington (2006)
17. Sadahiro, Y.: Analysis of the relationship among spatial tessellations. J. Geogr. Syst. **13**(4), 373–391 (2011)
18. Manning, C.D., Raghavan, P., Schutze, H.: Introduction to Information Retrieval. Cambridge University Press, Cambridge (2008)

# VW16E: A Robust Video Watermarking Technique Using Simulated Blocks

**Farnaz Arab and Mazdak Zamani**

**Abstract** The basic idea for video watermarking technique is concealing information in the video host for different purposes including authentication and tamper detection. The most common approach for concealing information in the video host is spatial domain. This paper focuses on video watermarking; particularly with respect to the Audio Video Interleaved (AVI) form of video file format in spatial domain by simulating blocks. It proposes a new watermarking technique that gives a high imperceptibility and efficient tamper detection compared to the other similar schemes.

**Keywords** Watermarking · Tamper detection · Block simulation · Robustness

## 1 Introduction

A digital watermark is a kind of indication, which is accommodated in the host medium such as digital image, audio, text, software or video. It can be commonly used for ownership protection. Watermarking is a technique of covering digital information in the carrier signal (host). The hidden data is not necessarily related to the content of the host [1]. Particularly for video files, in order to solve the problem of unlawful manipulation and dishonest distribution, video watermarking is applied [2].

Digital Video plays a major role of forensic evidence in court [3]. Hence the video files should be authenticable with ability to detect the tamper, thus a technique like watermarking is applied for the purpose. The watermark must not have any effect on visual information and must not reduce the ability for compromise on the video evidence. Therefore, high imperceptible watermark has responded to the

---

F. Arab (✉) · M. Zamani
Kean University, Union, NJ, USA
e-mail: faarab@kean.edu

M. Zamani
e-mail: mzamani@kean.edu



A.E. Hassanien et al. (eds.), *Multimedia Forensics and Security*,
Intelligent Systems Reference Library 115, DOI 10.1007/978-3-319-44270-9_9

mentioned necessity [4–7]. Video tamper detection is the challenge of today's researchers in the field of multimedia security [8, 9].

Although video watermarking has many properties, the main three properties are imperceptibility, robustness and payload or capacity which are closely related to each other for example when the robustness increases, imperceptibility would be decrease and vice versa [10, 11]. The correct balance between these conflicting requirements of watermarking should be found for any application and techniques [10, 12].

Nowadays cameras in many circumstance has been installed, even these cameras mounted on the streets for fights, drug deals and other improper activities in an environment. The police might see the crime as it was happening or use the video to help in any consequent investigation. Digital multimedia content can easily be duplicated and stored and even without losing fidelity. In Digital Video System (DVS) video file is very vital, because it can be used as a piece of evidence, on the other hand; manipulating the video file by many editing video software in the market is like a piece of cake, so easy and simple with low cost [13].

By growth of communication network, due to the characteristics of digital products such as easy to transform and easy to copy, digital tamper detection has been critical issues which need to be solved. Techniques used for video watermarking tamper detection compared to digital image are stagnant [10]. Ascribable to the natural redundancy between the video frames, proposed techniques for image tamper detection are not appropriate for digital video watermarking which are not presented for attacks including frame dropping, frame inserting, frame shifting and etc. Beside these attacks, techniques are restricted in ability to detect the tamper areas [14].

The tamper detection technique has to be designed to ensure the verification of video content and preventing forgery. Researchers have proposed digital watermarking to verify integrity of content for digital video [15–18]. A wide range of modifications in any domain could be utilized for watermarking techniques [19]. On the other hand video market is become more and more popular; the cameras' information results have a major role in safety of environment and people. In order not to change the concept of visual information, the embedded data should be imperceptible and robust. Hence, in addition to robustness and imperceptibility the constraint of computational is imposed to video watermarking [20].

Video application requires a large quantity of sequences to be processed. Watermarking techniques can also be applied in the frequency domain. In these techniques higher imperceptibility can be obtained as well as better robustness. The disadvantage of frequency domain methods is that they are computationally expensive when compared with spatial. Spatial domain techniques are best suit for video watermarking than other watermarking domains. Watermark can also be embedded in the frequency domains [21]. In transform domain, first the host is converted to the frequency domain then the watermark is added and then the inverse frequency transform is applied. One of the common transform methods is the Discrete Cosine Transform (DCT) which divides the image into low, middle and high frequency bands. In the aspect of imperceptibility the middle band is best chosen rather than two other frequency bands. If the watermark is embedded in high frequency band, the details of the edges and other information would be affected [22–24]. On the other hand, when the watermark is embedded into the low frequency, the imperceptibility is influenced negatively. The DCT is not more efficient than spatial domain when it comes to transparency and also it has intensive computation relatively [11]. Another common transform method is Discrete Wavelet Transform (DWT) which decompose the image into four sub bands that are low resolution approximation (LL), horizontal (HL), vertical (LH) and diagonal (HH) of detail components. The edge and texture patterns are located in high resolution sub bands. The watermark cannot embed in LL because the smoother part of the image is in this part and also the watermark cannot embed in HH because major details of the image will be lost. That is why the HL and LH are normally selected for watermarking [21]. The DWT also is not more efficient than spatial domain in aspect of transparency and also have more computation compared to DCT [25].

The most common approach for concealing information in the video host is spatial domain. The robustness of spatial domain techniques is not as high as other techniques. In order to improve the robustness of spatial domain, a watermark can be embedded several times repeatedly. As a result, if a single copy of that watermark can survive after attacks, that can be retrieved and the techniques passes the robustness test. Moreover, although spatial domain technique is easy to implement, sometimes adding noise entirely demolish the watermark and could be noticeable for attacker by comparing the anticipated sample with the received signal [10, 26, 27].

In order for spatial domain techniques to be as efficient as other techniques, more payload is needed to embed additional information. The additional information would include the redundant watermarks to ensure the achievable robustness and more metadata of pixels to ensure achievable efficiency to detect more attacks. All these required additional information will degrade the quality (imperceptibility) [9].

In this Paper VW16E technique is proposed technique and the design and implementation of the proposed technique is discussed. The technique is implemented in spatial domain. Each pixel is represented by 2 bytes in the AVI video file format. The block-wised technique is used to determine which block is exactly altered.

## 2 Block Simulation

The simulation of each $2 \times 2$ block consists of four consequent pixels in video stream as illustrated in Fig. 1. In simulation after reading first two pixels there is no need to skip to the next row for reading the last two pixels. In block simulation video data consider as a coherent data and video frame does not make a limitation in building the block. One block may build from the pixel of two different frames, for example the first three pixels come from the end of one block and last pixel from next frame. The technique will be able to watermark more pixels (almost all pixels)

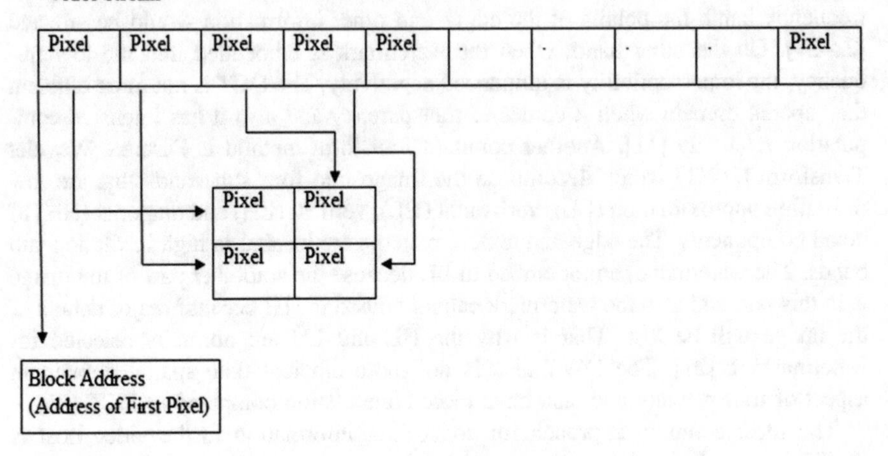


**Fig. 1** Block simulation

rather than common techniques which create real 2 × 2 blocks and exclude the pixels left out of blocks around the border.

The first pixel's address in each block represents the address of the block. The presented techniques divide the video stream into 2 × 2 non-overlapping simulated blocks. Thus, as shown in Fig. 2, there are 2 bytes in each cell of the block.

The presented watermark which is VW16E technique has 16 bits. The watermark is substituted with the 4 Least Significant Bits (LSB) of each pixel. The watermark is a combination of integrity bits and confidential bits. The watermark is consisting of 12-bits for integrity bits and 4-bits for confidential bits. Four bits of block's address and eight bit of pixels' data are involved in integrity bits. The 16-bits watermark is the signature for each simulated block with the size of 2 × 2 pixels.

Figure 3 shows, all pixels associate in creating the 16-bits watermark. Two MSB bits in each pixel are used as integrity information. So far, 8 bits of 16-bits are created. As Fig. 4 clarifies, the block's address is shown with 4 bytes and in the

**Fig. 2** Inside each block

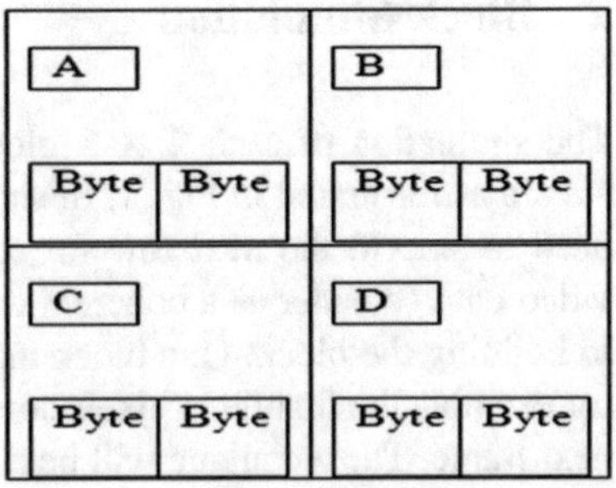

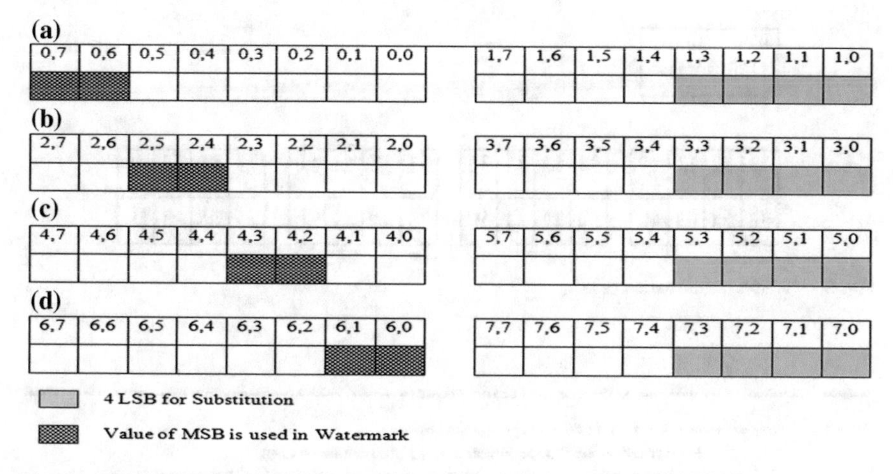


**Fig. 3** Chosen Bits as a Pixels' data for 16 Bits watermark

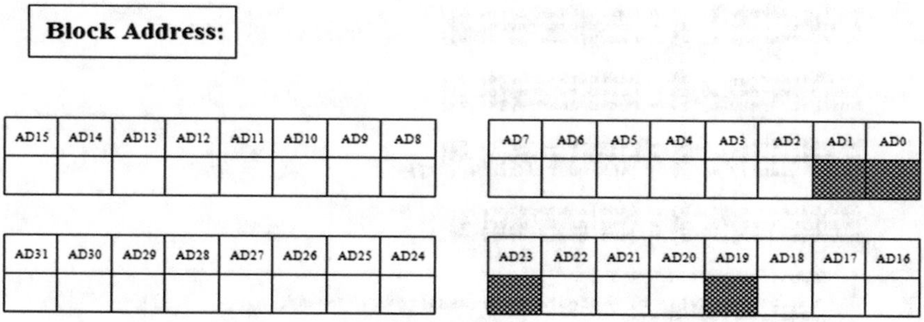

**Block Address:**

| AD15 | AD14 | AD13 | AD12 | AD11 | AD10 | AD9 | AD8 | AD7 | AD6 | AD5 | AD4 | AD3 | AD2 | AD1 | AD0 |
|---|---|---|---|---|---|---|---|---|---|---|---|---|---|---|---|

| AD31 | AD30 | AD29 | AD28 | AD27 | AD26 | AD25 | AD24 | AD23 | AD22 | AD21 | AD20 | AD19 | AD18 | AD17 | AD16 |
|---|---|---|---|---|---|---|---|---|---|---|---|---|---|---|---|

**Fig. 4** Block address Bytes for 16 Bits watermark

watermark just 4 bits are used. Therefore, 12 bits of 16-bits are introduced. The remained 4 bits of 16-bits watermark are reserved for confidential bits.

The integrity and confidential bits are set together and create the 16-bits watermark. As illustrated in Fig. 5, all the confidential bits are set at the first layer of LSB. The technique with 16 bits for watermark in which the confidential bits at the end layer of LSB is called Video Watermarking 16 at End and in short as "VW16E".

Figure 6 shows part of software cods which write the VW16E. First of all four pixels which are related to one block are red and expand. Then the integrity data is substituted with the mentioned order. So far 8 bits of watermark is written. In this program the size of message is stored it will use during the extraction. For writing four remains bits of watermark, in the "if", first is checked if the value of message size is finished writing confidential data will be started.

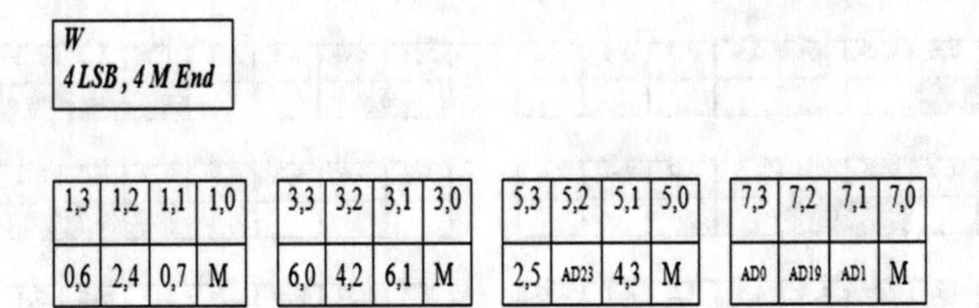


**Fig. 5** VW16E watermark

```
========================= LOOP For Writting VW16E=========================================

        temporarycurrenthostbyte = currenthostbyte;

        for (loopcounter6 = 0; loopcounter6 < 8; loopcounter6++)
        {
            hostbitsarray[loopcounter5, loopcounter6] = temporarycurrenthostbyte % 2;
            temporarycurrenthostbyte = temporarycurrenthostbyte / 2;
        }

    }

    hostbitsarray[1, 3] = hostbitsarray[0, 6];
    hostbitsarray[1, 1] = hostbitsarray[0, 7];
    hostbitsarray[1, 2] = hostbitsarray[2, 4];

    hostbitsarray[3, 3] = hostbitsarray[6, 0];
    hostbitsarray[3, 1] = hostbitsarray[6, 1];
    hostbitsarray[3, 2] = hostbitsarray[4, 2];

    hostbitsarray[5, 3] = hostbitsarray[2, 5];
    hostbitsarray[5, 1] = hostbitsarray[4, 3];
    hostbitsarray[5, 2] = (int)hostaddsarray[23];

    hostbitsarray[7, 3] = (int)hostaddsarray[0];
    hostbitsarray[7, 1] = (int)hostaddsarray[1];
    hostbitsarray[7, 2] = (int)hostaddsarray[19];

    if (msgsizearrayindex < 32)
    {
        hostbitsarray[1, 0] = msgsizearray[msgsizearrayindex];
        msgsizearrayindex++;
        hostbitsarray[3, 0] = msgsizearray[msgsizearrayindex];
        msgsizearrayindex++;
        hostbitsarray[5, 0] = msgsizearray[msgsizearrayindex];
        msgsizearrayindex++;
        hostbitsarray[7, 0] = msgsizearray[msgsizearrayindex];
        msgsizearrayindex++;

    }
    else
    {
        if (msgbitsarrayindex == 0 & loopcounter11 < msgsize)
        {
            currentmsgbyte = msgfile.ReadByte();
            temporarycurrentmsgbyte = currentmsgbyte;
            loopcounter11++;

            for (loopcounter12 = 0; loopcounter12 < 8; loopcounter12++)
            {
                msgbitsarray[loopcounter12] = temporarycurrentmsgbyte % 2;
                temporarycurrentmsgbyte = temporarycurrentmsgbyte / 2;
            }
        }
        hostbitsarray[1, 3] = msgbitsarray[msgbitsarrayindex];
        msgbitsarrayindex = (msgbitsarrayindex + 1) % 8;
        hostbitsarray[3, 3] = msgbitsarray[msgbitsarrayindex];
        msgbitsarrayindex = (msgbitsarrayindex + 1) % 8;
        hostbitsarray[5, 3] = msgbitsarray[msgbitsarrayindex];
        msgbitsarrayindex = (msgbitsarrayindex + 1) % 8;
        hostbitsarray[7, 3] = msgbitsarray[msgbitsarrayindex];
        msgbitsarrayindex = (msgbitsarrayindex + 1) % 8;
    }
```

**Fig. 6** VW16E technique codes for embedding watermark

# 3 Test the Imperceptibility

In the section, the result of implementation of the technique in details is presented. For testing the imperceptibility of the technique, 14 samples of AVI file format on recorded videos were tested. The parameters of each experimental video sequence are shown in the Table 1. They have been chosen based on the different length and different total frames. On the other hand, for confidential bits, four different size of text file have been chosen. The Message Parameters shows in Table 2.

In order for spatial domain watermarking to improve the robustness, a small watermark can be embedded several times. Thus, if a single copy of watermark survives, the method passes the robustness test. Therefore, all confidential messages which are embedded as confidential bits with the integrity bits are repetition of one RSA key. The confidential bits are referred as message in this research. Message No 1 is only one RSA key, message No 2 is 18 similar RSA key, message No 3 is 212 similar RSA key, and message No 4 is 1182 times repetition of same RSA key.

In this section, the experimental results of embedding four messages with the VW16E technique in all the video samples are explained. On the other hand, the Peak Signal to Noise Ratio (PSNR) of each original video with the embedded video one is calculated.

## 3.1 *Test Results of VW16E Technique for the Video Sample 1*

This section presents the result of embedding four messages into the Video Sample No 1. The size of messages is different. The size of M1 is 1,669 bytes, the size of M2 is 30,044 bytes, size of M3 is 353,828 bytes and size of M4 is 1,972,764 bytes while the size of Video Sample No 1 is 16,080,844 bytes. The maximum available bits which can participate for confidential message in sample No 1 is 5,356,800.

Because number of bits inside M4 is more than number of reserved bits for confidential message in Sample No 1, thus, the M4 is not appropriate to be substituted with the VW16E technique in the first video sample; therefore, there is no result for the last message. Each of remained three messages was embedded and the PSNR, which represent the loyalty of original video host to the watermarked video, is calculated. Table 3 shows the results of PSNRs. As expected, the result in Fig. 7 shows that, as the message size is increased, PSNR is decreased.

**Table 1** Specification of video samples

| Video sample no | Size (Bytes) | Length (Second) | Frame width (Pixel) | Frame height (Pixel) | Frame rate (Frame/Second) | Total frames | Total pixels | Integrity Bits (VW16E) |
|---|---|---|---|---|---|---|---|---|
| 1 | 16,080,844 | 0.00.03 | 320 | 180 | 29 | 93 | 5,356,800 | 16,070,400 |
| 2 | 29,272,060 | 0.00.04 | 320 | 240 | 29 | 127 | 9,753,600 | 29,260,800 |
| 3 | 39,239,260 | 0.00.07 | 320 | 180 | 29 | 227 | 13,075,200 | 39,225,600 |
| 4 | 57,212,956 | 0.00.11 | 320 | 180 | 29 | 331 | 19,065,600 | 57,196,800 |
| 5 | 68,964,988 | 0.00.13 | 320 | 180 | 29 | 399 | 22,982,400 | 68,947,200 |
| 6 | 78,813,220 | 0.00.11 | 320 | 240 | 29 | 342 | 26,265,600 | 78,796,800 |
| 7 | 125,128,444 | 0.00.18 | 320 | 240 | 29 | 543 | 41,702,400 | 125,107,200 |
| 8 | 129,042,292 | 0.00.14 | 320 | 320 | 29 | 420 | 43,008,000 | 129,024,000 |
| 9 | 243,690,052 | 0.00.47 | 320 | 180 | 29 | 1410 | 81,216,000 | 243,648,000 |
| 10 | 245,936,764 | 0.00.47 | 320 | 180 | 29 | 1423 | 81,964,800 | 245,894,400 |
| 11 | 259,417,036 | 0.00.50 | 320 | 180 | 29 | 1501 | 86,457,600 | 259,372,800 |
| 12 | 260,626,804 | 0.00.50 | 320 | 180 | 29 | 1508 | 86,860,800 | 260,582,400 |
| 13 | 296,064,676 | 0.00.53 | 320 | 192 | 29 | 1606 | 98,672,640 | 296,017,920 |
| 14 | 391,159,132 | 0.00.51 | 320 | 262 | 29 | 1555 | 130,371,200 | 391,113,600 |

**Table 2** Specification of message samples

| Name | File Type | Size (Bytes) | Confidential Bits |
|---|---|---|---|
| M1 | TXT | 1,669 | 13,352 |
| M2 | TXT | 30,044 | 240,352 |
| M3 | TXT | 353,828 | 2,830,624 |
| M4 | TXT | 1,972,764 | 15,782,112 |

**Table 3** Test results of VW16E technique for the video sample 1

| Host | Host size (Byte) | Confidential message | Confidential Bit | Integrity Bits for VW16E | PSNR |
|---|---|---|---|---|---|
| Video Sample 1 | 16,080,844 | M1 | 13,352 | 16,070,400 | 34.971 |
| Video Sample 1 | 16,080,844 | M2 | 240,352 | 16,070,400 | 34.965 |
| Video Sample 1 | 16,080,844 | M3 | 2,830,624 | 16,070,400 | 34.902 |
| Video Sample 1 | 16,080,844 | M4 | 15,782,112 | 16,070,400 | M4 too big |

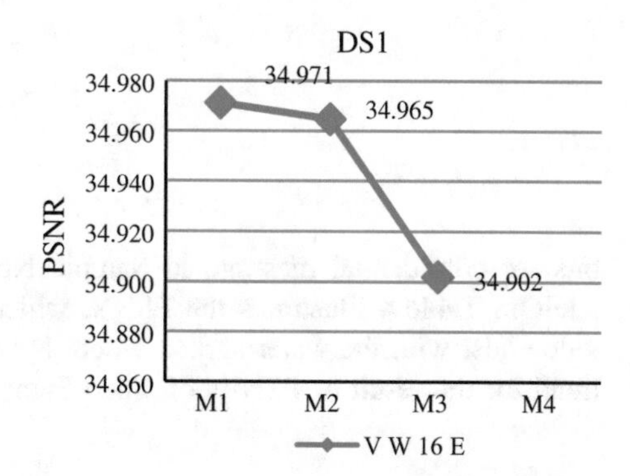


**Fig. 7** The test results' graph of VW16E technique for the video sample 1

## 3.2 *Test Results of VW16E Technique for the Video Sample 2*

In this section, the results of embedding four messages into the video sample No 2 with the size of 29,272,060 bytes have been presented. On the other hand, the size of M1 is 1,669 bytes, the size of M2 is 30,044 bytes, size of M3 is 353,828 bytes and size of M4 is 1,972,764 bytes. The maximum available bits which can participate for confidential message in sample No 2 is 9,753,600. Although the size of M4 is smaller than the size of Sample No 2, but M4 with the VW16E technique is not appropriate because number of bit inside M4 is more than number of reserved

**Table 4** Test results of VW16E technique for the video sample 2

| Host | Host size (Byte) | Confidential message | Confidential Bit | Integrity Bits for VW16E | PSNR |
|---|---|---|---|---|---|
| Video Sample 2 | 29,272,060 | M1 | 13,352 | 29,260,800 | 34.757 |
| Video Sample 2 | 29,272,060 | M2 | 240,352 | 29,260,800 | 34.754 |
| Video Sample 2 | 29,272,060 | M3 | 2,830,624 | 29,260,800 | 34.734 |
| Video Sample 2 | 29,272,060 | M4 | 15,782,112 | 29,260,800 | M4 too big |

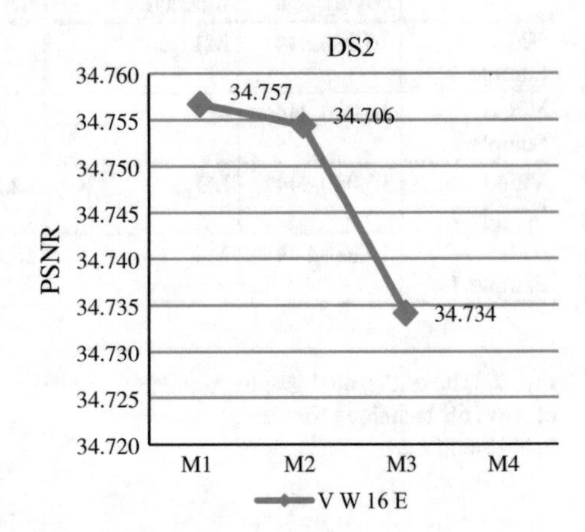


**Fig. 8** The test results' graph of VW16E technique for the video sample 2

bits for confidential message in Sample No 2 for substitution in this video. In addition Table 4 illustrates the PSNR, which presents the fidelity of the original video host with the watermarked video, is calculated. Figure 8 shows descending trend for the result of PSNRs of video Sample No 2 (Table 5).

**Table 5** Test results of VW16E technique for the video sample 3

| Host | Host size (Byte) | Confidential message | Confidential Bit | Integrity Bits for VW16E | PSNR |
|---|---|---|---|---|---|
| Video Sample 3 | 39,239,260 | M1 | 13,352 | 39,225,600 | 34.771 |
| Video Sample 3 | 39,239,260 | M2 | 240,352 | 39,225,600 | 34.769 |
| Video Sample 3 | 39,239,260 | M3 | 2,830,624 | 39,225,600 | 34.753 |
| Video Sample 3 | 39,239,260 | M4 | 15,782,112 | 39,225,600 | 34.667 |

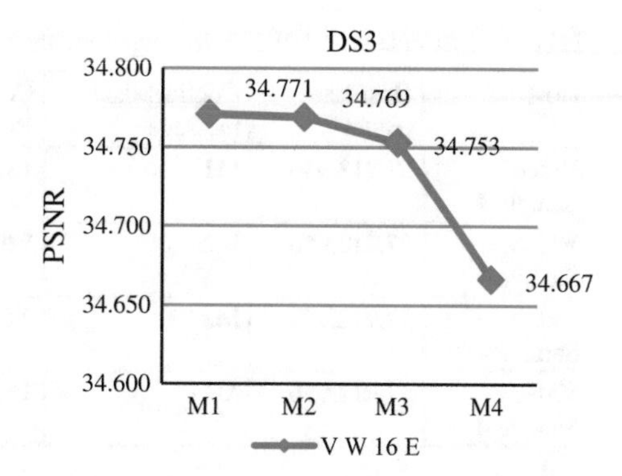


**Fig. 9** The test results' graph of VW16E technique for the video sample 3

## 3.3 Test Results of VW16E Technique for the Video Sample 3

The result of embedding four messages into the video sample No 3 is presented in this part. The size of four messages, which are embedded- in order- into the video sample No 3 is 1,669 bytes, 30,044 bytes, 353,828 bytes and 1,972,764 bytes. The third video stream has 39,239,260 bytes capacity. Each of these four messages was embedded and the PSNR, which signify the fidelity of original video host to the watermarked video, is calculated. Figure 9 depicts the trend of the PSNR; increasing size of message, decreasing the PSNRs.

## 3.4 Test Results of VW16E Technique for the Video Sample 4

This section presents the result of embedding four messages into the Video Sample No 4. Size of messages is different. The size of M1 is 1,669 bytes, the size of M2 is 30,044 bytes, size of M3 is 353,828 bytes and size of M4 is 1,972,764 bytes, while the size of Video Sample No 4 is 57,212,956 bytes. Each of these four messages was embedded and the PSNR, which represent the loyalty of original video host to the watermarked video, is calculated. Table 6 shows the results of PSNRs. Figure 10 shows the graph has the expected trend.

## 3.5 Test Results of VW16E Technique for the Video Sample 5

In this section, the results of embedding four messages into the video sample No 5 with the size of 68,964,988 bytes have been presented. On the other hand, the size

**Table 6** Test results of VW16E technique for the video sample 4

| Host | Host size (Byte) | Confidential message | Confidential Bit | Integrity Bits for VW16E | PSNR |
|---|---|---|---|---|---|
| Video Sample 4 | 57,212,956 | M1 | 13,352 | 57,196,800 | 34.864 |
| Video Sample 4 | 57,212,956 | M2 | 240,352 | 57,196,800 | 34.863 |
| Video Sample 4 | 57,212,956 | M3 | 2,830,624 | 57,196,800 | 34.852 |
| Video Sample 4 | 57,212,956 | M4 | 15,782,112 | 57,196,800 | 34.795 |

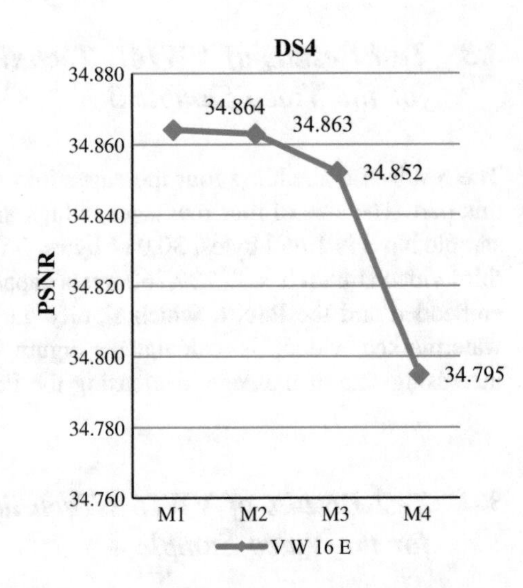


**Fig. 10** The test results' graph of VW16E technique for the video sample 4

of M1 is 1,669 bytes, the size of M2 is 30,044 bytes, size of M3 is 353,828 bytes and size of M4 is 1,972,764 bytes. Table 7 illustrates the PSNR which presents the fidelity of the original video host with the watermarked video is calculated. Figure 11 depicts by decreasing the size of message, PSNR is increased.

**Table 7** Test results of VW16E technique for the video sample 5

| Host | Host size (Byte) | Confidential message | Confidential Bit | Integrity Bits for VW16E | PSNR |
|---|---|---|---|---|---|
| Video Sample 5 | 68,964,988 | M1 | 13,352 | 68,947,200 | 35.001 |
| Video Sample 5 | 68,964,988 | M2 | 240,352 | 68,947,200 | 35.000 |
| Video Sample 5 | 68,964,988 | M3 | 2,830,624 | 68,947,200 | 34.992 |
| Video Sample 5 | 68,964,988 | M4 | 15,782,112 | 68,947,200 | 34.940 |

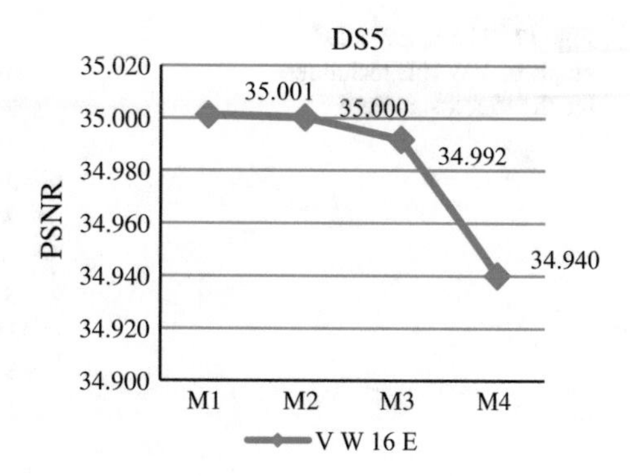


**Fig. 11** The test results' graph of VW16E technique for the video sample 5

## 3.6 *Test Results of VW16E Technique for the Video Sample 6*

The result of embedding four messages into the video sample No 6 is presented in this part. The size of four messages, which are embedded- in order- into the video sample No 6 is 1,669 bytes, 30,044 bytes, 353,828 bytes and 1,972,764 bytes. The sixth video stream has 78,813,220 bytes capacity. Table 8 shows the result that each of these four messages was embedded and the PSNR, which signify the fidelity of original video host to the watermarked video, is calculated. Figure 12 depicts the trend of the PSNR; it follows the expected trend. PSNRs are decreased by increasing size of message.

## 3.7 *Test Results of VW16E Technique for the Video Sample 7*

In this section the results of embedding four messages with VW16E technique into the video sample No 7 with the size of 125,128,444 bytes have been presented.

**Table 8** Test results of VW16E technique for the video sample 6

| Host | Host size (Byte) | Confidential message | Confidential Bit | Integrity Bits for VW16E | PSNR |
|---|---|---|---|---|---|
| Video Sample 6 | 78,813,220 | M1 | 13,352 | 78,796,800 | 34.784 |
| Video Sample 6 | 78,813,220 | M2 | 240,352 | 78,796,800 | 34.784 |
| Video Sample 6 | 78,813,220 | M3 | 2,830,624 | 78,796,800 | 34.776 |
| Video Sample 6 | 78,813,220 | M4 | 15,782,112 | 78,796,800 | 34.735 |

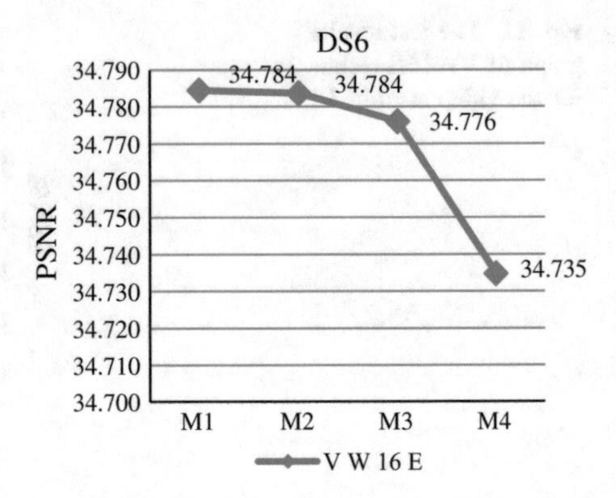


**Fig. 12** The test results' graph of VW16E technique for the video sample 6

The size of M1 is 1,669 bytes, the size of M2 is 30,044 bytes, size of M3 is 353,828 bytes and size of M4 is 1,972,764 bytes. In addition Table 9 illustrates the PSNR, which presents the fidelity of the original video host with the watermarked video, is calculated. As in Fig. 13 Illustrates by increasing the size of message the PSNR decreases.

## 3.8 Test Results of VW16E Technique for the Video Sample 8

This section presents the result of embedding four messages into the Video Sample No 8. The size of messages is different. The size of M1 is 1,669 bytes, the size of M2 is 30,044 bytes, size of M3 is 353,828 bytes and size of M4 is 1,972,764 bytes while the size of Video Sample No 8 is 129,042,292 bytes. Each of these four messages was embedded and the PSNR, which represent the loyalty of original video host to the watermarked video, is calculated. Table 10 shows the result of PSNRs and Fig. 14 depicts that the graph pursues the expected trend.

**Table 9** Test results of VW16E technique for the video sample 7

| Host | Host size (Byte) | Confidential message | Confidential Bit | Integrity Bits for VW16E | PSNR |
|---|---|---|---|---|---|
| Video Sample 7 | 125,128,444 | M1 | 13,352 | 125,107,200 | 34.917 |
| Video Sample 7 | 125,128,444 | M2 | 240,352 | 125,107,200 | 34.917 |
| Video Sample 7 | 125,128,444 | M3 | 2,830,624 | 125,107,200 | 34.912 |
| Video Sample 7 | 125,128,444 | M4 | 15,782,112 | 125,107,200 | 34.886 |

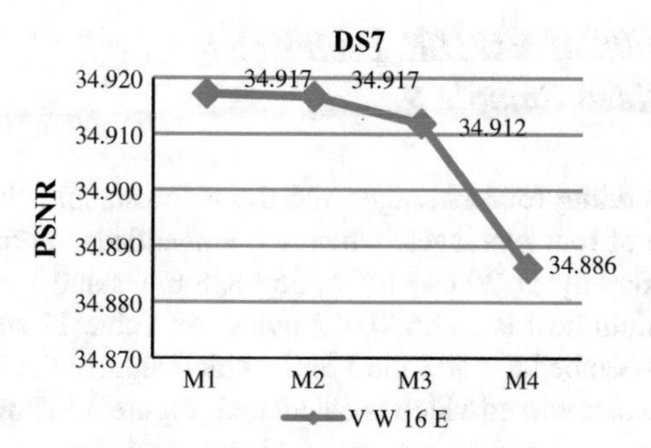


**Fig. 13** The test results' graph of VW16E technique for the video sample 7

**Table 10** Test results of VW16E technique for the video sample 8

| Host | Host size (Byte) | Confidential message | Confidential Bit | Integrity Bits for VW16E | PSNR |
|---|---|---|---|---|---|
| Video Sample 8 | 129,042,292 | M1 | 13,352 | 129,024,000 | 34.926 |
| Video Sample 8 | 129,042,292 | M2 | 240,352 | 129,024,000 | 34.924 |
| Video Sample 8 | 129,042,292 | M3 | 2,830,624 | 129,024,000 | 34.909 |
| Video Sample 8 | 129,042,292 | M4 | 15,782,112 | 129,024,000 | 34.840 |

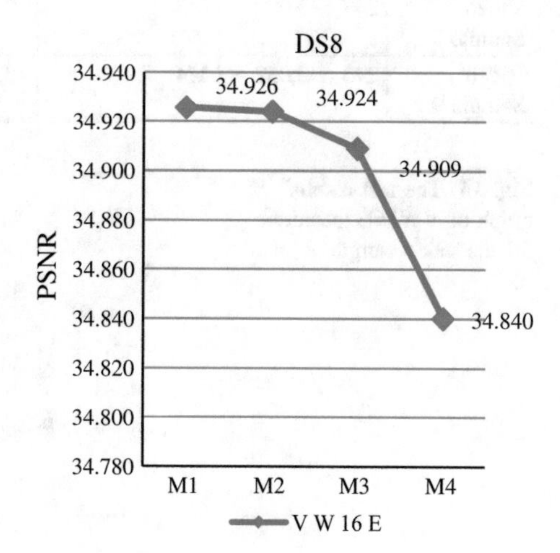


**Fig. 14** The test results' graph of VW16E technique for the video sample 8

### 3.9 Test Results of VW16E Technique for the Video Sample 9

The result of embedding four messages into the video sample No 9 is presented in this part. The size of four messages, which are embedded- in order- into the video sample No 9 is 1,669 bytes, 30,044 bytes, 353,828 bytes and 1,972,764 bytes. The total capacity of ninth host is 243,690,052 bytes. As Table 11 shows each of these four messages was embedded and the PSNR which signify the fidelity of original video host to the watermarked video is calculated. Figure 15 shows the trend of the PSNR; PSNRs are decreased by increasing size of message.

### 3.10 Test Results of VW16E Technique for the Video Sample 10

This section presents the result of embedding four messages into the Video Sample No 10. The size of messages is different. The size of M1 is 1,669 bytes, the size of

**Table 11** Test results of VW16E technique for the video sample 9

| Host | Host size (Byte) | Confidential message | Confidential Bit | Integrity Bits for VW16E | PSNR |
|---|---|---|---|---|---|
| Video Sample 9 | 243,690,052 | M1 | 13,352 | 243,648,000 | 34.702 |
| Video Sample 9 | 243,690,052 | M2 | 240,352 | 243,648,000 | 34.701 |
| Video Sample 9 | 243,690,052 | M3 | 2,830,624 | 243,648,000 | 34.700 |
| Video Sample 9 | 243,690,052 | M4 | 15,782,112 | 243,648,000 | 34.688 |

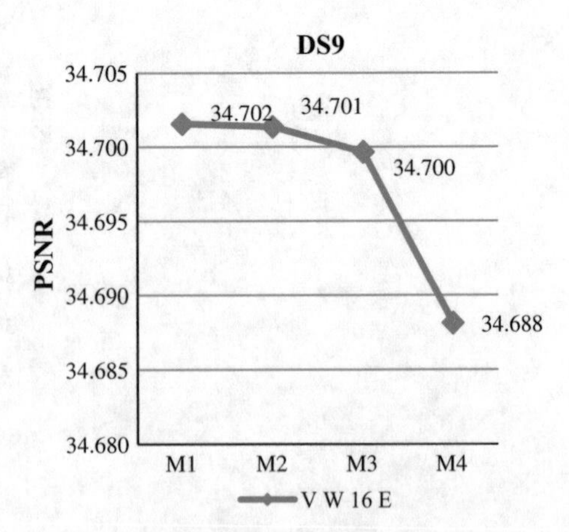


**Fig. 15** The test results' graph of VW16E technique for the video sample 9

**Table 12** Test results of VW16E technique for the video sample 10

| Host | Host size (Byte) | Confidential message | Confidential Bit | Integrity Bits for VW16E | PSNR |
|---|---|---|---|---|---|
| Video Sample 10 | 245,936,764 | M1 | 13,352 | 245,894,400 | 34.881 |
| Video Sample 10 | 245,936,764 | M2 | 240,352 | 245,894,400 | 34.880 |
| Video Sample 10 | 245,936,764 | M3 | 2,830,624 | 245,894,400 | 34.870 |
| Video Sample 10 | 245,936,764 | M4 | 15,782,112 | 245,894,400 | 34.831 |

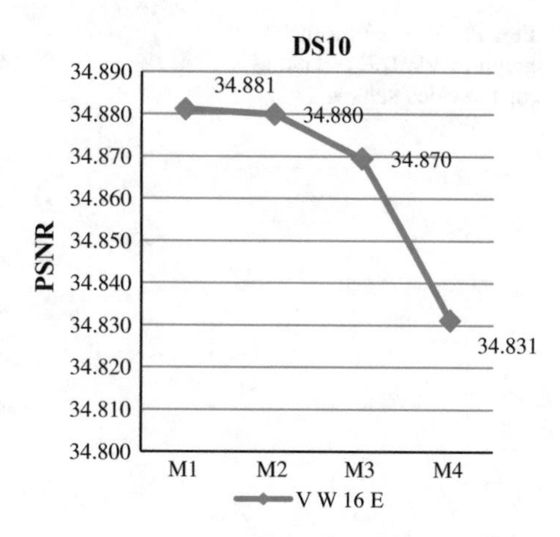


**Fig. 16** The test results' graph of VW16E technique for the video sample 10

M2 is 30,044 bytes, size of M3 is 353,828 bytes and size of M4 is 1,972,764 bytes while the size of Video Sample No 10 is 245,936,764 bytes. Each of these four messages was embedded and the PSNR which represent the loyalty of original video host to the watermarked video is calculated. Table 12 shows the results of PSNRs. As Fig. 16 depicts the PSNR has the expected trend.

## *3.11 Test Results of VW16E Technique for the Video Sample 11*

The result of embedding four messages into the video sample No 11 is presented in this part. The size of four messages, which are embedded- in order- into the video sample No 11 is 1,669 bytes, 30,044 bytes, 353,828 bytes and 1,972,764 bytes. The eleventh video stream has 259,417,036 bytes capacity. Each of these four messages was embedded and the PSNR which signify the fidelity of original video host to the

**Table 13** Test results of VW16E technique for the video sample 11

| Host | Host Size (Byte) | Confidential Message | Confidential Bit | Integrity Bits for VW16E | PSNR |
|---|---|---|---|---|---|
| Video Sample 11 | 259,417,036 | M1 | 13,352 | 259,372,800 | 35.010 |
| Video Sample 11 | 259,417,036 | M2 | 240,352 | 259,372,800 | 35.010 |
| Video Sample 11 | 259,417,036 | M3 | 2,830,624 | 259,372,800 | 35.007 |
| Video Sample 11 | 259,417,036 | M4 | 15,782,112 | 259,372,800 | 34.991 |

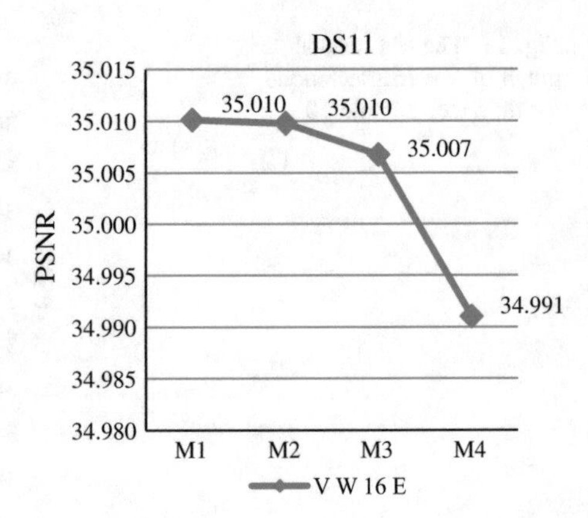


**Fig. 17** The test results' graph of VW16E technique for the video sample 11

watermarked video is calculated and have been shown in Table 13. Figure 17 depicts the trend of the PSNR; The PSNR is dropped with growth of the size of message.

## 3.12 *Test Results of VW16E Technique for the Video Sample 12*

This section presents the result of embedding four messages into the Video Sample No 12. The size of messages is different. The size of M1 is 1,669 bytes, the size of M2 is 30,044 bytes, size of M3 is 353,828 bytes and size of M4 is 1,972,764 bytes while the size of Video Sample No 12 is 260,626,804 bytes. Each of these four messages was embedded and the PSNR which represent the loyalty of original video host to the watermarked video is calculated. Table 14 shows the results of PSNRs. As Fig. 18 shows the PSNR is reduced by growth of message size.

**Table 14** Test results of VW16E technique for the video sample 12

| Host | Host size (Byte) | Confidential message | Confidential Bit | Integrity Bits for VW16E | PSNR |
|---|---|---|---|---|---|
| Video Sample 12 | 260,626,804 | M1 | 13,352 | 260,582,400 | 34.720 |
| Video Sample 12 | 260,626,804 | M2 | 240,352 | 260,582,400 | 34.719 |
| Video Sample 12 | 260,626,804 | M3 | 2,830,624 | 260,582,400 | 34.717 |
| Video Sample 12 | 260,626,804 | M4 | 15,782,112 | 260,582,400 | 34.705 |

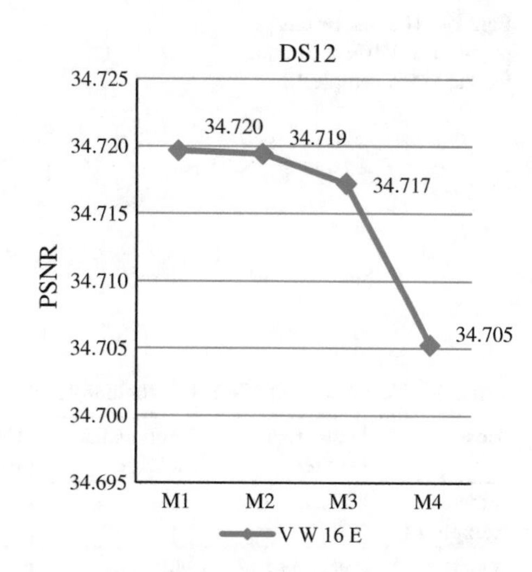


**Fig. 18** The test results' graph of VW16E technique for the video sample 12

## 3.13 *Test Results of VW16E Technique for the Video Sample 13*

In this section, the results of embedding four messages into the video sample No 13 with the size of 296,064,676 bytes have been presented. On the other hand, the size of M1 is 1,669 bytes, the size of M2 is 30,044 bytes, size of M3 is 353,828 bytes and size of M4 is 1,972,764 bytes. Table 15 illustrates the PSNR which presents the fidelity of the original video host with the watermarked video is calculated. Figure 19 shows PSNR is decreased by increasing the size of messages (Table 16).

**Table 15** Test results of VW16E technique for the video sample 13

| Host | Host size (Byte) | Confidential message | Confidential Bit | Integrity Bits for VW16E | PSNR |
|---|---|---|---|---|---|
| Video Sample 13 | 296,064,676 | M1 | 13,352 | 296,017,920 | 34.788 |
| Video Sample 13 | 296,064,676 | M2 | 240,352 | 296,017,920 | 34.788 |
| Video Sample 13 | 296,064,676 | M3 | 2,830,624 | 296,017,920 | 34.785 |
| Video Sample 13 | 296,064,676 | M4 | 15,782,112 | 296,017,920 | 34.770 |

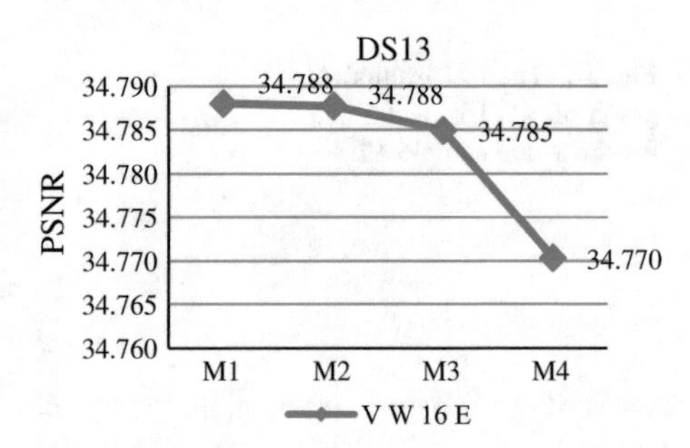


**Fig. 19** The test results' graph of VW16E technique for the video sample 13

**Table 16** Test results of VW16E technique for the video sample 14

| Host | Host size (Byte) | Confidential message | Confidential Bit | Integrity Bits for VW16E | PSNR |
|---|---|---|---|---|---|
| Video Sample 14 | 391,159,132 | M1 | 13,352 | 391,113,600 | 34.989 |
| Video Sample 14 | 391,159,132 | M2 | 240,352 | 391,113,600 | 34.989 |
| Video Sample 14 | 391,159,132 | M3 | 2,830,624 | 391,113,600 | 34.986 |
| Video Sample 14 | 391,159,132 | M4 | 15,782,112 | 391,113,600 | 34.973 |

### *3.14 Test Results of VW16E Technique for the Video Sample 14*

The result of embedding four messages into the video sample No 14 is presented in this part. The size of four messages which are embedded into the video sample No 14 in order is 1,669 bytes, 30,044 bytes, 353,828 bytes and 1,972,764 bytes. The fourteenth video stream has 391,159,132 bytes capacity. Each of these four

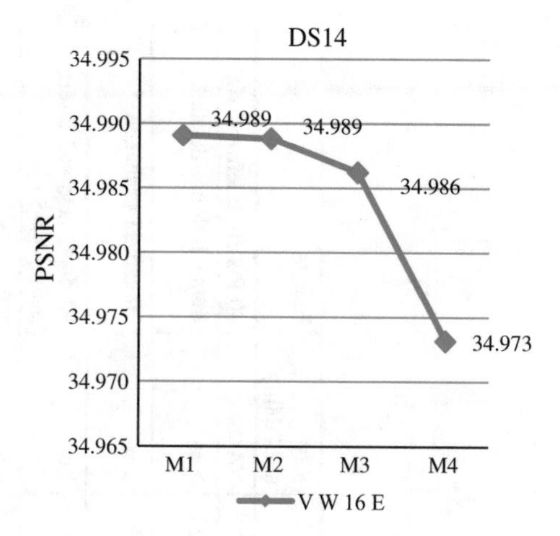


**Fig. 20** The test results' graph of VW16E technique for the video sample 14

messages was embedded and the PSNR which signify the fidelity of original video host to the watermarked video is calculated. In Fig. 20 the PSNR follows the expected trend.

## 4 Discussion and Analysis on Test and Evaluation

A novel spatial scheme was proposed for tamper detection to improve the imperceptibility and efficiency of tamper detection. VW16E technique could perfectly embed confidential and integrity information into the hosts, while the efficiency to detect the tamper is improved but the PSNR has not been improved considerably. The average PSNR was 34.84 dB in this phase. In this section, the test result is discussed, and the efficiency of the proposed technique is tested. The watermark only refers to the confidential message.

In this part, nine attacks applied on one watermarked video stream which has embedded with the VW16E technique the results are analyzed. In this section, some attacks including Frame Insert, Frame Exchange, Frame Deletion, Crop, Rotate, and Reverse Rotate, Frame Shifting, Salt & Pepper and Superimpose attacks are applied on the embedded video stream. Detecting the tampers is the next step to analyze on attacked videos. For this testing and evaluation we have chosen those videos that embedded with the techniques for largest message. This is because most portions of videos have the payload and it is obvious the entire videos have integrity bits for tamper detection.

**Table 17** Attack results on video sample no 1

| Sample No 1 | Attack | Address of first tampered pixel (Hexadecimal) | Tamper detect | Watermark extract | NC (%) | Key availability | Effects |
|---|---|---|---|---|---|---|---|
| 1 | Crop | 6081B | Yes | No | NA | NA | 40 Pix from left and right |
| **2** | Frame deletion | B6113B | Yes | Yes | 0.9999 | Yes | frame 70th has deleted |
| **3** | Frame exchange | 3239AB | Yes | Yes | 0.9999 | Yes | 20 and 93 exchange |
| **4** | Frame insert | 4C97FB | Yes | Yes | 0.9999 | Yes | 93 duplicated and insert before 30–50 |
| **5** | Reverse rotate | 2013 | Yes | No | NA | NA | 10° |
| **6** | Rotate | 2013 | Yes | No | NA | NA | 10° |
| **7** | Salt and Pepper | 17F6BB | Yes | Yes | 0.9999 | Yes | 10–12 applied |
| **8** | Frame shifting | 55CFB | Yes | Yes | 0.8085 | Yes | 1 shift left |
| **9** | Superimpose | 1828B | Yes | Yes | 0.9576 | Yes | 1–5 |

## *4.1 Attacks on Video Sample No 1*

The second message with the VW16E technique has been embedded in the sample No.1. Nine mentioned attacks applied on the first host. The nine attacks are Frame Insert, Frame Exchange, Frame Deletion, Crop, Rotate, and Rivers Rotate, Shift, Salt and Pepper and Superimpose attacks. These attacks are applied on the embedded video stream and subsequently the attacked video are analysed to know the video is tampered or not. Table 17 shows the tamper detection result of attacks. If the watermark can be extracted, the NC for the extracted watermark is calculated and as described in the section II, message used are the repetition of a key. According to the Table 17 in terms of specifying robustness the availability of even a single key is checked.

### 4.1.1 Crop Attack on Video Sample No 1

Forty (40) Pixels from top, bottom, left and right for all frames of the sample number one with the size of 16,080,844 bytes and with the message of 30,044 bytes have been cropped. The result shows modifying is detected and the address 6081B is returned as a first block which is tampered. Figure 21 shows the embedded frame, and then the result of tamper detection has shown in Fig. 22 for the same frame.

### 4.1.2 Frame Deletion Attack on Video Sample No 1

Frame number 70 has been deleted in sample number one with the size of 16,080,844 bytes with the message of 30,044 bytes. The result shows modifying is detected and the address B6113B is returned as a first block which is tampered. The

**Fig. 21** Watermarked frame for crop attack on video sample no 1

**Fig. 22** Result of tamper detection for crop attack on video sample no 1

**Fig. 23** Result of tamper detection for frame deletion attack on video sample no 1

result of tamper detection for the 71th frame has depicted in Fig. 23. The extracted watermark NC is 99.98 %.

#### 4.1.3 Frame Exchange Attack on Video Sample No 1

Frame number 20th exchange to 93th in sample number one with the size of 16,080,844 bytes with the message of 30,044 bytes. The result shows modifying is detected and the address 3239AB is returned as a first block which is tampered. The NC of the extracted watermark is 99.98\,%.

### 4.1.4 Frame Insert Attack on Video Sample No 1

Frame number 93 has been duplicated in the host number one with the size of 16,080,844 bytes and with the message of 30,044 bytes. The result shows modifying is detected and the address 4C97FB is returned as a first block which is tampered. The NC of the extracted watermark is 99.98 %.

### 4.1.5 Rotate Attack on Video Sample No 1

All the frames of the sample number one with the size of 16,080,844 bytes and with the message of 30,044 bytes have been rotated by 10°. The result shows modifying is detected the address 2013 is returned as a first block which is tampered. Figure 24 shows the embedded frame.

### 4.1.6 Reverse Rotate Attack on Video Sample No 1

All the frames of the sample number one with the size of 16,080,844 bytes and with the message of 30,044 bytes have been rotated by 10° and then have been reversed. The result shows modifying is not detected. As Fig. 25 illustrates the embedded frame, Fig. 26 shows the tampered frame.

**Fig. 24** Watermarked frame for rotate attack on video sample no 1

**Fig. 25** Watermarked frame for reverse rotate attack on video sample no 1

**Fig. 26** Tampered frame for reverse rotate attack on video sample no 1

### 4.1.7 Salt and Pepper Attack on Video Sample No 1

For the 10th, 11th, and 12th frames, salt and pepper have been applied in the sample number one with the size of 16,080,844 bytes and with the message of 30,044 bytes. The result shows modifying is detected and the address 17F6BB is returned as a first block which is tampered. Figure 27 shows the result of tamper detection for the same frame. The NC of the extracted watermark is 99.98 %.

**Fig. 27** Result of tamper detection for salt and pepper attack on video sample no 1

### 4.1.8 Shift Frames Attack on Video Sample No 1

One frame shift left for the sample number one with the size of 16,080,844 bytes and with the message of 30,044 bytes. The result shows modifying is detected and the address 55CFB is returned as a first block which is tampered. The NC of the extracted watermark is 80.85 %.

### 4.1.9 Superimpose Attack on Video Sample No 1

For the frame numbers 1, 2, 3, 4 and 5 in the sample number one with the size of 16,080,844 bytes and with the message of 30,044 bytes, Superimpose has been applied. The result shows modifying is detected and the address 1828B is returned as a first block which is tampered. The NC of the extracted watermark is 95.76 %.

## 5 Conclusion

Concealing information in the video host for different purposes including authentication and tamper detection is the basic idea for video watermarking technique. Spatial domain techniques are the most common approach for concealing information in the video host. We focused on video watermarking; particularly with respect to the Audio Video Interleaved (AVI) form of video file format in spatial domain by simulating blocks. We proposed a new watermarking technique that gives a high imperceptibility and efficient tamper detection compared to the other similar schemes.

## References


1. Arab, F., Daud, S.M., Hashim, S.Z.M., Manaf, A.A., Shubli, S.: Evaluation of video watermarking techniques. JDCTA: Int. J. Digit. Content Technol. its Appl. **7**(5), 484–491 (2013)
2. Zamani, M., Manaf, A.A., Abdullah, S., Chaeikar, S.S.: Mazdak Technique for PSNR Estimation in Audio Steganography Applied Mechanics and Materials, vol. 229, pp. 2798–2803. Trans Tech Publications Inc (2012)
3. Zamani, M., Manaf, A.A., Ahmad, R., Jaryani, F., Chaeikar, S.S., Zeidanloo, H.R.: (2010). Genetic Audio Watermarking Communications in Computer and Information Science, vol. 70, pp. 514–517. Springer, Berlin
4. Zamani, M., Manaf, A.A.: Genetic algorithm for fragile audio watermarking. Telecommun. Syst. (2015)
5. Zeki, A., Manaf, A.A., Ibrahim, A.A., Zamani, M.: A robust watermark embedding in smooth areas. Res. J. Inf. Technol. **3**(2), 123–131 (2011)
6. Zamani, M., Manaf, A.A., Ahmad, R., Zeki, A., Magalingam, P.: A novel approach for audio watermarking. Paper presented at the 5th international conference on information assurance and security, Xian (2009)
7. Zamani, M., Ahmad, R., Manaf, A.A., Zeki, A.: An approach to improve the robustness of substitution techniques of audio steganography. Paper presented at the international conference on computer science and information technology, Beijing (2009)
8. A Jabbar Altaay, A., Bin Sahib, S., Zamani, M. (2013). Correlation Analysis of the Four Photo Themes in Five Layers Communications in Computer and Information Science (Vol. 398, pp. 211–222): Springer Berlin Heidelberg
9. Hassanien, A.E.: A copyright protection using watermarking algorithm. Informatica, Lith. Acad. Sci. **17**(2), 187–198 (2006)
10. Agarwal, H., Ahuja, R., Bedi, S.: Highly robust and imperceptible luminance based hybrid digital video watermarking scheme for ownership protection. Int. J. Image, Graph. Sign. Process. **4**(11) (2012)
11. Yu, P.F., Yu, P.C., Xu, D.: Palmprint authentication based on DCT-based watermarking. Appl. Mech. Mater. **457**, 893–898 (2014)
12. Ishtiaq, M., Jaffar, M.A., Khan, M.A., Jan, Z., Mirza, A.M.: Robust and imperceptible watermarking of video streams for low power devices. In Signal Processing, Image Processing and Pattern Recognition. Springer (2009)
13. Zamani, M., Taherdoost, H., Manaf, A.A., Ahmad, R., Zeki, A.: Robust Audio Steganography via Genetic Algorithm. In: 3rd International Conference on Information & Communication Technologies ICICT2009, pp.149–153. Karachi, Pakistan, 15–16 Aug 2009
14. Sinha, S., Bardhan, P., Pramanick, S., Jagatramka, A., Kole, D.K., Chakraborty, A.: Digital video watermarking using discrete wavelet transform and principal component analysis. Int. J. Wisdom Based Comput. **1**(2), 7–12 (2011)
15. Zamani, M., Abdul Manaf, A., Abdullah, S.: An overview on audio steganography techniques. Int. J. Digit. Content Technol. Its Appl. **6**(13), 107–122 (2012)
16. Zamani, M., Abdul Manaf, A., Rouhani Zeidanloo, H., Shojae Chaeikar, S.: Genetic substitution-based audio steganography for high-capacity applications. Int. J. Internet Technol. Secured Trans. **3**(1), 97–110 (2011)
17. Jabbar Altaay, A.A., Bin Sahib, S., Zamani, M.: Pixel Correlation Behavior in Different Themes Communications in Computer and Information Science, vol. 398. Springer, Berlin (2013)
18. Zamani, M., Abdul Manaf, A., Ahmad, R., Jaryani, F., Taherdoost, H., Zeki, A. (2009). A Secure Audio Steganography Approach. 4th International Conference for Internet Technology and Secured Transactions, London. 1–6

19. Jabbar Altaay, A.A., Bin Sahib, S., Zamani, M.: An Introduction to Image Steganography Techniques. International Conference on Advanced Computer Science Applications and Technologies, pp.122–126. Kuala Lumpur (2012)
20. Hasnaoui, M., Mitrea, M.: Semi-fragile watermarking for video surveillance applications. In: Signal Processing Conference (EUSIPCO), 2012 Proceedings of the 20th European, pp. 1782–1786 (2012)
21. Chimanna, M.A., Khot, S.: Robustness of video watermarking against various attacks using Wavelet Transform techniques and Principle Component Analysis. In: 2013 International Conference on Information Communication and Embedded Systems (ICICES), pp. 613–618 (2013)
22. Zamani, M., Manaf, A.A., Ahmad, R.: Knots of substitution techniques of audio steganography. Paper presented at the international conference on telecom technology and applications, Manila (2009)
23. Jabbar Altaay, A.A., Bin Sahib, S., Zamani, M.: Statistical steganalysis of four photo themes before embedding. Paper presented at the international conference on advanced computer science applications and technologies, Sarawak (2013)
24. Abdullah, S., Manaf, A.A., Zamani, M.: Capacity and quality improvement in reversible image watermarking approach. Paper presented at the international conference on networked computing and advanced information management, Seoul (2010)
25. Arab, F., Daud, S.M., Hashim, S.Z.: Discrete wavelet transform domain techniques. In: International Conference on Informatics and Creative Multimedia (ICICM), pp. 340–345 (2013). Zamani, M., Manaf, A.A., Ahmad, R.: Current problems of substitution technique of audio steganography. Paper presented at the international conference on artificial intelligence and pattern recognition, Orlando (2009)
26. Zamani, M., Manaf, A.A., Daruis, R.: Azizah technique for efficiency measurement in steganography. Paper presented at the 8th international conference on information science and digital content technology, Jeju (2012)
27. Arab, F., Abdullah, S.M., Hashim, S.Z.M., Manaf, A.A., Zamani, M.: A robust video watermarking technique for the tamper detection of surveillance systems. Multimedia Tools Appl. (2015)

# A Robust and Computationally Efficient Digital Watermarking Technique Using Inter Block Pixel Differencing

**Shabir A. Parah, Javaid A. Sheikh, Nazir A. Loan and G.M. Bhat**

**Abstract** The growth of internet coupled with the rise in networked infrastructure has resulted in exponential increase in the multimedia content being shared over the communication networks. The advancement in technology has resulted in increase in multimedia piracy. This is due to the fact that it is very easy to copy, duplicate and distributes multimedia content using current day technology. In such a scenario Digital Rights Management is one of the prominent issues to be dealt with and tremendous work is going on in this direction round the globe. Digital watermarking and fingerprinting have emerged as fundamental technologies to cater to DRM issues. These technologies have been found to be of prominent use in content authentication, copy protection, copyright control, broadcast monitoring and forensic applications. Various requirements of a digital watermarking system include Imperceptibility, Robustness, Security, Payload and Computational complexity. The main requirement of real time DRM systems is lesser computational complexity and high robustness. This chapter proposes and analyses a robust and computational efficient Image watermarking technique in spatial domain based on Inter Block Pixel Difference (IBPD). The cover image is divided into 8 × 8 non overlapping blocks and difference between intensities of two pixels of adjacent blocks at predefined positions is calculated. Depending upon the watermark bit to be embedded; both the pixels are modified to bring the difference in a predefined zone. The experimental results reveal that the proposed scheme is capable of providing high quality watermarked images in addition to being robust to various singular and hybrid image processing and geometrical attacks like Salt and Pepper noise, Gaussian noise, Sharpening, Compression, Rotation and Cropping etc. Further the implementation of the scheme in pixel domain reduces the computational complexity drastically and makes the proposed scheme an ideal candidate for real time applications.

---

S.A. Parah (✉) · J.A. Sheikh · N.A. Loan · G.M. Bhat
Department of Electronics and Instrumentation Technology, University of Kashmir, Srinagar 190006, India
e-mail: shabireltr@gmail.com





A.E. Hassanien et al. (eds.), *Multimedia Forensics and Security*,
Intelligent Systems Reference Library 115, DOI 10.1007/978-3-319-44270-9_10

**Keywords** Blind watermarking • Spatial domain • Forensics • Fingerprinting • Computational complexity • Inter block pixel difference (IBPD)

## 1 Introduction

The exponential growth of internet has boosted the area of information technology to a large extent. This has resulted into sharing of abundant data over the networked infrastructure. The ease of information flow has resulted in various e-facilities creeping in our day to day lives and making life easy. Some of the prominent electronic facilities include e-banking, e-commerce, e-governance and e-healthcare etc. This however, poses various challenges in terms of security of data and copyright issues of the multimedia owners. This is due to the fact that digital data being shared via internet is available for both the authentic users as well as unauthentic users and hence can be easily copied, altered and redistributed. For such a situation, the major concern for the authors, owners, buyers and suppliers of data (which need to be addressed) are the copyright protection, copy control and authenticity of the content [1]. Digital watermarking has evolved as an effective solution for such a situation.

Digital watermarking is a technique of hiding an important information into a host message which can be retrieved at a later stage for proof of copyright or against any illegal effort to either reproduce or manipulate it. The embedded information can be anything, for example plain text, image or a serial number [2]. For the case of medical images, a watermark may consist of hospital logo and patient's information (Electronic Health Record). In case of fingerprint images, it could be the authentication mark to prove that the fingerprint is genuine. It is pertinent to mention that to secure the watermark, a key is usually used to scramble the watermark before embedding it into cover media and same is used at extraction stage. A digital watermarking algorithm is mainly characterized by Robustness, Security, Capacity and Imperceptivity [3]. Depending on different applications and requirements, watermarking methods can be broadly classified into two broad categories, fragile and robust watermarking. Fragile watermarks are mainly used for content protection, authentication and tamper detection, while robust watermarks are used for ownership verification and copyright protection [4, 5]. For medical image watermarking, authentication matters a lot as the image contains the critical information about the patient. For general images robustness is of major concern while for fingerprint watermarking both authentication and robustness are important. A digital watermark is generally embedded into a cover media in spatial or transform domain [6, 7]. The spatial domain embedding methods are easy to implement as they directly operate on pixels. This results in less computational overhead [8]. The earliest and computationally efficient watermarking method in spatial domain is Least Significant Bit (LSB) substitution in which watermark bits are directly placed in the LSBs of the cover image [9, 10]. Transform domain methods like Discrete Cosine Transform (DCT) and Discrete Wavelet Transform (DWT) manipulate transform coefficients to embed watermark. Though these type of watermarking algorithms are robust but are computationally complex [11].

Development of robust watermarking algorithm in spatial domain is a challenging task. In this chapter we present a robust and computationally efficient digital image watermarking technique using IBPD for grey scale and color images. The scheme has been tested for various standard grey scale and color test images. We have also used some medical and fingerprint images to test the algorithm. It has been observed that in addition to computational efficiency and robustness to various singular attacks the proposed system is robust to various simultaneous attacks as well. Rest of the chapter has been organized as follows. Section 2 presents a detailed account of related literature. Proposed scheme is presented in Sect. 3. Experimental results and discussions are presented in Sect. 4. The chapter concludes in Sect. 5.

## 2 Related Literature

With an aim to meet various challenging requirements of multimedia watermarking systems a lot of research work is being carried out round the globe. Most of the work involves optimization of robustness, imperceptibility, payload, security and computational complexity. For DRM, robustness is utmost important. A robust watermarking technique based on error correction codes and dual encryption along with DWT-DCT has been proposed by Kalra et al. [12]. Though the scheme is robust but, it is computationally complex and has low capacity. A blind image watermarking technique based on Fractional Fourier transform has been proposed by Jun et al. [13]. The authors demonstrate that the scheme is highly robust against JPEG compression and other image processing operations but, has higher computational overhead. A DCT domain based blind watermarking technique has been reported by Ma et al. in [14]. Arnold transform has been used to scramble the watermark to reduce the correlation between pixels and increase the security of watermark. This scheme however, has low capacity and poor imperceptivity. A spatial domain watermarking scheme has been proposed in [15]. The scheme has been shown to be robust to various geometric and signal processing attacks but its capacity is very low. A color image watermarking scheme based on geometric correction using Quaternion Exponent Moments (QEM) has been reported in [16]. The secret watermark bits are embedded into the host image by modifying the Quaternion Fourier Transform (QFT) coefficients. The scheme is highly robust against rotation and scaling attacks but its capacity is low. A robust watermarking scheme for gray scale images is presented in [17]. The host image as well as watermark is encrypted and then encrypted watermark is embedded into the encrypted host image. The security of this scheme is high but the imperceptivity is low. A novel watermarking approach for color images based on Discrete Wavelet Transform (DWT) using bio-inspired optimization principle is presented in [18]. A good perceptual quality watermarked image is yielded by this technique but its capacity is low. A watermarking technique for gray scale images in DWT domain based on dynamic block selection has been reported in [19]. Binary watermark of size 8 × 8 is embedded into the significant wavelets in the dynamic blocks with strong edges because the change in blocks with strong edges is less noticeable to human eye. Since its capacity depends on the size of $HL_n$ and $LH_n$ subbands of n-level DWT of an image therefore, its capacity is very low and hence cannot be used for the applications where large data is to be embedded. For copyright protection and ownership verification of color images a robust watermarking scheme is presented in [20]. It takes the advantage of Gray Level Co-occurrence Matrix (GLCM) and Principal Component Analysis (PCA) for watermark embedding. This scheme achieves high transparency, imperceptivity and robustness but its capacity is very low because the payload is strictly image dependent. A watermarking scheme based on partial calculation of radial moments and interval phase modulation is proposed in [21] for gray scale images which is robust against geometric attack. This scheme has very high imperceptivity and is highly robust against rotation attack but the capacity of this scheme is very low, only 127 bits are embedded in a cover image of size 512 × 512. Further, the scheme is not tested for cropping attack which is one of the crucial geometric attacks that needs to be taken into account. A blind color image watermarking technique has been proposed in [22]. The scheme is robust to signal processing and geometric attacks and is computationally more efficient. Local quaternion exponent moments, based color image watermarking technique is proposed in [23]. The scheme is robust against geometric attacks and other image processing operations but offers low payload. For copyright protection and authentication of gray scale images, Chinese Remainder Theorem-based watermarking scheme has been presented in [24]. The scheme provides good quality watermarked images and has been reported to be robust against various attacks. However, without applying any image processing operation on the watermarked images, this scheme reports a bit error rate of 0.5. Based on the concept of mathematical remainder, a watermarking technique is reported in [25] in which watermark is embedded into the low-frequency DCT coefficients of an image. The scheme is robust against JPEG compression and other image processing operations but it provides watermarked images of poor quality and also its embedding capacity is low.

With an aim to achieve high degree of robustness and imperceptivity Agarwal et al. [26] proposed a watermarking technique based on Fuzzy-BPN architecture. The reported performance of the scheme has been found good on both fronts but is its main drawback is high computational complexity. An attempt to reduce the computational complexity has been made by Verma et al. [27]. The scheme utilizes concepts of significant region (SR) based image watermarking technique using lifting wavelet transform (LWT). This scheme however has reduced payload ability.

A novel robust watermarking technique presented based on correlation of corresponding elements in two successive 8 × 8 DCT blocks has been reported in [28]. The cover image (512 × 512) has been converted in 8 × 8 blocks and DCT of each block is computed. Watermark embedding is carried out based on the correlation of the two corresponding DCT coefficients in adjacent blocks. For embedding watermark bits 1 and 0 the coefficient difference in selected adjacent blocks are brought in predefined zones using a modification factor. The scheme has been shown robust to various singular and simultaneous attacks. However, it fails to utilize all 64 × 64 DCT blocks to embed watermark as such payload capability is low and the performance of the scheme to noise attacks like White Gaussian Noise and salt and pepper noise is not up to the mark. Further as the scheme has been implemented in transform domain, computational complexity is high. Our work has been inspired by this work [28]; we however implement the scheme in spatial domain with some changes in embedding approach, thus reducing computational cost drastically. Further we have been able to use all the blocks for embedding watermark thus enhancing payload.

## 3 Proposed Watermarking Algorithm

The block diagram of the proposed system is shown in Fig. 1. The original image I is divided into non-overlapping blocks each of size $8 \times 8$. The block pre-processing has been discussed in Sect. 3.1.

In each block one bit of watermark is embedded, therefore the total number of watermark bits that can be embedded equals the total number of blocks i.e., a watermark of $64 \times 64$ can be embedded in a $512 \times 512$ grey scale cover image while as three $64 \times 64$ watermarks can be embedded in a color image of same size. Watermark embedding has been discussed in detail in Sect. 3.2. The proposed algorithm uses blind extraction. Section 3.3 gives a detailed account of extraction process.

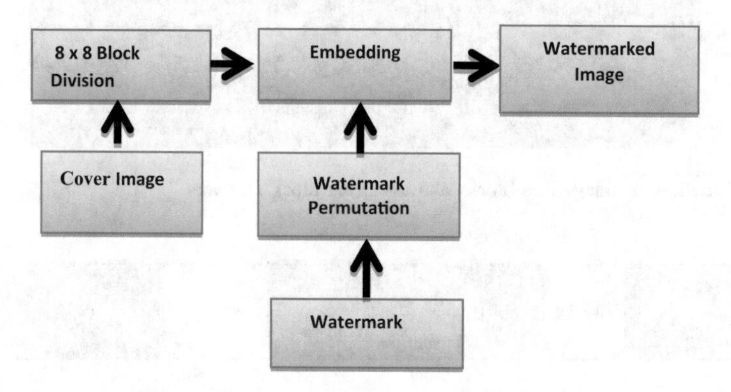


**Fig. 1** Block diagram of proposed scheme

## 3.1 Block Pre-processing

The cover image I of size $M \times N$ is divided into non-overlapping blocks each of size $8 \times 8$. The total number of blocks is equal to PQ where $P = M/8$ and $Q = N/8$. A unique block number is assigned to each block according to Fig. 2.

If an arbitrary block of the cover image is represented by $B_x$, $1 \leq x \leq P \cdot Q$. Then the blocks are arranged in a single row with increasing order of x as shown in Fig. 3.

The pixel of xth block $B_x$ at position (i, j) is denoted by $B_x(i,j)$ where $1 \leq i,j \leq 8$. In each block one bit of watermark is embedded, therefore the total number of watermark bits that can be embedded into a grayscale image of size $M \times N$ is equal to the total number of blocks i.e. $P \times Q$, while for color image of same size, it is equal to three times the grayscale image. The binary watermark $W$ to be embedded is permuted with a key before embedding and is represented by $Wp$. The watermark bit $Wp(x)$ is embedded into the block $B_x$ by modifying the difference between two predetermined pixels of two adjacent blocks. The difference 'D' between the two pixels of adjacent blocks is calculated and is given as:

$$D = B_x(i,j) - B_{x+1}(k,l) \tag{1}$$

| $B_1$ | $B_2$ | $B_3$ | $B_4$ | $B_5$ | ..... | $B_{(Q)}$ |
|---|---|---|---|---|---|---|
| $B_{(2Q)}$ | $B_{(2Q-1)}$ | $B_{(2Q-2)}$ | $B_{(2Q-3)}$ | $B_{(2Q-4)}$ | ..... | $B_{(Q+1)}$ |
| $B_{(2Q+1)}$ | $B_{(2Q+2)}$ | $B_{(2Q+3)}$ | $B_{(2Q+4)}$ | $B_{(2Q+5)}$ | ..... | $B_{(3Q)}$ |
| $B_{(4Q)}$ | $B_{(4Q-1)}$ | $B_{(4Q-2)}$ | $B_{(4Q-3)}$ | $B_{(4Q-4)}$ | ..... | $B_{(3Q+1)}$ |
| $B_{(4Q+1)}$ | $B_{(4Q+2)}$ | $B_{(4Q+3)}$ | $B_{(4Q+4)}$ | $B_{(4Q+5)}$ | ..... | $B_{(5Q)}$ |
| ⋮ | ⋮ | ⋮ | ⋮ | ⋮ | ⋮ | ⋮ |
| $B_{(P*Q)}$ | $B_{(P*Q-1)}$ | $B_{(P*Q-1)}$ | $B_{(P*Q-2)}$ | $B_{(P*Q-3)}$ | ..... | $B_{(P*Q)-(Q-1)}$ |

**Fig. 2** Dividing of image into blocks and assigning block numbers

| $B_1$ | $B_2$ | ..... | $B_{(Q)}$ | $B_{(Q+1)}$ | $B_{(Q+2)}$ | ..... | $B_{(P*Q-1)}$ | $B_{(P*Q)}$ |
|---|---|---|---|---|---|---|---|---|

**Fig. 3** Arranging blocks in a row

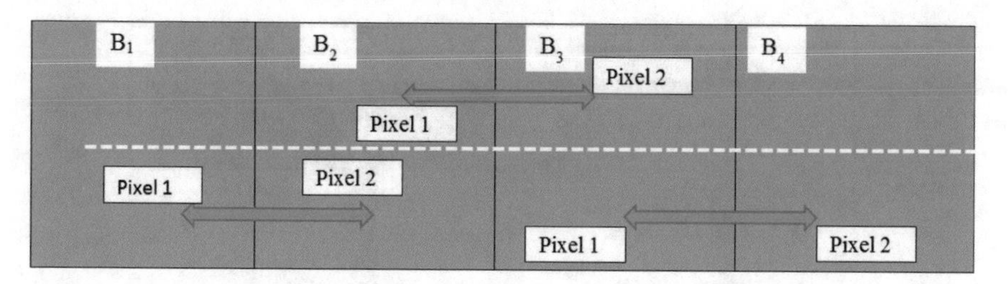


**Fig. 4** Pixel selection for even and odd numbered blocks

where $B_x(i,j)$ is pixel of block $B_x$ and $B_{x+1}(k,l)$ is pixel of block $B_{x+1}$. When x is even, the two pixels one from upper half of the block $B_x$ and another from the upper half of the block $B_{x+1}$ are randomly chosen for embedding purpose as depicted from Fig. 4. While for odd x, the two pixels are taken from lower half of the blocks $B_x$ and $B_{x+1}$.

This is done to ensure correct watermark extraction. The two pixels chosen may have same or different indices. Depending upon the watermark bit to be embedded both the pixels $B_x(i,j)$ and $B_{x+1}(k,l)$ are modified by a modification parameter 'm' so that the difference '*D*' is made to lie in a specific zone. Figure 5 shows various difference Zones for embedding binary 0 and 1.

The modification factor is given as

$$m = q \times abs(\frac{Av(B_x) - Med(B_x)}{Av(B_x)}) \tag{2}$$

where $Av(B_x)$ is the average of pixels of block $B_x$, $Med(B_x)$ is the median of 64 pixels of block $B_x$ and q is a scaling variable, we set q = 0.1.

## 3.2 Watermark Embedding

The watermark embedding algorithm for grayscale images is summarized as under:

Step 1 Read the original image I.

Step 2 Divide I into $8 \times 8$. non-overlapping blocks and arrange them into a row vector as shown in Fig. 3, xth block of I is denoted by $B_x$

Step 3 Read the watermark '*w*.

Step 4 Permute the watermark with a key, the permuted watermark is denoted by *Wp*.

Step 5 Initialize the threshold value *Th* to 80 and Embedding strength *Em* to 20. (we have set values of *Th* and *Em* as per [28]).

Step 6 Compute the difference '*D*' using Eq. 1 according to the Fig. 4.

Step 7 Compute the modification parameter m using Eq. 2.

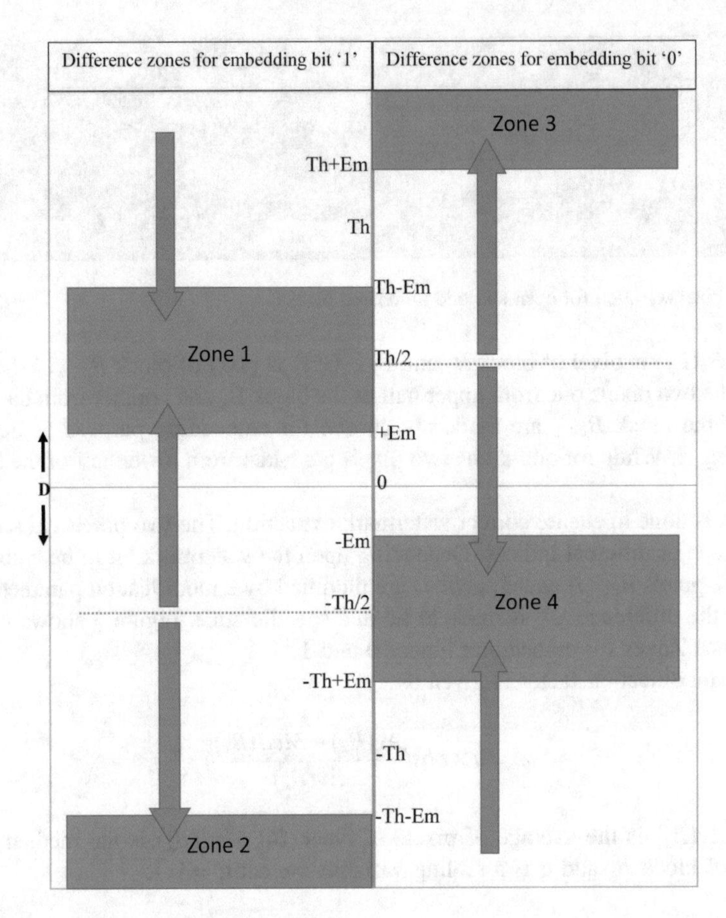


**Fig. 5** Various difference zones for embedding

***For embedding watermark bit Wp(x) = 1; do the following:***

*If {D ≥ (Th-Em) then subtract m/2 from pixel $B_x$ (i, j) and add m/2 to pixel $B_{x+1}$ (i, j) until the difference 'D' becomes less than (Th-Em)}.*

*Else if {D ≤ Em and D ≥* (−Th/2) then add m/2 to pixel $B_x$ *(i, j) and subtract m/2 from pixel $B_{x+1}$ (i, j) until difference 'D' becomes greater than Em}.*

*Else if {D ≤* (−Th/2) then subtract m/2 from pixel $B_x$ *(i, j) and add m/2 to pixel $B_{x+1}$ (i, j) until the difference D becomes less than (–Th-Em)}.*

*End of embedding of bit 1*

***For embedding watermark bit Wp(x) = 0 do the following:***

*If {D ≥ Th/2 then add m/2 to pixel $B_x$ (i, j) and subtract m/2 from pixel $B_{x+1}$ (i, j) until difference 'D' becomes greater than (Th + Em)}.*

*Else if {D ≥ (-Em) and D < Th/2 then subtract m/2 from pixel $B_x$ (i, j) and add m/2 to pixel $B_{x+1}$ (i, j) until the difference 'D' becomes less than (–Em)}.*

*Else if {D ≤ (-Th + Em) then add m/2 to pixel $B_x$ (i, j) and subtract m/2 from pixel $B_{x+1}$ (i, j) until difference 'D' becomes greater than (-Th + Em)}.*
*End of embedding of bit 0*

**Step 8**. Repeat the steps from step 5 to step 7 until the whole watermark *Wp* is embedded into the image I. After complete embedding the blocks are brought rearranged to obtain watermarked image. It is evident from the embedding process that for embedding binary 1 difference is adjusted to lie in zone 1 or zone 2 while as for watermark bit zero it is modified to lie in zone 3 or zone 4.

The watermark embedding algorithm for color images is same as for grayscale image where each plane (Red, Green, Blue) of the image is treated as grayscale image and a watermark is embedded in each plane (i.e. three logos are embedded into a color image) by following the above mentioned steps.

### 3.3 Watermark Extraction

For extraction of watermark from the watermarked image, the watermarked image is divided into $8 \times 8$ blocks and the blocks are arranged in a single row according to Fig. 3. The difference between the pixels $B_x(i,j)$ and $B_{x+1}(k,l)$ of adjacent blocks is calculated using Eq. 1 according to Fig. 4. If the difference is less than $(-Th - Em)$ or falls between *Em* and $(Th - Em)$, i.e. difference zone 1 or zone 2, a bit 1 is extracted and if the difference is greater than $(Th + Em)$ or falls between $-Th + Em)$ and $-Em$, i.e. zone 3 or zone 4, a bit 0 is extracted. Since we embedded a permuted watermark therefore, the inverse permutation of bits is required to get the extracted watermark.

## 4 Experimental Results and Discussions

The experimental results have been carried out on various test images including commonly used gray scale test images, medical images and fingerprint images. Both gray scale and color images used for testing purpose are of size $512 \times 512$. The grayscale images and various binary logos (i.e. Logo 1, logo 2 and Logo 3) used as watermarks are shown in Fig. 6 while as the color images have been depicted in Fig. 7. It is pertinent to mention here that one logo has been embedded in grey scale images while as three in a color image with Logo 1 is embedded into red plane, Logo 2 in Green plane and Logo 3 in Blue plane.

The proposed scheme has been evaluated using various image quality metrics like Normalized cross correlation (NCC), Bit Error Rate (BER) Peak Signal to Noise Ratio (PSNR) and Mean Square Error [1, 9, 28]. The system robustness has

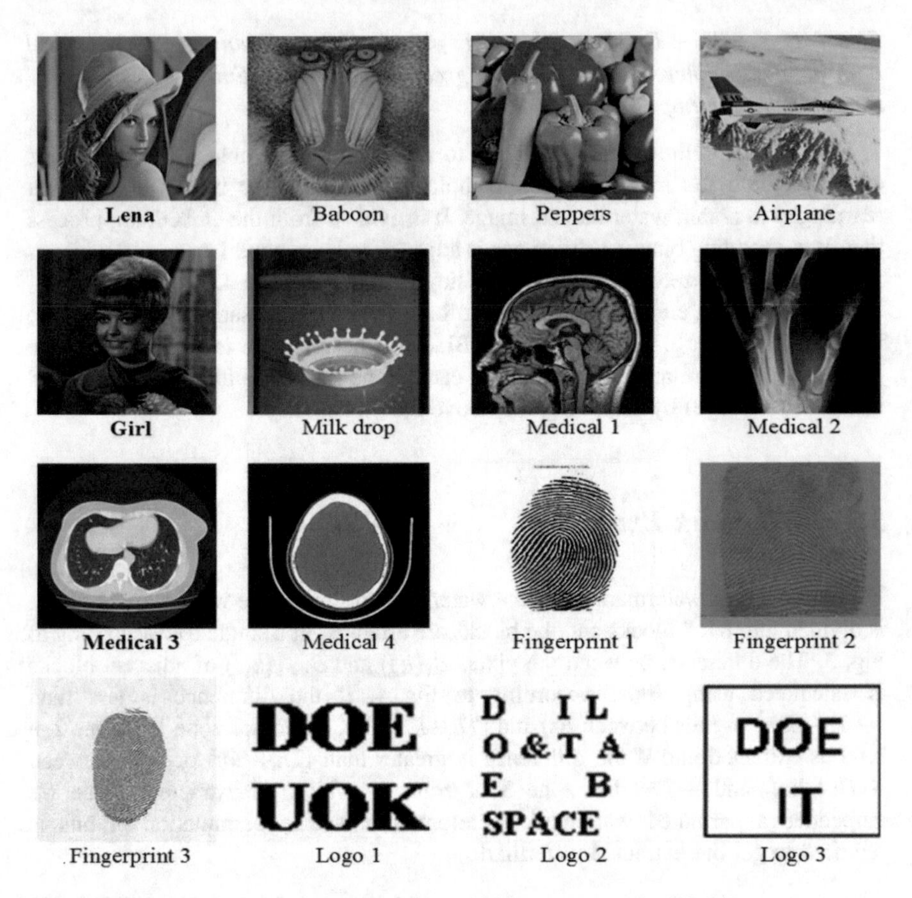


**Fig. 6** 512 × 512 Grayscale images and 64 × 64 binary watermarks [29–31]

been evaluated in terms of quality metrics like Normalized cross correlation (NCC) and Bit Error Rate (BER). NCC gives the similarity between original watermark and extracted watermark while as the dissimilarity between the two is judged by BER. Lesser BER and higher NCC values indicate that the algorithm is robust to the attacks. If W is the original watermark and $W^*$ is the extracted watermark then BER and NCC are given as

$$BER = \frac{1}{RS}\left[\sum_{i=1}^{R}\sum_{j=1}^{s} W(i,j) \oplus W^*(i,j)\right] \tag{3}$$

$$C = \frac{\sum_i \sum_j W(i,j) * W^*(i,j))}{\sum_i \sum_j [W(i,j)]^2} \tag{4}$$

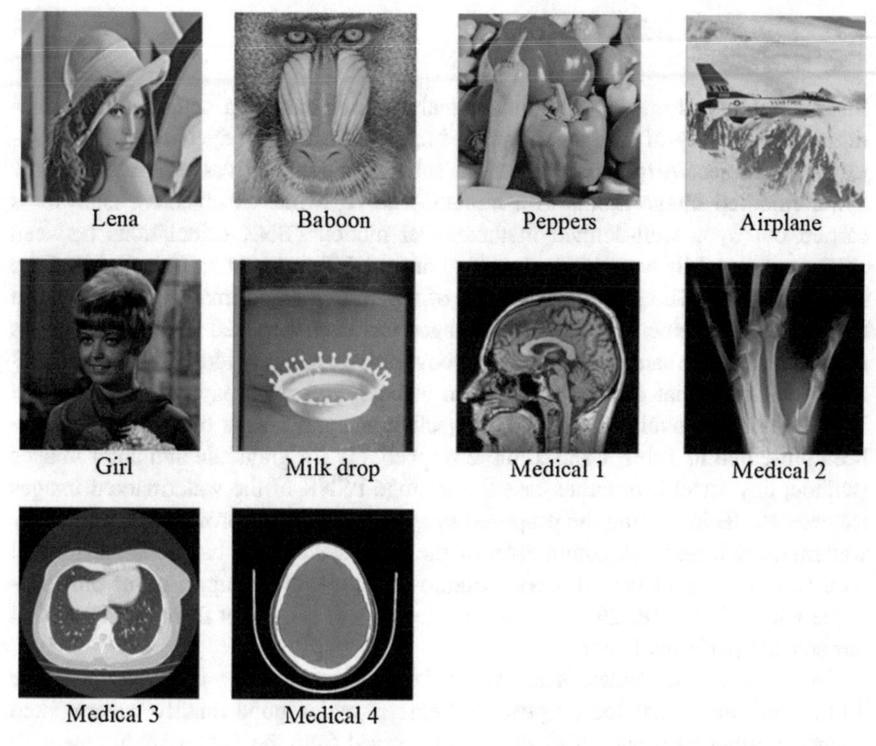


**Fig. 7** 512 × 512 color images [29, 30]

where W(i, j) is original watermark bit at position (i, j) and $W^{*}(i,j)$ is the extracted watermark bit at position (i, j) and R × S is the size of watermark. Peak Signal to Noise Ratio (PSNR) is used to measure the imperceptivity of watermarked image. If 'I' represents the cover image and 'Iw' represents the watermarked image then PSNR is given as

$$PSNR(dB) = 10\log_{10}\frac{255^2}{MSE} \tag{5}$$

where MSE is the mean square error and is given as:

$$MSE = \frac{1}{MN}\sum_{i=1}^{M}\sum_{j=1}^{N}[I(i,j) - I_w(i,j)]^2 \tag{6}$$

Here M and N are the dimensions of the images. It is in place to mention here that the computations have been carried out using MATLAB R2010b on windows 7 platform with Intel i3-3110 M processor working at 2.40 GHz and with 4 GB RAM.

## *4.1 Imperceptivity Analysis*

Imperceptibility is one of the fundamental requirements of a watermarking algorithm. The quality of the watermarked images is usually judged from their subjective and objective quality indices. The subjective quality gives an idea about how a watermarked image seems to a human observer while as objective analysis is carried out by a well-defined mathematical model. PSNR calculations between cover image and its watermarked version are used to arrive at an objectivity of the imperceptibility. The subjective quality of the watermarked images as obtained in the proposed scheme for gray scale images has been depicted in Fig. 8 while as Fig. 9 presents the same for color based cover images. It is evident from the Figs. 8 and 9 that no visual distortions are seen in both case for a payload of 4096, and 12288 bits respectively. Further, the objective analysis results of the scheme have been presented in Table 1 and Table 2 respectively for grayscale and color images (without any attack). In either case the average PSNR of the watermarked images exceeds 40 dB indicating the proposed system is capable of providing high quality watermarked images. A comparison of the proposed system has been carried out with various state of the art works. Figure 10 shows the comparison of our technique with [13, 14, 18, 26] for cover image Lena. It is evident from the figure that our scheme performs better.

From the above Tables, it is observed that average value of PSNR is above 40 dB, indicating that the proposed scheme provides good quality watermarked images. Further BER and NCC values as observed from the Tables are in tune with expected values (under no attack).

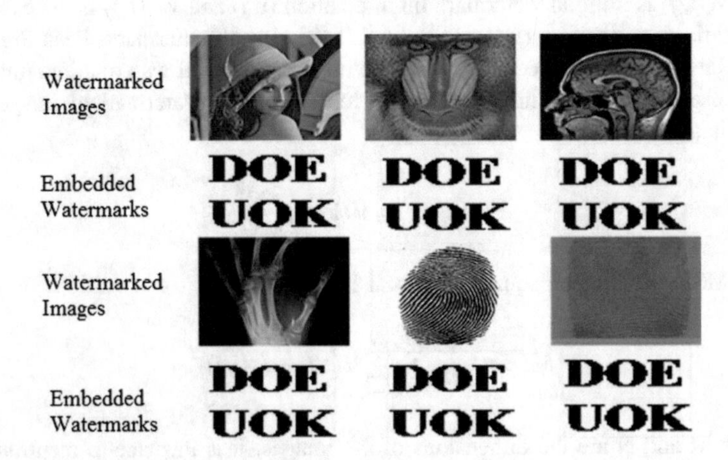


**Fig. 8** Watermarked images and embedded watermarks

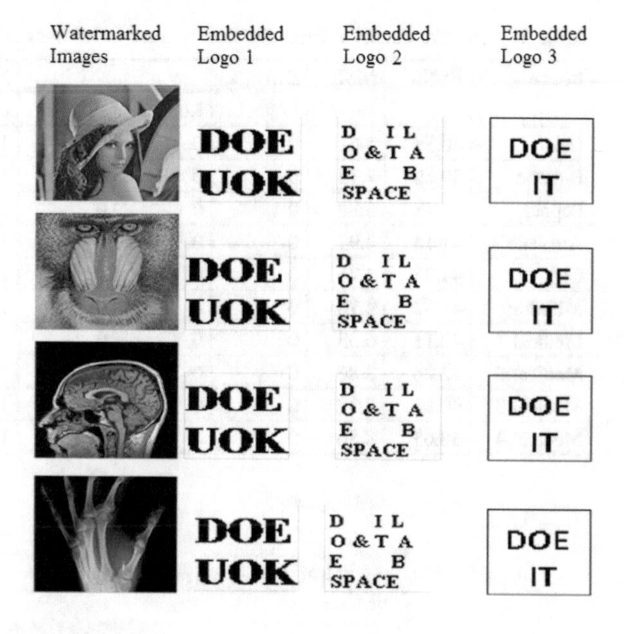


**Fig. 9** Watermarked color images and various embedded watermarks

**Table 1** Objective quality metrics for watermarked grayscale images without attack

| Image | PSNR | MSE | BER (%) | NCC |
|---|---|---|---|---|
| Lena | 40.53 | 5.75 | 0 | 1 |
| Baboon | 40.08 | 6.38 | 0 | 1 |
| Pepper | 40.706 | 5.52 | 0 | 1 |
| Airplane | 40.99 | 5.16 | 0 | 1 |
| Girl | 41.47 | 4.63 | 0 | 1 |
| Milkdrop | 41.42 | 4.67 | 0 | 1 |
| Medical 1 | 40.09 | 6.36 | 0 | 1 |
| Medical 2 | 41.46 | 4.63 | 0 | 1 |
| Medical 3 | 40.75 | 5.46 | 0 | 1 |
| Medical 4 | 38.52 | 9.14 | 0 | 1 |
| Fingerprint 1 | 31.65 | 44.42 | 0 | 1 |
| Fingerprint 2 | 41.9 4 | 4.15 | 0 | 1 |
| Fingerprint 3 | 41.70 | 4.38 | 0 | 1 |

## 4.2 *Payload Analysis*

In the proposed watermarking scheme one bit of watermark is embedded into each block of the cover image. Therefore, for a gray scale image of size $512 \times 512$ total number of 4096 bits, are embedded while as for a color image of size $512 \times 512$, 4096 bits have been embedded into each color plane resulting into a payload of $4096 \times 3 = 12288$ bits. A comparison of the proposed technique with some state of art techniques in the field has been carried out and presented here. Figure 11a

**Table 2** Objective quality metrics for watermarked color images without attack

| Image | PSNR | MSE | BER (%) | | | NCC | | |
|---|---|---|---|---|---|---|---|---|
| | | | Logo 1 | Logo 2 | Logo 3 | Logo 1 | Logo 2 | Logo 3 |
| Lena | 40.58 | 5.68 | 0 | 0 | 0 | 1 | 1 | 1 |
| Baboon | 39.60 | 7.12 | 0 | 0 | 0 | 1 | 1 | 1 |
| Pepper | 40.43 | 5.88 | 0 | 0 | 0 | 1 | 1 | 1 |
| Airplane | 41.18 | 4.95 | 0 | 0 | 0 | 1 | 1 | 1 |
| Girl | 41.35 | 4.75 | 0 | 0 | 0 | 1 | 1 | 1 |
| Milkdrop | 41.32 | 4.78 | 0 | 0 | 0 | 1 | 1 | 1 |
| Medical 1 | 40.11 | 6.33 | 0 | 0 | 0 | 1 | 1 | 1 |
| Medical 2 | 42.26 | 3.86 | 0 | 0 | 0 | 1 | 1 | 1 |
| Medical 3 | 40.82 | 5.37 | 0 | 0 | 0 | 1 | 1 | 1 |
| Medical 4 | 38.65 | 8.86 | 0 | 0 | 0 | 1 | 1 | 1 |

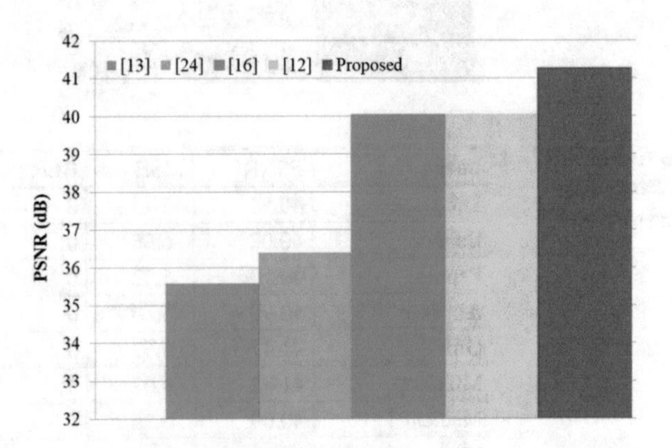


**Fig. 10** Comparison of PSNR for Lena image

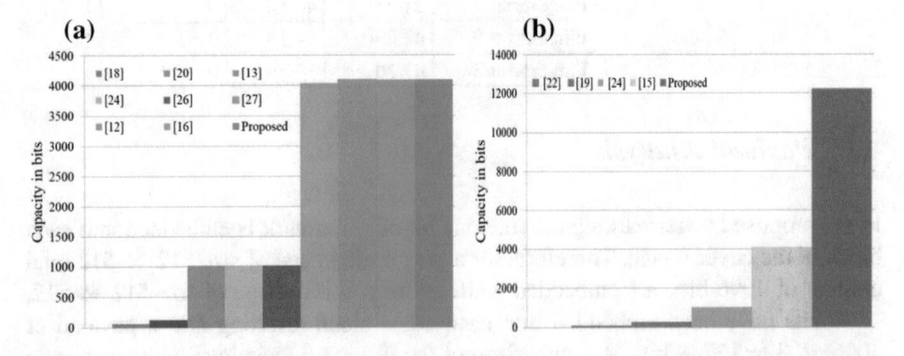


**Fig. 11** Comparison of capacity for (**a**) *Gray* scale images and (**b**) Color images

presents payload comparison of our technique for same size gray scale images while as Fig. 11b depicts the same when color images have been used as cover images. It is evident from the mentioned Figures that proposed technique offers better payload irrespective of nature of the image compared to existing techniques under comparison.

## *4.3 Robustness Analysis*

The ability to resist an attack is one of the premier requirements of a DRM system. The proposed system has been evaluated for robustness to various singular and hybrid attacks. Various attacks carried out on the implemented system include Salt and Pepper noise, Gaussian noise, Sharpening, Compression, Cropping and Rotation etc. The detailed analysis for each attack is presented below:

### 4.3.1 Salt and Pepper Noise

The Salt and Pepper noise is added to the watermarked images with noise density 0.01. The images distorted by Salt and Pepper noise and the watermarks extracted from them are shown in Fig. 12 and Fig. 13 respectively for grayscale images and color images. The objective metrics for the distorted images are reported in Table 3 and Table 4 respectively for grayscale and color images.

The small BER values and near unity NCC values as evident from Tables 3 and 4 indicate that the proposed scheme is robust to this attack. Further the extracted watermarks are easily recognizable and hence subjective quality of the extracted watermarks is good. The subjective and objective results show that the proposed scheme is robust to this attack. Table 5 reports the average values of NCC and BER for Color and grayscale images for varying density of salt and pepper noise. It is evident from the Table that robustness decreases as noise density increases.

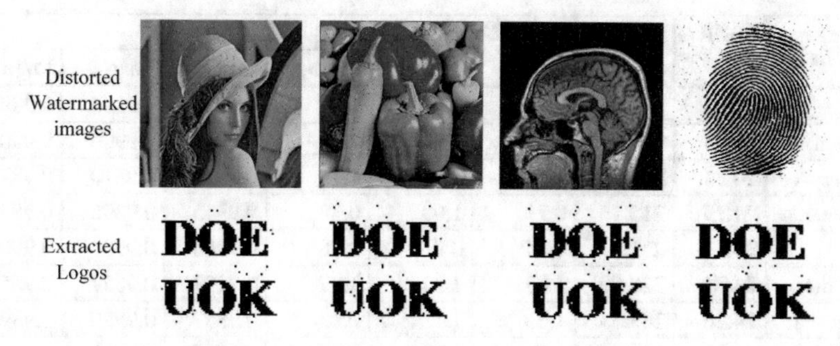


**Fig. 12** Distorted watermarked *gray* scale images by Salt and pepper noise (0.01) and extracted logos

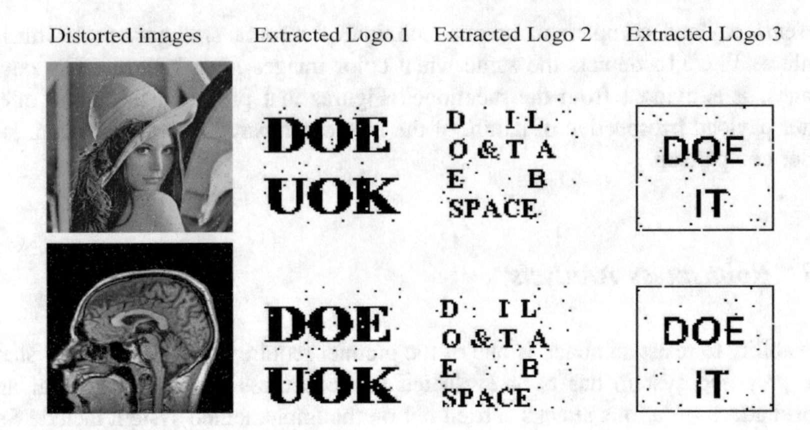


**Fig. 13** Distorted watermarked color images by Salt and pepper noise (0.01) and extracted logos

**Table 3** Objective quality metrics for watermarked grayscale images distorted with Salt and Pepper noise (0.01)

| Image | PSNR | MSE | BER (%) | NCC |
|---|---|---|---|---|
| Lena | 25.2328 | 194.8957 | 1.12 | 0.9893 |
| Baboon | 25.5975 | 179.1958 | 0.95 | 0.9907 |
| Pepper | 25.1180 | 200.1173 | 1.34 | 0.9883 |
| Airplane | 25.0970 | 201.0871 | 1.05 | 0.9872 |
| Girl | 24.5359 | 228.8176 | 1.39 | 0.9862 |
| Milk drop | 25.1603 | 198.1752 | 1.07 | 0.9890 |
| Medical 1 | 24.3231 | 240.3118 | 1.29 | 0.9866 |
| Medical 2 | 23.6526 | 280.4273 | 0.93 | 0.9900 |
| Fingerprint 1 | 22.4224 | 372.2572 | 0.71 | 0.9917 |
| Fingerprint 2 | 25.8542 | 168.9134 | 1.07 | 0.9883 |

**Table 4** Objective quality metrics for watermarked color images for Salt and Pepper noise (0.01)

| Image | PSNR | MSE | BER (%) | | | NCC | | |
|---|---|---|---|---|---|---|---|---|
| | | | Logo 1 | Logo 2 | Logo 3 | Logo 1 | Logo 2 | Logo 3 |
| Lena | 25.06 | 202.65 | 0.98 | 1 | 1.2 | 0.9897 | 0.9896 | 0.9896 |
| Baboon | 25.03 | 204.15 | 0.95 | 0.9 | 1 | 0.9897 | 0.9909 | 0.9909 |
| Pepper | 24.75 | 217.78 | 1 | 1.7 | 1.15 | 0.9903 | 0.9810 | 0.9810 |
| Airplane | 24.79 | 215.76 | 0.71 | 1.03 | 0.88 | 0.9945 | 0.9896 | 0.9896 |
| Girl | 24.47 | 231.97 | 1.34 | 0.9 | 0.88 | 0.9866 | 0.9909 | 0.9909 |
| Milk drop | 24.54 | 228.50 | 1.03 | 1.32 | 1.07 | 0.9903 | 0.9857 | 0.9857 |
| Medical 1 | 24.26 | 243.33 | 1.05 | 1.1 | 0.93 | 0.9897 | 0.9890 | 0.9890 |
| Medical 2 | 23.75 | 273.89 | 1.07 | 1.12 | 1.15 | 0.9890 | 0.9882 | 0.9882 |

**Table 5** Average BER and NCC against Salt and Pepper noise with varying noise density

| Noise density | 0.01 | | 0.05 | | 0.1 | |
|---|---|---|---|---|---|---|
| | Average BER | Average NCC | Average BER | Average NCC | Average BER | Average NCC |
| Grayscale Images | 1.09 | 0.9887 | 5.1517 | 0.9461 | 10.0917 | 0.8959 |
| Color images | 0.85 | 0.7910 | 5.1715 | 0.9475 | 10.0350 | 0.8985 |

### 4.3.2 Addition of Gaussian Noise

The watermarked images have been distorted with Gaussian noise with zero mean and different variance values. The distorted images and extracted logos from them are shown in Fig. 14 and Fig. 15 respectively for Grayscale and color images. The objective metrics obtained for this attack are reported in Tables 6 and 7.

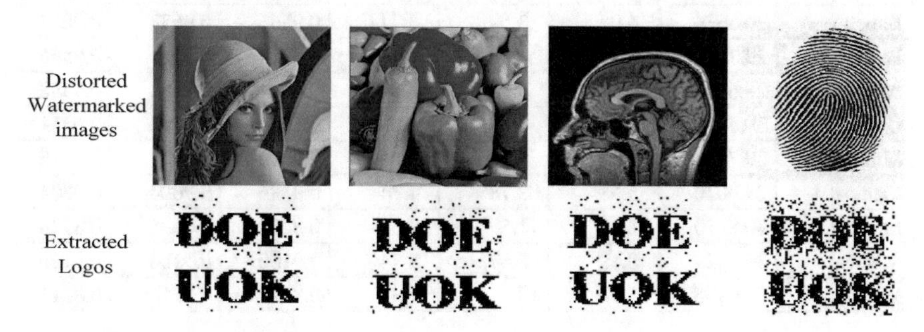


**Fig. 14** Distorted watermarked *gray* scale images by Gaussian noise and extracted Logos (Variance 0.001)

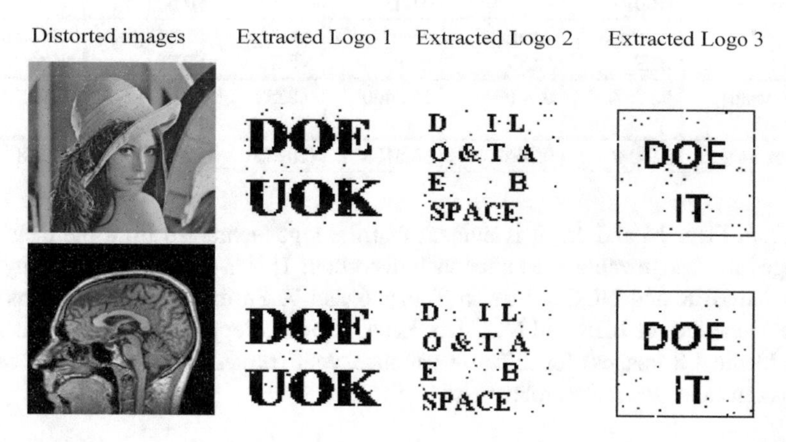


**Fig. 15** Distorted watermarked color images by Gaussian noise and extracted logos (variance 0.001)

**Table 6** Objective analysis for watermarked grayscale images for Gaussian noise (Variance 0.001)

| Image | PSNR | BER(%) | NCC | Image | PSNR | BER (%) | NCC |
|---|---|---|---|---|---|---|---|
| Lena | 29.61 | 3.47 | 0.9666 | Medical 2 | 30.83 | 8.5 | 0.9073 |
| Baboon | 29.59 | 3.22 | 0.9679 | Medical 3 | 30.12 | 8.3 | 0.9131 |
| Pepper | 29.63 | 3.69 | 0.9607 | Medical 4 | 31.08 | 15.3 | 0.8301 |
| Airplane | 29.66 | 3.88 | 0.9600 | Fingerprint 1 | 29.24 | 16.3 | 0.8197 |
| Girl | 29.77 | 3.98 | 0.961 | Fingerprint 2 | 29.70 | 3.08 | 0.9704 |
| Milkdrop | 29.75 | 4.10 | 0.9576 | Fingerprint 3 | 30.79 | 4.10 | 0.9590 |
| Medical 1 | 29.89 | 8.0 | 0.9155 | | | | |

**Table 7** Objective analysis for watermarked color images for Gaussian noise (Variance = 0.001)

| Image | PSNR | BER (%) | | | NCC | | |
|---|---|---|---|---|---|---|---|
| | | Logo 1 | Logo 2 | Logo 3 | Logo 1 | Logo 2 | Logo 3 |
| Lena | 29.6466 | 3.71 | 4.13 | 3.64 | 0.9590 | 0.9560 | 0.9560 |
| Baboon | 29.5634 | 4.05 | 3.59 | 3.27 | 0.9566 | 0.9637 | 0.9637 |
| Pepper | 29.7305 | 3.56 | 4.32 | 4.79 | 0.9621 | 0.9566 | 0.9566 |
| Airplane | 29.6767 | 2.81 | 3.96 | 3.98 | 0.9704 | 0.9588 | 0.9588 |
| Girl | 29.8839 | 3.88 | 4.69 | 5.0 | 0.9597 | 0.9511 | 0.9511 |
| Milk drop | 29.7496 | 3.66 | 5.71 | 3.61 | 0.9610 | 0.9415 | 0.9415 |
| Medical 1 | 29.9036 | 3.98 | 3.78 | 4.74 | 0.9586 | 0.9604 | 0.9604 |
| Medical 2 | 30.8930 | 12.08 | 11.5 | 12.52 | 0.8725 | 0.8788 | 0.8788 |
| Medical 3 | 30.1429 | 8.18 | 8.47 | 8.74 | 0.9100 | 0.9126 | 0.9126 |
| Medical 4 | 31.0955 | 15.04 | 17.21 | 17.31 | 0.8328 | 0.8211 | 0.8211 |

**Table 8** Average BER and NCC against Gaussian noise with different variance

| Variance AWGN | 0.001 | | 0.003 | | 0.005 | |
|---|---|---|---|---|---|---|
| | Average BER | Average NCC | Average BER | Average NCC | Average BER | Average NCC |
| Grayscale images | 5.1267 | 0.9463 | 15.8400 | 0.8352 | 22.8217 | 0.7635 |
| Color images | 5.2450 | 0.9460 | 16.5183 | 0.8280 | 23.4050 | 0.7618 |

From Figs. 14 and 15, it is evident that the logos extracted from the distorted images are recognizable even after such distortion. This is also substantiated by the reported BER and NCC values in Tables 6 and 7. Further Table 8 contains the Average values of BER and NCC (for Lena, Baboon, Pepper, Airplane, Medical 1 and Medical 2 images) for different variance. As expected increase in variance of Gaussian noise decreases robustness.

### 4.3.3 Sharpening

The watermarked images are sharpened using a 3 × 3 filter kernel. The sharpened images and the watermarks extracted from them are shown in Fig. 16 for Grayscale images.

The image quality metrics obtained for sharpening attack are reported in Tables 9 and 10 for grayscale and color images respectively. It could be observed from the Tables 9 and 10 that BER values are less and NCC near unity indicating that proposed scheme is robust to this attack. From Fig. 16 it is clear that recognizable watermarks are extracted from the sharpened images which reveal that the proposed technique is also robust to sharpening.

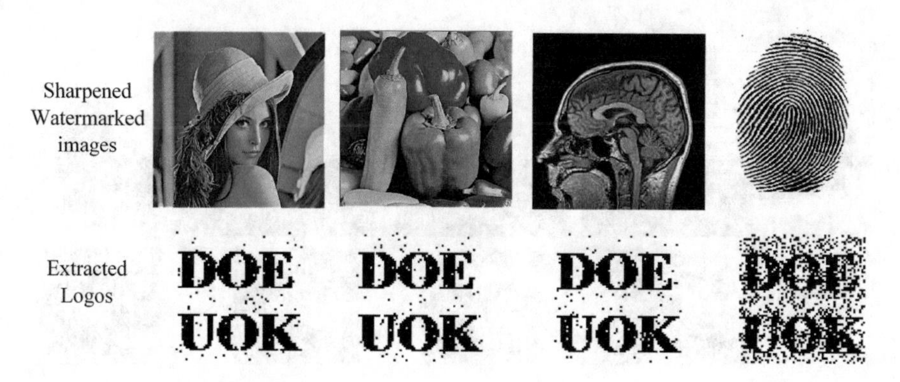


**Fig. 16** Sharpened images and extracted logos

**Table 9** Objective metrics obtained for Grayscale images for Sharpening attack

| Image | PSNR | BER (%) | NCC | Image | PSNR | BER (%) | NCC |
|---|---|---|---|---|---|---|---|
| Lena | 25.9972 | 2.56 | 0.9724 | Medical 1 | 27.6069 | 2.54 | 0.9741 |
| Baboon | 21.5270 | 9.16 | 0.9083 | Medical 2 | 32.2859 | 0.6 | 0.9928 |
| Pepper | 26.4964 | 2.44 | 0.9741 | Fingerprint 1 | 29.7130 | 21 | 0.7201 |
| Airplane | 24.6192 | 3.0 | 0.9686 | Fingerprint 2 | 28.8284 | 0 | 1 |

**Table 10** Objective metrics obtained for color images for Sharpening attack

| Image | PSNR | BER (%) | | | NCC | | |
|---|---|---|---|---|---|---|---|
| | | Logo 1 | Logo 2 | Logo 3 | Logo 1 | Logo 2 | Logo 3 |
| Lena | 26.2386 | 2.17 | 3.56 | 2.88 | 0.9762 | 0.9629 | 0.9629 |
| Baboon | 20.3016 | 1.01 | 1.33 | 1.36 | 0.9004 | 0.8656 | 0.8656 |
| Pepper | 26.3050 | 2.54 | 5.13 | 2.37 | 0.9741 | 0.9467 | 0.9467 |
| Airplane | 25.1438 | 3.27 | 4.13 | 2.15 | 0.9655 | 0.9571 | 0.9571 |
| Medical 1 | 27.6192 | 2.93 | 2.91 | 2.34 | 0.9738 | 0.9717 | 0.9717 |
| Medical 2 | 32.6174 | 0.4 | 0.56 | 0.56 | 0.9955 | 0.9942 | 0.9942 |

### 4.3.4 Cropping

Cropping is one the important intentional attacks that must be taken into account while developing a watermarking algorithm for images. The watermarked images are cropped by 25 % at top left corner (TLC), top right corner (TRC), bottom left corner (BLC) and bottom right corner (BRC). Various 25 % cropped images at top left corner and the logos extracted from them are shown in Fig. 17 and Fig. 18 respectively.

The average image indices obtained for gray scale and color images used for the analysis (as shown in Figs. 17 and 18) cropped at various corners have been reported in Table 11. It is clear from the subjective and objective analysis that our technique could withstand this attack as well.

**Fig. 17** Cropped images and Logos extracted

**Fig. 18** Cropped Lena image and Logos extracted

**Table 11** Average BER and NCC values for Cropping at different corners

| 25 % cropping at | Grayscale image | | Color image | |
|---|---|---|---|---|
| | BER (%) | NCC | Average BER (%) | Average NCC |
| Top left corner | 17.48 | 0.7546 | 22.72 | 0.7481 |
| Top right corner | 18.14 | 0.7456 | 20.41 | 0.7526 |
| Bottom left corner | 17.65 | 0.7525 | 20.6 | 0.7460 |
| Bottom right corner | 17.63 | 0.7525 | 20.47 | 0.7508 |

### 4.3.5 Rotation Attack

Rotation is one of the important intentional attacks carried out on the watermarked images. To check robustness of our scheme to this attack we have rotated the watermarked images by 10°, 20° and 30°. Figure 19 shows some gray scale watermarked images and extracted watermarks after a rotation of 30° while as Fig. 20 shows the color Lena image rotated by 10° and logos extracted from it.

The BER and NCC for the extracted logos for various angles of rotation for grayscale and color images are presented in Table 12. It is pertinent to mention here that BER and NCC values reported are average values obtained after logo extraction from R, G and B planes. The Figs. 19, 20 and Table 12-show that the recognizable watermarks are extracted from the rotated images even after 30° rotation. This implies that the proposed scheme is robust to rotation attack, however the robustness decreases as the angle of rotation is increased.

### 4.3.6 JPEG Compression

JPEG is one of the standard compression techniques used to compress the images. The watermarked images have been subjected to JPEG compression with of

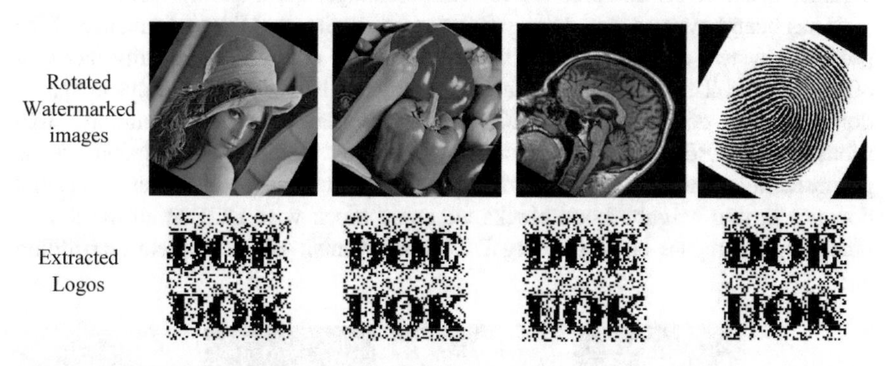


**Fig. 19** Rotated images and extracted logos

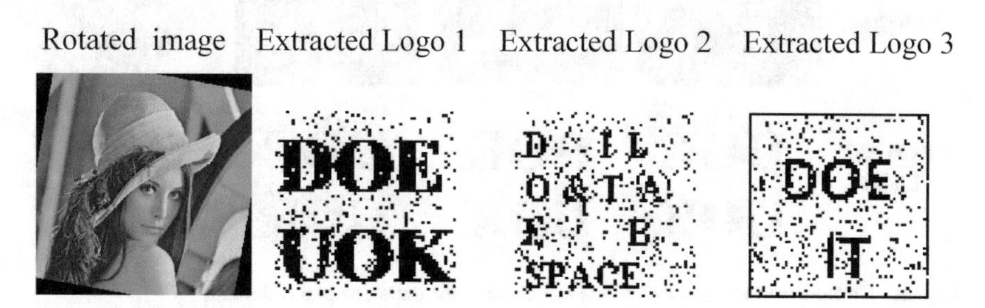


**Fig. 20** Rotated images and Logos extracted

**Table 12** BER and NCC for extracted logos against varying rotation angles

| Angle | Image | Lena | Baboon | Pepper | Airplane | Medical 1 | Medical 2 |
|---|---|---|---|---|---|---|---|
| For Grayscale images | | | | | | | |
| 10 | BER (%) | 8.1 | 8.8 | 8.13 | 7.93 | 7.9 | 8.64 |
| | NCC | 0.9007 | 0.8825 | 0.8959 | 0.8949 | 0.8973 | 0.8842 |
| 20 | BER (%) | 12.9 | 15.28 | 12.53 | 12.45 | 13.40 | 14.18 |
| | NCC | 0.8349 | 0.8077 | 0.8404 | 0.8407 | 0.8239 | 0.8090 |
| 30 | BER (%) | 18.07 | 18.14 | 18.46 | 18.6 | 17.92 | 18.68 |
| | NCC | 0.7725 | 0.07752 | 0.7704 | 0.7632 | 0.7663 | 0.7504 |
| For Color images | | | | | | | |
| 10 | Average BER (%) | 8.71 | 10.43 | 9.30 | 8.42 | 9.38 | 10.17 |
| | Average NCC | 0.8995 | 0.8825 | 0.8911 | 0.9026 | 0.8907 | 0.8777 |
| 20 | Average BER (%) | 15.15 | 17.76 | 14.84 | 14.21 | 15.54 | 16.85 |
| | Average NCC | 0.8260 | 0.8006 | 0.8284 | 0.8348 | 0.8195 | 0.7964 |
| 30 | Average BER(%) | 19.37 | 20.7 | 19.64 | 19.09 | 20.11 | 21.14 |
| | Average NCC | 0.7789 | 0.7663 | 0.7729 | 0.7773 | 0.7662 | 0.7494 |

different quality factors. The compressed images for a quality factor of 90 and the logos extracted from them are shown in Fig. 21. Table 13 shows the performance parameters for JPEG compression for various images for a quality factor of 90.

It has been however seen during the experimentation that the performance of the proposed system deteriorates significantly for JPEG compression quality factor of 80 and below. This is due to the fact that the work has been carried out in spatial domain which does not take JPEG quantization Table into consideration. High robustness to JPEG compression was observed by making slight change in the proposed algorithm. Instead of comparing two predefined pixel values (in spatial domain) in two neighborhood blocks we apply block wise DCT to all the 8 × 8 blocks and compare corresponding DCT coefficients. After complete embedding

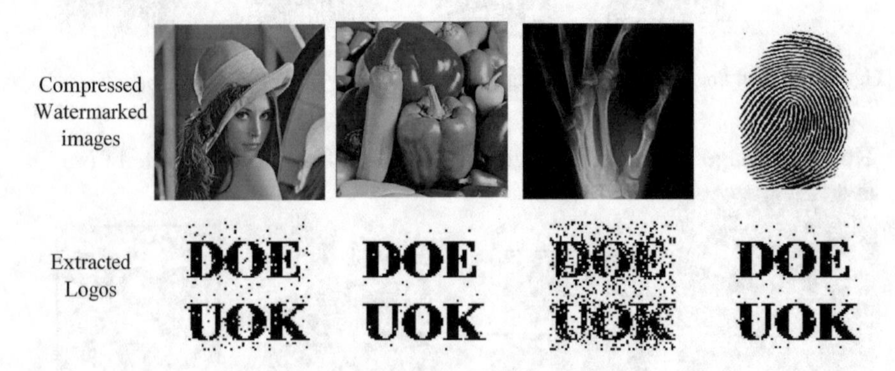


**Fig. 21** Compressed images and logos extracted from them in spatial domain

**Table 13** Performance parameters for JPEG compression at quality 90

| Image | PSNR | BER (%) | NCC |
|---|---|---|---|
| Lena | 38.3636 | 4.35 | 0.9528 |
| Baboon | 35.6552 | 0.52 | 0.9928 |
| Pepper | 36.9842 | 0.4 | 0.9952 |
| Airplane | 39.3694 | 3.2 | 0.9576 |
| Fingerprint 1 | 31.49 | 1.54 | 0.9817 |
| Medical 1 | 41.237 | 20.1 | 0.7542 |
| Medical 2 | 44.6909 | 16.2 | 0.8194 |

**Table 14** Image quality metrics for compressed Lena image

| Quality Factor | | 90 | 80 | 70 | 60 | 50 | 40 |
|---|---|---|---|---|---|---|---|
| For Grayscale image | *Spatial Domain* | | | | | | |
| | PSNR (dB) | 38.39 | 37.6 | 36.7 | 36.07 | 35 | 34.5 |
| | BER (%) | 4 | 27.7 | 33.6 | 40 | 42 | 46 |
| | NCC | 0.9566 | 0.6915 | 0.6329 | 0.5839 | 0.5574 | 0.5302 |
| | *DCT Domain* | | | | | | |
| | PSNR (dB) | 37.10 | 36 | 35.26 | 34.7 | 34.3 | 33.7 |
| | BER (%) | 0 | 0 | 0 | 0.05 | 0.4 | 2.23 |
| | NCC | 1 | 1 | 1 | 0.9993 | 0.9972 | 0.9799 |
| For Color image | *Spatial Domain* | | | | | | |
| | PSNR | 34.125 | 33.35 | 32.7 | 32.1 | 31.9 | 30.7 |
| | Average BER (%) | 23.72 | 36 | 44 | 46 | 48 | 55 |
| | Average NCC | 0.8378 | 0.6460 | 0.5585 | 0.5367 | 0.5125 | 0.4713 |
| | *DCT Domain* | | | | | | |
| | PSNR | 33.38 | 32.52 | 31.96 | 31.57 | 31.2 | 30.7 |
| | Average BER (%) | 2.57 | 2.41 | 2.49 | 2.57 | 3.64 | 4.56 |
| | Average NCC | 0.9727 | 0.9744 | 0.9738 | 0.9730 | 0.9617 | 0.9520 |

IDCT of the modified blocks is computed to get the watermarked image. The results obtained for JPEG compression for greyscale and color image Lena for varying quality factors in Spatial and Transform domain (DCT) image are reported in Table 14. It is evident from the data presented in mentioned Tables that the scheme is highly robust in case of transform domain embedding compared to spatial domain embedding. Figure 22 shows various watermarked images subjected to JPEG compression of Quality factor 40. The extracted watermarks are easily recognizable, which justifies our argument. However, one has to pay more computational cost if robustness against JPEG is of prime concern.

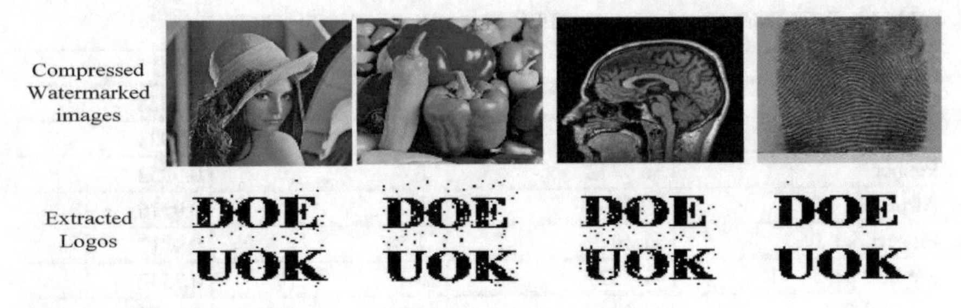


**Fig. 22** Compressed images with quality 40 and extracted logos

### 4.3.7 Rotation + Cropping

The robustness of the proposed technique is also examined for hybrid attack of rotation and cropping. The watermarked images are first cropped 25 % and then rotated by 15°. Figure 23 shows various watermarked gray scale images cropped by 25 % at TLC and subjected to 15° rotation. The subjective quality of the image shows that recognizable watermarks are extracted from the watermarked image using the proposed technique.

Figure 24 shows watermarked color image 'Lena' and the extracted watermarks for the above mentioned attack specifications. It is observed that logos extracted from Red and Blue planes are recognizable, however that extracted from green plane is not. To conclude, it was observed during experimentation that after cropping the image, as the angle of rotation is increased the robustness to this hybrid attack becomes less. Table 15 shows BER and NCC values obtained using this attack.

### 4.3.8 Salt and Pepper Noise + Gaussian Noise

The watermarked images are subjected to Salt and pepper noise with noise density 0.01 and then the Gaussian noise (variance = 0.001) is added to the distorted

**Fig. 23** Watermarked images after cropping + rotation and logos extracted

**Fig. 24** Watermarked Lena image after Cropping + Rotation attack and extracted logos

**Table 15** BER and NCC for extracted logos against Cropping + Rotation attack

| Image | Grayscale | | Color | |
|---|---|---|---|---|
| | BER (%) | NCC | Average BER (%) | Average NCC |
| Lena | 30 | 0.6515 | 30 | 0.6440 |
| Baboon | 30 | 0.6315 | 30 | 0.6325 |
| Pepper | 30 | 0.6412 | 30 | 0.6460 |
| Airplane | 30 | 0.6456 | 30 | 0.6450 |
| Girl | 30 | 0.6505 | 30 | 0.6386 |
| Medical 1 | 30 | 0.6446 | 30 | 0.6480 |
| Medical 2 | 30 | 0.6301 | 30 | 0.6217 |

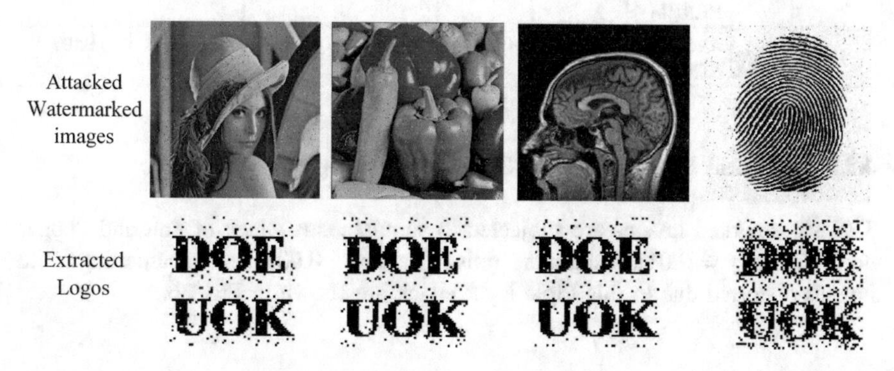


**Fig. 25** Watermarked images after addition of Salt and Pepper noise and Gaussian noise

images. The resultant images obtained after such a combined attack are shown in Fig. 25.

The objective quality indices for this simultaneous attack are presented by Table 16 for both the grayscale and color images. The above figure and values of BER and NCC presented in Table 16 reveals that the proposed technique is robust to this combined attack.

**Table 16** BER and NCC for extracted logos for hybrid attack of Salt and Pepper noise, Gaussian noise and Sharpening

| Attack | Quality metric | Lena | Baboon | Pepper | Airplane | Medical 1 | Medical 2 |
|---|---|---|---|---|---|---|---|
| Salt and Pepper noise + Gaussian noise | *For Grayscale images* | | | | | | |
| | PSNR (dB) | 24.11 | 24.48 | 24.04 | 24.06 | 23.58 | 23.18 |
| | BER (%) | 4.8 | 3.9 | 5.1 | 4.4 | 4.7 | 8.7 |
| | NCC | 0.9490 | 0.9607 | 0.9469 | 0.9555 | 0.9517 | 0.9080 |
| | *For Color images* | | | | | | |
| | PSNR | 24.00 | 24.09 | 23.82 | 23.75 | 23.39 | 23.16 |
| | Average BER(%) | 5.13 | 4.46 | 5.34 | 5.16 | 4.9 | 12.8 |
| | Average NCC | 0.9496 | 0.9562 | 0.9498 | 0.9528 | 0.9513 | 0.8635 |
| Salt and Pepper noise + Gaussian noise + Sharpening | *For Grayscale images* | | | | | | |
| | PSNR (dB) | 21.97 | 19.79 | 22.16 | 21.35 | 21.99 | 22.50 |
| | BER (%) | 4.8 | 10.6 | 5.15 | 5.3 | 5.71 | 3.39 |
| | NCC | 0.9500 | 0.8928 | 0.9431 | 0.9442 | 0.9421 | 0.9617 |
| | *For Color images* | | | | | | |
| | PSNR | 21.97 | 18.86 | 22.00 | 21.45 | 22.17 | 22.66 |
| | Average BER(%) | 5.45 | 14.70 | 5.73 | 4.99 | 5.54 | 4.64 |
| | Average NCC | 0.9439 | 0.8510 | 0.9324 | 0.9390 | 0.9470 | 0.9507 |

### 4.3.9 Salt and Pepper Noise + Gaussian Noise + Sharpening

The watermarked images are subjected to simultaneous attack of Salt and Pepper noise (density = 0.01), Gaussians noise (variance 0.001) and sharpening. The images distorted due to this triple layer attack are shown in Fig. 26.

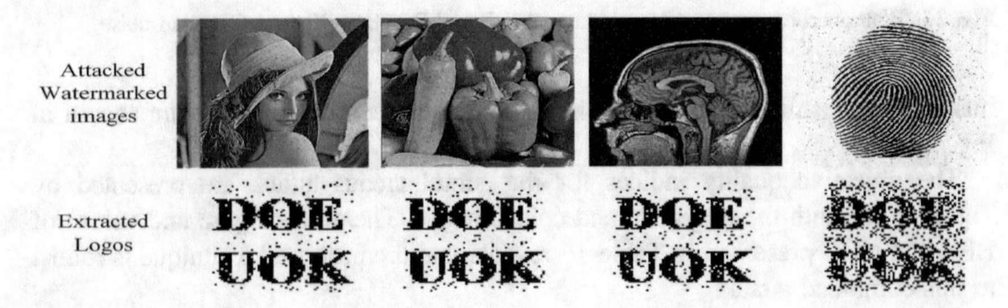


**Fig. 26** Watermarked images after addition of Salt and Pepper noise, Gaussian noise and Sharpening

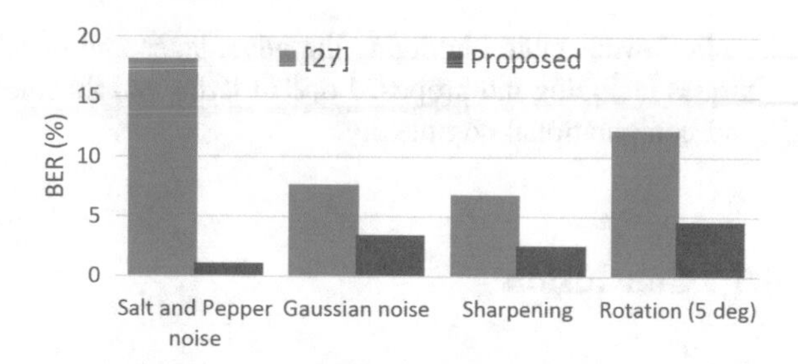


**Fig. 27** BER Comparison for various attacks

The NCC and BER obtained after watermark extraction in this case are reported in Table 16. It is pertinent to mention here that BER and NCC values for color images represent average values computed for three different planes. It is evident from Fig. 26 and Table 16 that the proposed algorithm is robust to this attack as well.

A comparison of the proposed scheme has been carried out with that of Das et al. Figure 27 shows the comparison of BER for Lena image for various attacks like Salt and Pepper noise (density = 0.01), Gaussian noise (variance = 0.001), Sharpening and Rotation (5°). It is evident from Fig. 27 that the proposed technique performs better in comparison to [28].

## *4.4 Computational Complexity Analysis*

The computational complexity of a watermarking system is one of the most important factors. For real time applications this factor assumes most important significance. Lesser the computational complexity better the algorithm. Total computational time is sum of embedding and extraction time. We have compared the computational time of the proposed scheme with some state of art works [22, 26, 27] and the comparison results have been reported in Table 17. From the Table it is clear that our technique outperforms the techniques under comparison. The lesser complexity in our case is due to the fact that our technique involves simple algebraic manipulations.

**Table 17** Computational time in seconds

| Execution time | Image size | Embedding time (sec) | Extraction time (sec) | Total time (sec) |
|---|---|---|---|---|
| [26] | 512 × 512 | NA | NA | 8.5 |
| [22] | 512 × 512 | 0.5244 | 0.3701 | 0.8945 |
| [27] | 512 × 512 | 0.5173 | 0.5989 | 1.1162 |
| Proposed | 128 × 128 | 0.50 | 0.0003 | 0.5003 |
| | 256 × 256 | 0.55 | 0.0156 | 0.5656 |
| | 512 × 512 | 0.59 | 0.0624 | 0.6524 |

Following table highlights the advantages and disadvantages of different techniques including the proposed one in terms of robustness, imperceptivity, capacity and computational complexity.

## 5 Conclusion

A blind watermarking algorithm for grayscale and color images is proposed and analyzed in this chapter. The data has been embedded using Inter Block Pixel Difference (IBPD). The cover image is divided in various 8 × 8 blocks and one bit of data has been embedded in each block. The relative difference between two predefined elements in a pair of adjacent blocks is modified and brought in a specified region depending upon the nature of data bit (whether 0 or 1) to be embedded. The scheme has been extensively tested and has been not only found robust to various singular image processing operations like cropping, rotation, salt & pepper noise, Gaussian noise and sharpening but also to various hybrid attacks. The proposed scheme has been compared to a state of art watermarking techniques and the comparison results show that our technique besides being robust to various image processing and geometric attacks is capable of providing high quality watermarked images. Further, the use of IBPD in spatial domain makes the scheme computationally efficient. This has been testified by comparing the computational complexity of our scheme with some state of art works in the area. The robust nature and computational efficiency of the scheme makes the system an ideal candidate for real time DRM applications (Table 18).

**Table 18** Advantages and disadvantages of different technique

| Technique | Advantages | Disadvantages |
|---|---|---|
| [13] | 1. Robustness against JPEG compression and other signal processing operations is high<br>2. Capacity is high | 1. Imperceptivity is low<br>2. Computational complexity is high as embedding is carried out in transform domain |
| [14] | 1. Robustness against JPEG compression and other signal processing operations is high | 1. Very low capacity<br>2. Very low imperceptivity<br>3. Computational complexity is high due to transform domain embedding |
| [17] | 1. Robustness against JPEG compression and other signal processing operations is high<br>2. High capacity | 1. Imperceptivity is low<br>2. Computational complexity is high |
| [25] | 1. Robustness against JPEG compression and other signal processing operations is high | 1. Very low capacity<br>2. Very low imperceptivity<br>3. High computational complexity |

(continued)

**Table 18** (continued)

| Technique | Advantages | Disadvantages |
|---|---|---|
| [26] | 1. Highly robust to common image processing operations and JPEG compression<br>2. Imperceptivity is very high<br>3. High capacity | 1. High computational complexity |
| [27] | 1. Robustness and imperceptivity are high | 1. Very low capacity<br>2. Computational complexity is high; as embedding is carried out in transform domain |
| [28] | 1. Robustness against JPEG compression is high<br>2. High capacity | 1. Robustness against some common image processing operations is very low<br>2. Low capacity<br>3. Computational complexity is high |
| Proposed | 1. Highly robust to common image processing operations<br>2. Robust to various simultaneous attacks<br>3. Imperceptivity is high<br>4. High payload<br>5. Computationally efficient, due to spatial domain embedding | 1. Robustness against JPEG compression is poor |

**Acknowledgments** The authors would like to thank the Department of Science and Technology (DST) New Delhi for providing financial support under DST inspire fellowship scheme.

# References

1. Parah, S.A., Sheikh, J.A., Loan, N.A., Bhat, G.M.: A blind watermarking technique in spatial domain using inter-block pixel value differencing. In: 2015 International Conference on Advances in Computer, Communication and Electronic Engineering (COMMUNE), pp. 321–326 (2015)
2. Mustafa, O., Rameshwar, R.: Digital image watermarking basics, and hardware implementation. Int. J. Model. Optim. **2**, 19–24 (2012)
3. Parah, S.A., Sheikh, J.A., Hafiz, A.M., Bhat, G.M.: A secure and robust information hiding technique for covert communication. Int. J. Electron. Taylor & Francis (2014). doi:10.1080/00207217.2014.954635
4. Sami, E., Lala, Z., Thawar, A., Zyad, S.: Watermarking of digital images in frequency domain. Int. J. Autom. Comput. Springer **7**, 17–22 (2010)
5. Hassanien, A.E.: A copyright protection using watermarking algorithm. Informatica Lith. Acad. Sci. **17**(2), 187–198 (2006)
6. Namita, C., Jaspal, B.: Performance comparison of digital image watermarking technique: A survey. Int. J. Comput. Appl. Technol. Res. **2**, 126–130 (2013)
7. Radhika, V., Kalpan, B.: Comparative analysis of watermarking in digital image using DCT & DWT. Int. J. Sci. Res Publ. **3**, 1–4 (2013)

8. Parah, S.A., Sheikh, J.A., Ahad, F., Loan, N.A., Bhat, G.M.: Information hiding in medical images: a robust medical image watermarking system for E-healthcare. Multimed. Tools Appl. Springer (2015). doi:10.1007/s11042-015-3127-y
9. Parah, S.A., Sheikh, J.A., Bhat, G.M.: A secure and efficient spatial domain data hiding technique based on pixel adjustment. Am. J. Eng. Technol. Res. **14**, 33–39 (2014)
10. Samesh, O., Adnane, C., Bassel, S.: Multiple binary images in spatial and frequency domain. Int. J. Comput. Theory Eng. **5**, 598–602 (2013)
11. Parah, S.A., Sheikh, J.A., Bhat, G.M., Hafiz, A.M.: Data hiding in scrambled images: a new double layer security data hiding technique. Comput. Electr. Eng. Elsevier **40**, 70–82 (2014)
12. Kalra, G., Talwar, R., Sadawarti, H.: Digital image watermarking in frequency domain using ecc and dual encryption technique. Res. J. Appl. Sci. Eng. Technol. Maxwell Scientific Organization **6**(18), 3365–3371 (2013)
13. Lang, J., Zhang, Z.: Blind digital watermarking method in the fractional Fourier transform domain. Opt. Lasers Eng. Elsevier **53**, 112–121 (2014)
14. Ma, F., Zhang, J., Zhang, W.: A blind watermarking technology based on dct domain. In: Proceedings of the IEEE International Conference on Computer Science and Service System (CSSS), pp 398–401. IEEE Computer Society (2012)
15. Lei, D., Long, G.B.: Localized image watermarking in spatial domain resistant to geometric attacks. Int. J. Electron. Commun. (AEU). Elsevier **63**, 123–131(2009)
16. Yang, H., Zhang, Y., Wang, P., Wang, X., Wang, C.: A geometric correction based robust color image watermarking scheme using quaternion Exponent moments. Optik. Elsevier **125**, 4456–4469 (2014)
17. Guo, J., Zheng, P., Huang, J.: Secure watermarking scheme against watermark attacks in the encrypted domain. J. Vis. Commun. Image R. Elsevier **30**, 125–135 (2015)
18. Vahedi, E., Zoroofi, R.A., Shiva, M.: Toward a new wavelet-based watermarking approach for color images using bio-inspired optimization principles. Digit. Sig. Process. Elsevier **22**, 153–162 (2012)
19. Keyvanpour, M.R., Bayat, F.M.: Robust dynamic block-based image watermarking in DWT domain. Procedia. Comput. Sci. Elsevier **3**, 238–242 (2011)
20. Prathap, I., Natarajan, V., Anitha, R.: Hybrid robust watermarking for color images. Comput. Electr. Eng. Elsevier **40**, 920–930 (2014)
21. Golabi, S., Helfroush, M.S., Danyali, H., Owjimehr, M.: Robust watermarking against geometric attacks using partial calculation of radial moments and interval phase modulation. Inf. Sci. Elsevier **269**, 94–105 (2014)
22. Su, Q., Niub, Y., Wanga, Q., Sheng, G.: A blind color image watermarking based on DC component in the spatial domain. Optik. Elsevier **124**, 6255–6260 (2013)
23. Yang, W.X., Pan, N., Ying, Y.H., Peng, W.C., Long, W.A.: A new robust color image watermarking using local quaternion exponent moments. Inf. Sci. Elsevier **277**, 731–754 (2014)
24. Patra, J.C., Phua, J.E., Bornand, C.: A novel DCT domain CRT-based watermarking scheme for image authentication surviving JPEG compression. Digit. Sign. Proces. Elsevier **20**, 1597–1611 (2010)
25. Lin, S., Shie, S., Guo, J.Y.: Improving the robustness of DCT-based image watermarking against JPEG compression. Comput. Stan. Interfaces. Elsevier **32**, 54–60 (2010)
26. Agarwal C, Mishra A. Sharma A (2015) A novel gray-scale image watermarking using hybrid Fuzzy-BPN architecture. Egyptian Informatics Journal. Elsevier, Vol. 16, pp 83–102
27. Verma, V.S., Jha, R.K., Ojha, A.: Significant region based robust watermarking scheme in lifting wavelet transform domain. Expert Syst. Appl. Elsevier **42**, 8184–8197 (2015)
28. Das, C., Panigrahi, S., Sharma, V.K., Mahapatra, K.K.: A novel blind robust image watermarking in DCT domain using inter-block coefficient correlation. Int. J. Electron. Commun. Elsevier **68**, 244– 253 (2014)
29. http://sipi.usc.edu/database/database.php?volume=misc
30. https://openi.nlm.nih.gov/index.php
31. http://www.shutterstock.com/s/fingerprint/search.html

# JPEG2000 Compatible Layered Block Cipher

**Qurban A. Memon**

**Abstract** Multimedia security is ever demanding area of research covering different aspects of electrical engineering and computer science. In this chapter, our main focus is encryption of JPEG2000 compatible images. Though both stream and block cipher have been investigated in the literature, but this chapter provides a detailed study of block cipher as applied to images, since JPEG2000 generates various subband sizes as blocks. In the first section, we briefly define various encryption components like wavelet transform, bit plane decomposition, XOR operation, artificial neural network, seed key generator and chaotic map functions, for interest of the reader. Later in Sect. 2, we present literature review of various encryption techniques from two perspectives: applications to highlight scope of research in this domain; and approaches to provide overall view of multimedia encryption. The section three provides a new two-layer encryption technique for JPEG2000 compatible images. The first step provides a single layer of encryption using a neural network to generate a pseudo-random sequence with a 128-bit key, which XORs with bit planes obtained from image subbands to generate encrypted sequences. The second step develops another layer of encryption using a cellular neural network with a different 128-bit key to develop sequences with hyper chaotic behavior. These sequences XOR with selected encrypted bit planes (obtained in step 1) to generate doubly-encrypted bit planes. Finally, these processed bit planes go through reverse process, followed by inverse wavelet transform to generate encrypted image. In order to test this approach, the section four presents commonly adopted testing criteria like 0/1 balancedness, NIST statistical test, correlation and histogram tests done on seed generator and encrypted images to demonstrate robustness of the proposed approach. It is also shown that the key size is above 256 bits.

Q.A. Memon (✉)
UAE University, 15551 Al-Ain, United Arab Emirates
e-mail: qurban.memon@uaeu.ac.ae



A.E. Hassanien et al. (eds.), *Multimedia Forensics and Security*,
Intelligent Systems Reference Library 115, DOI 10.1007/978-3-319-44270-9_11

**Keywords** JPEG2000 · Multimedia security · Feed forward neural network · Random sequence · Cellular neural network · Block cipher · Encryption

# 1 Introduction

In current multimedia environment, security and protection of data is essential to fulfil vendor rights and client requirements. As Internet is evolving so are the interests, applications and risks. People are fascinated by recent applications to simplify their professional work [1], and secure their data access [2]. With respect to storage or transferring of images and videos, organizations prefer to use encryption of multimedia as an alternative to other approaches. Nowadays, a great deal of money is being invested to increase level of security for image data transmitted or stored over public channels, and a lot of research works are being reported in this field.

As encryption technology is growing to protect data in storage, so are the challenges for forensic investigators since tools need decryption key to collect evidence. As an example, if a hard drive is fully encrypted, then investigator's options are limited to decrypt it. In recent developments, corporate organizations, like the Blackberry tend to embed encryption on their proprietary data to avoid their misuse, and in fact drive for effective encryption is coming from corporate sector, not from individual users.

More recently, images are wavelet transformed before encryption in order to enable selective compression and/or encryption. Since encryption is typically done on binary images; and operation on bits is done through XOR operation, they need to be explained first. Typically pseudo-random sequences are generated through artificial neural networks and are claimed to be best suited for this purpose as they possess features like high security, no distortion and its ability to perform for nonlinear input-output characteristics. Likewise, seed for artificial neural network is generated through a nonlinear chaotic function. All of these preliminary functions and/or tools are a prerequisite for most of the encryption algorithms used today. These are described below for interest of the reader.

## 1.1 Background

This section presents background information about some needed steps, which are used in any encryption algorithm in general, and in the proposed approach in particular. Each of these is briefly discussed below.

*Discrete Wavelet Transform:* For any JPEG2000 compatible operation, the 2-D version of Wavelet Transform (2-D DWT) is applied first. The mathematical representation to obtain wavelet coefficients from image pixels is given by [3]:

$$W_{\varphi}(j_o, m, n) = \frac{1}{\sqrt{MN}} \sum_{x=0}^{M-1} \sum_{y=0}^{N-1} p(x, y)\, \varphi_{j_o, m, n}(x, y)$$
$$W_{\psi}^{i}(j, m, n) = \frac{1}{\sqrt{MN}} \sum_{x=0}^{M-1} \sum_{y=0}^{N-1} p(x, y)\, \psi_{j, m, n}^{i}(x, y) \qquad , i = \{H, V, D\} \tag{1}$$

Similarly, for reconstruction of the image, the inverse wavelet transform is used at the receiver from the received wavelet coefficients. The inverse wavelet transform is described mathematically as follows [3]:

$$p(x, y) = \frac{1}{\sqrt{MN}} \sum_{m} \sum_{n} W_{\varphi}(j_o, m, n)\, \varphi_{j_o, m, n}(x, y) + \frac{1}{\sqrt{MN}} \sum_{i=H,V,D} \sum_{j=j_0} \sum_{m} \sum_{n} W_{\psi}^{i}(j, m, n)\, \psi_{j, m, n}^{i}(x, y) \tag{2}$$

In Eqs. (1) and (2), $\phi_{j0, m, n}(x, y)$ is the approximation wavelet function, and $\psi_{j, m, n}^{i}(x, y)$ is the detail wavelet function. The superscript $i$ could be $H$ for the horizontal detail image, $V$ for vertical, and $D$ for diagonal. The subscripts $m$ and $n$ are the location indices of the wavelet coefficient in the detail image, $j$ is the level of wavelet decomposition (*sub-band*), and $M$ and $N$ are the number of rows and columns in the image respectively. In Eq. (2), the first term relates to approximation coefficients and the second term relates to horizontal, vertical and diagonal coefficients.

In JPEG2000, typically images are decomposed into three to five wavelet levels to opt for greater number of subband representations. Each time, subbands are wavelet-decomposed to next level; the size of new subbands reduces to quarter of the original. The wavelet functions used in JPEG2000 typically belong to the first member of the Cohen-Daubechies-Feauveauthe family [4].

*Bit Plane Decomposition*: By bit plane decomposition, it is meant that an image *p*(*x*, *y*) of size NxN with G = 256 gray levels is decomposed into eight (~log $_2$ (G)) binary images. Thus, an 8-bit data will have eight binary images. Each of these binary images is called a bit plane with bits corresponding to a given bit position in each of the binary numbers representing the pixel gray level. The first bit plane will contain the set of most significant bit of each gray level value; likewise the last bit plane will contain the set of least significant bits. The first bit plane gives the most critical approximation of the pixel, whereas each later bit plane improves this approximation as we continue to add on successive bit planes.

*XOR operation*: This is one of the Boolean operations that can be done on binary images. Typically, these operations are known as masking functions, where *p*(*x*, *y*) is a binary image and a masking binary pattern is chosen to mask its bits. In case of XOR operator, it inverts bits. It produces output value of logical one, whenever *p* (*x*, *y*) and masking bit are different. In other words, it is used to highlight differences in a binary image. Likewise, XNOR is used to highlight similarities in a binary image. The XOR operation is commutative, associative and self-inverse. For example, we want to XOR 8-bit gray level value 205 with 180. 205 is 11001101 in binary and 180 is 10110100 in binary. The result of XOR in bitwise operation is 1111001 in binary or 121 in decimal. Interestingly, if this result is XORed with 180 again, we get original value 205 back. This reversible property of XOR is usually exploited in encryption, where masking is used in transmitter and unmasking is done on receiver side.

*Cellular Neural Networks*: Cellular neural networks (CNNs) provide both continuous time and local interconnection features. The basic circuit is called a cell, which contains linear and non-linear elements and sources. Cells can be characterized as multiple input-single output nonlinear processors all described by one, or one among several different, parametric functionals [5]. A state variable characterizes a cell itself, and the notion of distance implies that the network is intrinsically defined in space; and typically 1-, 2- or 3- dimensional space is considered. The neighborhood adjacent cells only connect directly to each other, but are indirectly affected by other cells due to propagation effects caused by continuous time dynamics of the system. The cell topology can be considered as rectangular, triangular, hexagonal or a 3-dimensional array realized as a stack of 2-dimensional layers. Cells may be identical or different otherwise with typically a small neighborhood. Though cells are characterized by adjacent neighbors due to local nature, yet they are assigned some global properties due to continuous time features. For illustrative purposes, a cellular neural network is shown in Fig. 1.

In recent works, the authors in [6] have explored criteria for the existence of unique equilibrium point and its global asymptotic stability of continuous delayed CNNs to offset oscillations caused by existence of time delays in CNNs. The authors in [7] propose two kinds of cellular neural networks based on mem-elements—MC-CNN lets a mem-capacitor replace the conventional linear

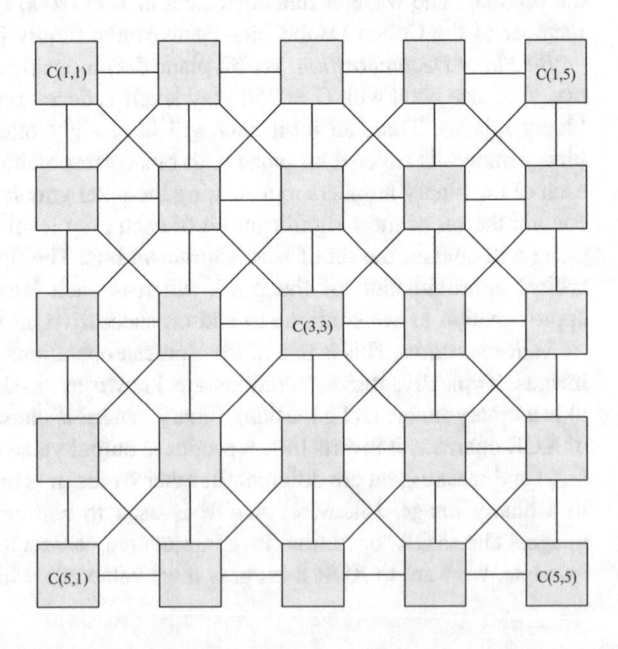


**Fig. 1** Sample 5 × 5 Cellular neural network

capacitor of a cellular neural network cell, while EM-CNN is an economical on fabricating cost for better implementation of CNN. A new model of CNNs with transient chaos is proposed in [8], where authors propose adding negative self-feedback once dynamic equations have been transformed into discrete time resulting in transient chaos.

Due to dynamics in CNNs, cellular neural networks have found applications in image processing, pattern recognition and classification, and combinatorial optimization amongst others. CNN has typically one type of unit processor in one layer; however, some applications require collaboration of distinct dynamics. As an example, the authors [9] propose randomly reconfigurable cellular neural network to mimic joint effort of distinct types of neurons. In biomedicine area, recently a new algorithm using fuzzy cellular neural network has been proposed [10] to automatically detect white blood cells by developing a complete contour around cell. Study by the authors [11] has shown that as the number of cells increases beyond four, a hyper-chaotic behavior is observed in the cellular neural network, thus requires more keys to describe the state of the system.

*Chaotic Map*: A number of chaotic maps exist like piecewise linear chaotic map (PWLCM) [12], standard map [13], cat map [14] and baker map [15] for block ciphers. Each of these maps can be used for block ciphers. The PWLCM is given as [12]:

$$x(n) = f(x(n-1)) = \begin{cases} \frac{x(n-1)}{k} & if\, x(n-1) \in [0, k] \\ \frac{x(n-1)-k}{0.5-k} & if\, x(n-1) \in [k, 0.5] \\ f(1-x(n-1)) & if\, x(n-1) \in [0.5, 1] \end{cases} \tag{3}$$

where $f$ is the transfer function, $k$ ε [0, 0.5] and x(n) ε [0, 1]. $x(0)$ and $k$ are used as secret keys. For a dynamical system to generate highest lyapunov exponent, $k$ is typically chosen to be 0.5. Similarly, standard map [13], cat map [14] and baker map [15] can be stated below in Eqs. (4), (5) and (6) respectively:

$$\begin{aligned} x_{j+1} &= (x_j + y_j) modN \\ y_{j+1} &= \left( y_j + ksin\frac{x_{j+1}N}{2\pi} \right) modN \end{aligned} \tag{4}$$

$$\begin{bmatrix} x_{j+1} \\ y_{J+1} \end{bmatrix} = \begin{bmatrix} 1u \\ vuv+1 \end{bmatrix} \begin{bmatrix} x_j \\ y_J \end{bmatrix} (modN) \tag{5}$$

$$\begin{cases} x_{j+1} = \frac{N}{k_i}(x_j - N_i) + y_j mod \frac{N}{k_i} \\ y_{j+1} = \frac{k_i}{N}\left( y_j - y_j mod \frac{N}{k_i} \right) + N_i \end{cases} with \begin{cases} k_1 + k_2 + k_3 + \ldots = N \\ N_i = k_1 + k_2 + \ldots + k_{i-1} \\ N_i \le x_j \le N_i + k_i \\ 0 \le y_j \le N \end{cases} \tag{6}$$

*Key Generator*: Many chaotic key generators exist but the one used in this research involves 1-D cubic map [16]. It takes 64-bit random key Key = [$Key_1$,

$Key_2$, $Key_3$, $Key_4$] to calculate initial condition based on its 16-bit component ($Key_i$) such that $y(0) = \left(\sum \frac{Key_i}{2^{16}}\right) mod(1)$ [16] of 1-D cubic map and returns values of the map using iterations. The states of the cubic map are written as [17]:

$$y(n+1) = \lambda y(n)(1 - y(n).y(n)) \tag{7}$$

where $\lambda$ is typically set at 2.59 as a control parameter, and state of equation is satisfied by $0 \le y(n) \le 1$. In order to generate initial conditions for the neural network, the Eq. (7) is first iterated 50 times and values are discarded; and then iterated again to initialize $w_0$, $w_1$, $w_2$, $w_3$, $A_0$, $B_0$, $C_0$, $D_0$, $K_0$, $K_1$, $K_2$, $K_3$, $n_0$, $n_1$, $n_2$, $n_3$. In order for Eq. (7) to provide randomness and reproducibility of same initial conditions each time it is run even on different computing machine, it should be ensured that values like Key, $\lambda$ set at 2.59, and initial iteration of 50 to discard values are used with same precision arithmetic. Further test analysis for robustness of sequences is discussed in performance analysis section.

*Reproducibility*: Reproducibility is the ability of an entire program or an experiment to yield same results every time it is run, by the same person or the different one working independently. This means that, every time code is run, it shall produce same results, with high degree of precision. Reproducibility is important for debugging and building confidence. Sometimes, lack of reproducibility is termed as a weakness of any random generator. Thus, there are general conditions for the random generator for reproducibility:

1. One should use the same random number generator seed.
2. The model of generating code shall not change.
3. The same initial values or conditions are used.
4. The precision of computation shall remain same as originally used.
5. Avoid running macros or custom visual basic for application (VBA) functions, which do not exactly generate same results from simulation to simulation.

There are some technical concerns with reproducibility like portability of code generator from one operating system to another, or use of parallelization involving number of threads on the platform. The first one relates to precision; and the second one to number of threads running for the same code. The only caution that should be exercised is that same precision is to be used across platforms, and that one iteration of random number generation model shall not refer directly or indirectly to another value in an independent thread. This same concern holds on multiple CPU's. That is, if code is run on multiple CPU's or a CPU with multiple threads, the order of generating random numbers shall remain as of generated using serial computation. With these constraints in mind, 100,000 bits of code sequence were generated on same machine twice and once on another machine, and were found to be same each time.

*Standardization*: Information Technology lab (ITL) at NIST develops technical, management, etc. standards and guidelines, like 800-series report, for cost-effective privacy and security of sensitive information. In 800-38E report [18] published in

January 2010, the author informs about approval of XTS-AES mode of Advanced Encryption Standard (AES) algorithm for protection of data on storage devices referenced by IEEE P1619 Task group. The implementation mode of this block cipher defines three elements [19]: a key; fixed length data; and implementation procedure. The sample implementation is reported in [20].

Another institution heavily involved in developing information security expertise is the European Union Agency for Network and Information Security Agency, which works with groups to develop recommendations on practice in information security. In [21], the author(s) collate a series of recommendations for algorithms, tools, parameters, etc. for cryptographic protocols in use. The important points to note in this document are [21]:

a. Recommended Block Ciphers
  i. *AES*: The Advanced Encryption Standard, or AES, is recommended as a block cipher for future applications [22, 23]. AES uses block length of 128 bits and supports three key lengths: 128, 192 and 256 bits. The one with longer uses more rounds and thus are slower by 40 %.
  ii. *Camellia*: The Camellia block cipher [24] cipher uses Feistel cipher design and supports a block length of 128 bits with three key lengths: 128, 192, and 256 [24]. The one with longer keys are slower by 33 %.
b. Legacy Block Ciphers
  i. *3DES*: This standard comes with two variants: a two-key version and a three-key version [25]. Both, two-key version and three-key version come with an effective key length of 112-bits, supporting a block length of 64-bits and thus suffers from a related key attack [21].
  ii. *Kasumi*: This cipher [26] has a 128-bit key, supports block length of 64-bits, and is typically used in 3GPP. This cipher suffers from various problems and is not recommended for use in further applications [21].
  iii. *Blowfish*: This cipher [27] has key size range of 32–448 bits and is recommended for legacy applications. It supports a block size of 64-bits, and typically used in some IPsec configurations.
c. Historical (not-recommended) Block Ciphers
  i. *DES*: DES supports a 56-bit key with block size of 64-bits, and thus not secure by today's standards, as it suffers from linear [28] and differential cryptanalysis [29].

The chapter is structured as follows. In the next section, literature review is presented from two perspectives: firstly multimedia security techniques are highlighted that target specific applications; secondly general overview of recent approaches is discussed that are claimed to be effective and robust. Section 3 presents a new two-layer block cipher approach for image encryption. In Sect. 4, we analyze the performance of the proposed approach with regard to key space, histogram, correlation coefficient, NIST statistical test and 0/1 balancedness. Section 5 presents discussion and conclusions, followed by references.

## 2 Literature Review

In this section, we intend to present survey of work done in the domain of image security approaches and applications, the details of which are described below:

### *2.1 Applications*

In literature, a lot of work can be found that presents applications addressed by multimedia security tools. For example, the authors in [30] address some key issues within financial domain, and propose use of recurrent neural networks (RNNs) in multimedia information processes such as coding/fusing and securing data. The authors claim that e-financial secure solution using neural network techniques has the potential to replace paper based cheque of traditional banking or the debit card to execute confidential transfers of funds over Internet. The solution offers four levels of security: coding, fusing, encryption, and speaker recognition. In [31], the authors propose use of an existing feature extraction algorithm developed for fingerprint recognition, to be adapted for vein recognition. In modification, the operation of false feature elimination is applied only once instead of three times. Though restricted to black and white image, the authors implement this algorithm, and realize it on a Field-Programmable Gate-Array and claim for a success rate of over 99 %. The compression of optically encrypted digital holograms using artificial neural networks is investigated in [32]. In fact, the authors encrypt real-world three-dimensional (3D) objects, suitable for secure 3D object storage and transmission applications, and then lossy compression is achieved by non-uniformly quantizing the complex-valued encrypted digital holograms using a unsupervised artificial neural network. The authors in [33] discuss a closed-set based text-dependent speech-and-speaker identification system based on spoken Arabic digit recognition from zero to ten. The speech signals of the Arabic digits are processed then graphically, in a way that the signal is treated as an object image for processing to recognize right words and identifying the speaker. For shape recognition, a reconfigurable hardware architecture involving specialized interconnected tiny neural networks has been presented in [34], where authors propose a cooperative working strategy amongst tiny neural networks with online training and run time weight modification to build parallel architecture for shape recognition. In [35], the authors provide a benchmark designed to reflect more realistic face processing applications, and describe pipeline for inferring facial attributes with a robust face alignment technique and use of dropout-SVM approach to linear SVM training. As an another application, visual cryptography for color transfer is investigated in [36], where authors propose encryption of color image into a number of noise-like binary share images such that whenever a number of shares images is collected, a colorful version of the secret image can be reconstructed with simple decryption computations. In another work [37], the authors integrate merits of semi-supervised learning and ensemble learning to design an approach for remotely sensed images to detect changes. An ensemble agreement is exploited to choose unlabeled patterns for training step. Finally, fusion of base classifiers (i.e., multilayer perceptron, elliptical basis function neural network and fuzzy k-nearest neighbor) is used based on a combination rule to assign each to a specific class. The authors in [38] examine the utility of the advanced wordlview-2 (WV-2) data for mapping using SVM and ANN classification algorithms, specifically the testing the advent of the additional WV-2 bands in mapping spatial extent of six endangered tree species to produce wall-to-wall thematic map to help design management plans and policies for assessing and monitoring natural resources. In [39], the authors propose a state of-the-art traffic based intrusion detection system built on a novel randomized data partitioned learning model using a compact network feature set within a time interval and feature selection techniques to design and develop botnet detection methodology for large scale network to make it immune to packet encryption.

## 2.2 Approaches

Though a lot of multimedia security approaches have been proposed, and work is still going on, but for sake of reader's interest, selective well known approaches are briefly mentioned here in this sub-section. As an example, the author [40] investigates neural network properties to propose low-cost authentication for images or videos. The author claims that the approach has the embedded ability to detect whether the data is modified maliciously. The author finally highlights open issues in this field like: which property of neural networks to be exploited for data protection; which neural network models are suitable for data protection; and learning ability of neural networks, etc. In another work [41], the authors use neural network in the receiver for the purpose of decryption by exploiting back propagation algorithm in the receiver to train it with a 12-bit cipher text as an input, and 8-bit plain text being the target output. The plain text at the input also includes some impurity based on some pre-determined key to mislead any possible eavesdropper. Similarly, the work in [42] targets securing image data transmission using randomness in encryption algorithm by introducing confusion in the data, and addition of impurities to misguide the cryptanalyst.

The neural network in combination with Chaos has also been investigated. As an example, a neural network is proposed by authors [43], which is composed of a chaotic neuron layer and a linear neuron layer. The network is then used to construct a block cipher that encrypts the plaintext into a cipher text using a key in order to construct a chaotic neural network based block cipher with good computing security. In that, the block cipher involves two processes: a diffusion process implemented by chaotic neuron layer and a confusion process implemented by a linear neuron layer. These processes are iterated a number of times to improve encryption complexity. In another work [44] that uses chaos and neural network together, the authors exploit neural network structure to process many media contents in a parallel manner. The scheme combines encryption and watermarking together. The encryption part uses random sequences generated from chaos system with the help of an encryption key. This key is then used to encrypt media contents with a neural network structure. However, the apparent disadvantage in this scheme is that more sub-keys need to be transmitted to the receiver. In another research [45], the authors propose an image encryption/decryption algorithm based on chaotic neural network. The employed network comprises two layers each with three layers: chaotic neuron layer (CNL) and permutation neuron layer (PNL). The approach uses a 160-bit-long authentication code to generate initial conditions and the parameters of both layers. The overall process is repeated several times to make it robust and increasingly complex. The proposed method uses two more keys where a slight mismatch in one of them fails in successfully decrypting an image. The authors in [46] propose an encryption method based on a hybrid chaos-based encryption algorithm. The algorithm employs permutation–diffusion architecture that uses chaotic control parameters for permutation. These control parameters for the permutation stage are generated by a logistic map. In the diffusion stage, another chaotic logistic map with different initial conditions and parameters is used to generate initial conditions for a hyper-chaotic Hopfield neural network to generate a key stream for image homogenization of the shuffled image. In order to provide summarized view, the authors in [47] present survey of chaos-based encryption in multimedia to address full and partial encryption for different applications and requirements. The authors conclude that multimedia security is not yet mature and additional efforts are needed for further developments to ensure multiple levels of security with low computational requirements and format compliance.

The biometric is also investigated together with cryptography to boost security. For example, the authors in [48] propose secure and blind biometric protocol based on asymmetric encryption of the biometric data to addresses concerns of user's privacy, template protection, and trust issues. In this approach, the authors integrate merits of both biometric (for example face, iris, hand geometry, finger print) authentication and public key cryptography. The classification of feature vector is achieved using generic classifiers like SVM and neural networks. In another work [49], the authors present a new image encryption technique involving two main operations based on neural chaotic generator. The two operations are permutation at pixel level and masking and permutation at bit level with perturbing orbit technique using piecewise linear chaotic map. A symmetric encryption-decryption block cipher with 160-bit key is also investigated in [50], where authors combine spatiotemporal chaotic system based on coupled map lattice to generate weight matrix, etc., and a chaotic neural network. Using theoretical analysis and experimental results, the authors claim that the approach is able to resist differential attack, brute-force attack, statistical attack, and plaintext attack. To satisfy security requirements of confidentiality, authentication, and non-repudiation, the authors in [51] couple artificial neural networks and chaos dynamics to compute a digital envelope consisting of the encrypted signal, the secret key and the associated hash value to develop smart energy system against illegal activities. The appropriate

ANN structure is then found to successfully model the dynamics of chaos on the basis of minimizing an appropriate continuously differentiable error function. The author in [52] propose a novel image encryption technique based on reversible 2nd order one-dimensional cellular automata to act like a block cipher in a fully parallel mode. The symmetric scheme generates pseudo-random permutations with 128-bit key and then injects it into a parallelizable encryption scheme acting on different image blocks. Another symmetric image encryption algorithm is proposed in [53], where spatiotemporal chaos of the mixed linear–nonlinear coupled map lattices is diffused to develop cryptography features in dynamics and is claimed to perform better than the logistic map or the system of coupled map lattices. The authors in [54] propose efficient symmetric encryption scheme based on cyclic elliptic curve and chaotic system to address delay in the system performance, improve information security, and limited key space. In that, the cipher encrypts 256-bit of plain image to 256-bit of cipher image within eight 32-bit registers by generating pseudorandom bit sequences for round keys based on a piecewise nonlinear chaotic map. Similarly in [55], the author proposes digital image parallel encryption structure using local spread of cellular neural network. In that, feistel framework is used, where a cell encrypts one image block and then is affected by the output of 8-neighboring cells due to diffusion effect. Additionally, in [56], the author discusses an overview of how traditional data encryption techniques can be revised to improve good performance for secured communications. With this perspective, the authors present a comprehensive survey of existing chaos-based encryption along with comparative tables to help pick a suitable technique for an application at hand.

As a summary, many research works have appeared in literature to address encryption of data either before transmission or for storage. The issues that are still being investigated are complexity, the robustness in presence of malicious attack, as well as compatibility with current standards. In this chapter, neural network structures are examined in combination with wavelet transform for JPEG2000 compatible image encryption and decryption. The motivation behind use of wavelets is that current image transmission and storage is mostly preferred using JPEG2000, which is an evolving standard for image coding and transmission. This is motivated by the fact that the JPEG2000 is better of about 20 % at compressing images, and that it can allow an image to be retained without any distortion or loss [57]. For further reading on this subject, the readers are encouraged to refer to [58, 59].

## 3 Proposed Approach

In this section, we present the proposed approach. Consider plain image $p(\mathrm{x}, \mathrm{y})$ of size N × N. The first step in JPEG2000 is to apply $n$-level wavelet transform to the image. For purpose of simplicity, we assume $n = 2$. In wavelet transform decomposition, each level produces four frequency subbands of the input image, where each is of quarter-size of the original. In the proposed approach, we do not apply encryption on these subbands directly; rather these subbands undergo bit plane decomposition to generate eight binary images for each subband. Depending upon complexity need, a set of these binary images is transformed into encrypted bit plane images. If desired, these encrypted subbands can then undergo next step of the JPEG2000 encoder, or otherwise inverse wavelet transform is applied to generate encrypted image. There are three variables that add complexity to the encryption process executed through chaotic sequences: one is the number of levels the image undergoes wavelet decomposition; another is the number of subbands that are bit plane decomposed; and the third is the set of bit planes for encryption.

The proposed approach is shown in Fig. 2, where we have a XOR operation between pseudo-random sequence generated by a chaotic neural network and the binary bits from subband image pixels. We call this first step as single layer encryption.

In order to generate pseudo-random sequence, a chaotic neural network, as shown in Fig. 3, is employed to introduce non-linearity in generating a sequence. A 64-bit input key i.e., $A = [A_1, A_2, A_3, \ldots, A_{64}]$ is applied at the input layer such that 8-bits enter at each node of the layer. The output of this layer may be written similar to [60], as:

$$B = f^{n0}(Aw_o + A_o, K_0) \tag{8}$$

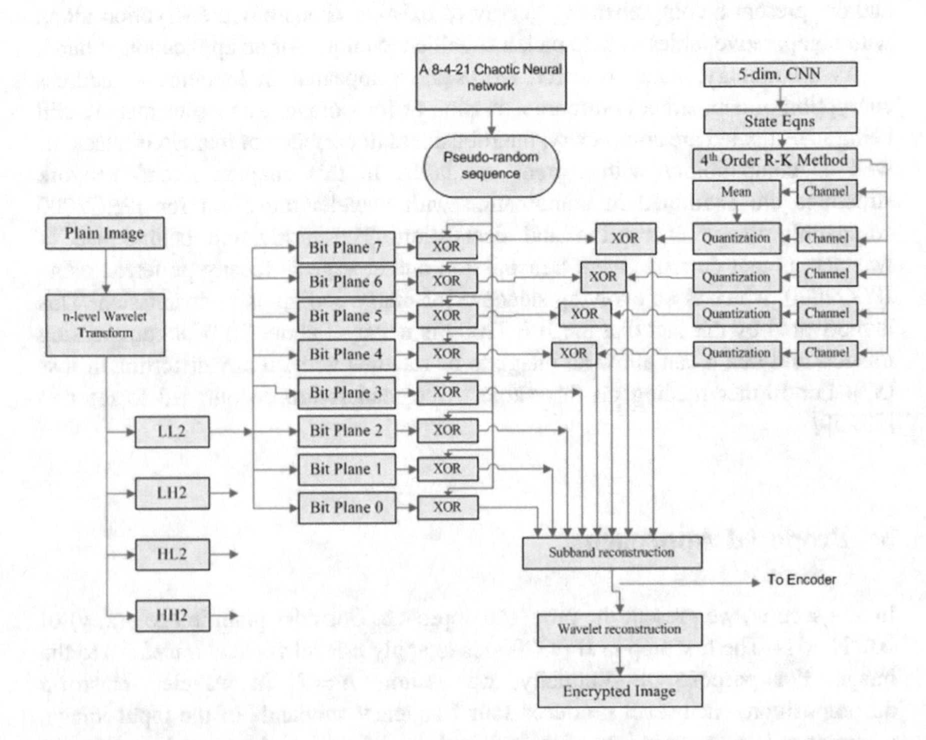


**Fig. 2** Proposed approach

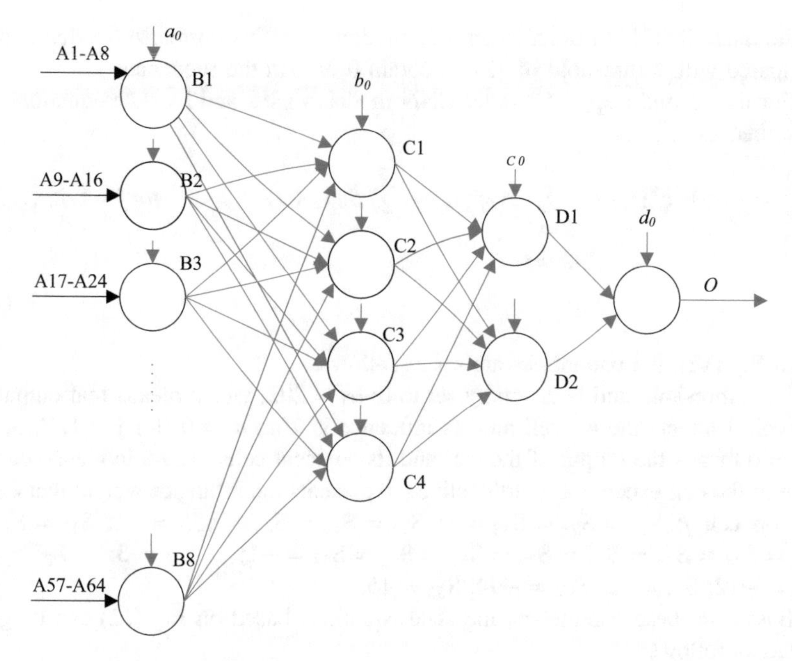


**Fig. 3** Generation of N-bit pseudorandom sequence for chaos

where $w_O$ is the matrix of size 8 × 8 i.e., $w_O$ = [$w_{0,0}$, $w_{0,1}$ $w_{0,2}$ $w_{0,3}$ $w_{0,4}$ $w_{0,5}$ $w_{0,6}$ $w_{0,7;}$ $w_{1,0}$,…,; $w_{7,7]}$, A is the input vector, the bias is $A_O$ = [$a_0$, $a_1$, $a_2$, $a_3$, $a_4$, $a_5$, $a_6$, $a_7$], $K_O$ is the control parameter [$k_0$, $k_1$, $k_2$, …, $k_7$] and $n0$ is random number generated by key generator in the range $1 \leq n0 \leq 10$.

The output of each layer becomes input to the next layer, apart from becoming input to that neuron itself. Continuing in the same fashion, the output of remaining layers is calculated as follows:

$$C = f^{n1}(Bw_1 + B_o, K_1) \tag{9}$$

$$D = f^{n2}(Cw_2 + C_o, K_2) \tag{10}$$

$$O = f^{n3}(Dw_3 + D_o, K_3) \tag{11}$$

where the matrices $w_1$, $w_2$, $w_3$ have sizes equivalent to 4 × 8, 2 × 4 and 1 × 2; $B_O$, $C_O$, $D_O$ with sizes 4 × 1, 2 × 1, and 1 × 1; $K_1$, $K_2$, $K_3$ with sizes 4 × 1, 2 × 1, and 1 × 1, respectively. During iterations at each layer, the control parameters are also adjusted using respective layer outputs in such a way that respective range lies in [0.4, 0.6], for example $K_O = 0.2 \times B + 0.4$ to get chaotic behavior. Like $n_O$, the values of $n_1$, $n_2$, and $n_3$ are obtained through key generation. Once the value of output is obtained between 0 and 1, then this value is normalized in the range 0–255. In order to enforce randomness, this normalized value is then compared with a threshold of 127 to obtain 0 or 1 in the sequence.

For the second step, a 5th order CNN model is used and its state equations are described as in [5]:

$$\frac{dx_j}{dt} = -x_j + a_j f(x_j) + \sum_{\substack{k=1 \\ k \neq j}}^{5} A_{jk} f(x_k) + \sum_{k=1}^{5} S_{jk} x_k + \breve{I}_j \qquad for\, j = 1, 2, \ldots, 5 \tag{12}$$

In Eq. (12), the parameters are set as follows:

Ĭ is a threshold and is generally set to 0; $a_4 = 202$, which means that output of the cell 4 affects the 4th cell and its influence is 202; $a_j = 0$ (for j = 1, 2, 3, 5); $A_{jk} = 0$ means the output of the cell and its adjacent cells has no influence on the state of the cell except the fourth cell; $S_{jk}$ represents the influence weight that $k$ cell has on cell $j$: $S_{11} = S_{23} = S_{33} = 1$; $S_{13} = S_{14} = S_{45} = S55 = -1$; $S_{12} = S_{15} = S_{21} = S_{24} = S_{25} = S_{34} = S_{35} = S_{51} = S_{52} = S_{54} = -1$; $S_{22} = 3$; $S_{31} = 11$; $S_{32} = -12$; $S_{41} = 92$; $S_{44} = -94$; $S_{53} = 15$.

Based on these parameters, the state equations based on Eq. (12) can be generated as follows:

$$\begin{aligned} \frac{dx_1}{dt} &= -x_3 - x_4 \\ \frac{dx_2}{dt} &= 2x_2 + x_3 \\ \frac{dx_3}{dt} &= 11x_1 - 12x_2 \\ \frac{dx_4}{dt} &= 92x_1 - 95x_4 - x_5 + 202f(x_4) \\ \frac{dx_5}{dt} &= 15x_3 - 2x_5 \end{aligned} \tag{13}$$

The lyapunov exponents (being important characteristics of a chaotic system) of Eq. (13) are: 0.2953, 0.5285, 0.1264, −3.9205, −17.4382, which proves that the system is hyper chaotic. In order to solve these state equations, a classical fourth-order Runga-Kutta method [61] is used, as:

$$y_{n+1} = y_n + \frac{h}{6}(k_1 + 2k_2 + 2k_3 + k_4) \tag{14}$$

where

$$k_1 = f(t_n, y_n);\; k_2 = f\left(t_n + \frac{1}{2}h, y_n + \frac{1}{2}hk_1\right),\; k_3 = f\left(t_n + \frac{1}{2}h, y_n + \frac{1}{2}hk_2\right),\; k_4 = f(t_n + h, y_n + hk_3)$$

In Eq. (14), the initial values are set as: step size $h = 0.005$, $x_1(0) = 0.1$, $x_2(0) = x_3(0) = x_4(0) = x_5(0) = 0.2$, and number of iterations μ is set at 16384 to generate μ hyper chaotic values. Thus five channels of hyper chaotic sequence are generated with each channel being 16384 values. The mean '$m$' of channel $x_1$ is calculated as follows:

$$x_{ij} = 0,\ if\ x_{ij} \leq m$$
$$x_{ij} = 1,\ if\ x_{ij} \geq m$$

where $i = 1, 2, 3, 4, 5$ and $j = 1, 2, 3, 4, \ldots., 16384$. Thus four channels ($x_1$, $x_2$, $x_3$, and $x_4$) are quantified into four binary sequences ($X_1$, $X_2$, $X_3$, $X_4$). These four binary sequences are XORed with four most significant sequences from step 1 (single layer encryption) to generate four most significant doubly-encrypted sequences.

## 4 Performance Analysis

In this section, we report on performance of the proposed approach. Specifically, different measurements and tests are explained to demonstrate complexity, effectiveness and robustness.

*Key space*: The key space of the proposed scheme can be derived from three parts: chaotic neural network key generator, $n$-level wavelet signal decomposition, and cellular neural network. There are two keys used in chaotic neural network: one is the 64-bit seed to neural network and another is the 64-bit to calculate initial conditions. The number of bits needed for a typical $n$-level wavelet transform does not exceed 3, and that how many of the bit planes are to be encrypted is also three. A 128-bit key to drive CNN hyper chaotic system parameters is used. Thus, the key size for this encryption is: 128 (from neural network) + 128 (from CNN) + 3 (wavelet decomposition level) + 3 (no. of bit planes) + 3 (no. of encrypted planes) ~>256, thus the key space is at least $2^{256} \sim 11.56 \times 10^{76}$.

*Statistical Test*: In literature, there are many suits found for examining the randomness of generated pseudo-random sequences, like NIST statistical test suit (STS) package [62], DIEHARD statistical suit [63], and ENT package [64]. Here, in this subsection, the generated pseudo-random sequences are tested by NIST package [62] to quantify and verify the randomness level. The suite includes statistical tests for individual sequences, and in turn generates a $p$-value. The criterion is that if this value compared to a significance level typically set at 0.01 (for confidence of 99 %) is determined to be equal to 1, then the sequence appears to be random, otherwise non-random. The results of all of these tests run on first 100,000 bits sequence are shown in Table 1, where we notice that the sequence generated passes the NIST statistical test for individual sequences.

**Table 1** NIST Statistical tests

| NIST statistical test | *p*-value | Pass rate |
|---|---|---|
| Frequency | 0.827319 | Pass |
| Block-frequency | 0.817450 | Pass |
| Cumulative sums (forward) | 0.954010 | Pass |
| Cumulative sums (reverse) | 0.808165 | Pass |
| Runs | 0.071589 | Pass |
| Longest runs of ones | 0.942786 | Pass |
| Rank | 0.161278 | Pass |
| FFT | 0.417645 | Pass |
| Linear complexity | 0.619830 | Pass |
| Serial | 0.287165 | Pass |
| Approximate entropy | 0.612450 | Pass |
| Lempel-Ziv Compression | 0.387341 | Pass |
| Overlapping templates | 0.608741 | Pass |

*0/1 Balancedness Test:* In the work [65], Golomb stated that the noise like sequence should look like an equality distribution. This means that the generated chaotic sequence should have equal distribution between 0 and 1 (i.e., equal number of 1's and 0's). In order to judge on the proposed approach by equality distribution, a number of tests were run on chaotic generator to produce the sequences of different length. These lengths were estimated to be 65536 (based on wavelet subband of size 256 × 256), 16384 (based on wavelet subband of size 128 × 128), 4096 (based on wavelet subband of size 64 × 64), and 1024 (based on wavelet subband of size 32 × 32). The results are depicted in Table 2 and Fig. 4. These results show that the numbers are quite close to 50 %.

*Histogram Analysis:* Generally, histogram of an image depicts pixel distribution density against intensity level. Here, we use it to analyze encrypted image pixel distribution. In order to test the proposed approach on this perspective, 512 × 512 "Camera-man" image was encrypted using $n = 2$ with four most significant bit planes. After encrypting these bit planes, the process is run in reverse to construct encrypted image and histogram calculated. The results are shown in Fig. 5, where $x$ axis shows gray level while $y$ axis shows count at that gray level. It can be clearly seen that histogram of the encrypted image is fairly uniform with statistical

**Table 2** Equality distribution within the chaotic sequence generated by chaotic neural network

| Sequence length | Count of 1's | Percentage |
|---|---|---|
| 1024 | 515 | 50.29 |
| 4096 | 2055 | 50.17 |
| 16384 | 8206 | 50.08 |
| 65536 | 32775 | 50.01 |

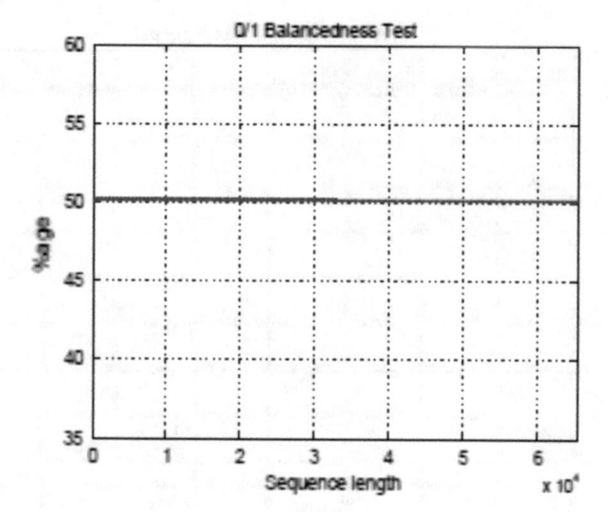


**Fig. 4** Percentage distribution of 1's in the sequence

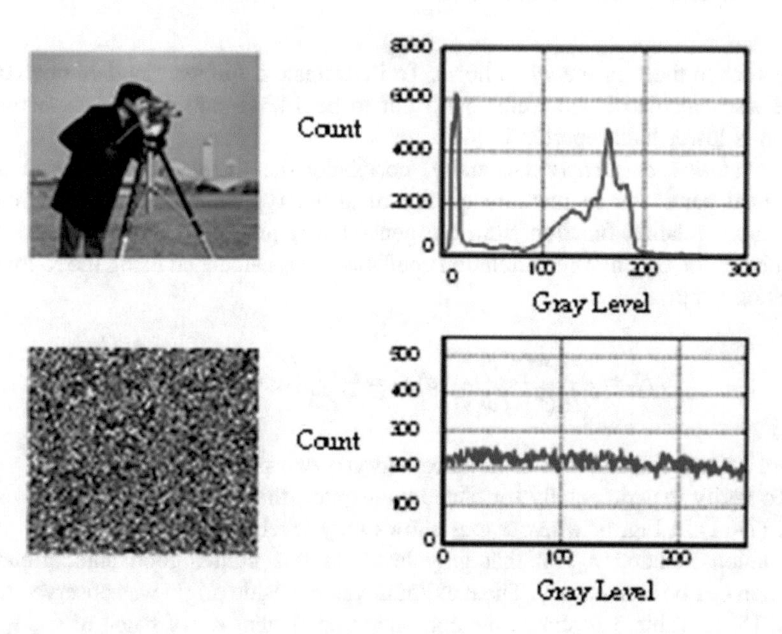


**Fig. 5** Histogram analysis of proposed approach

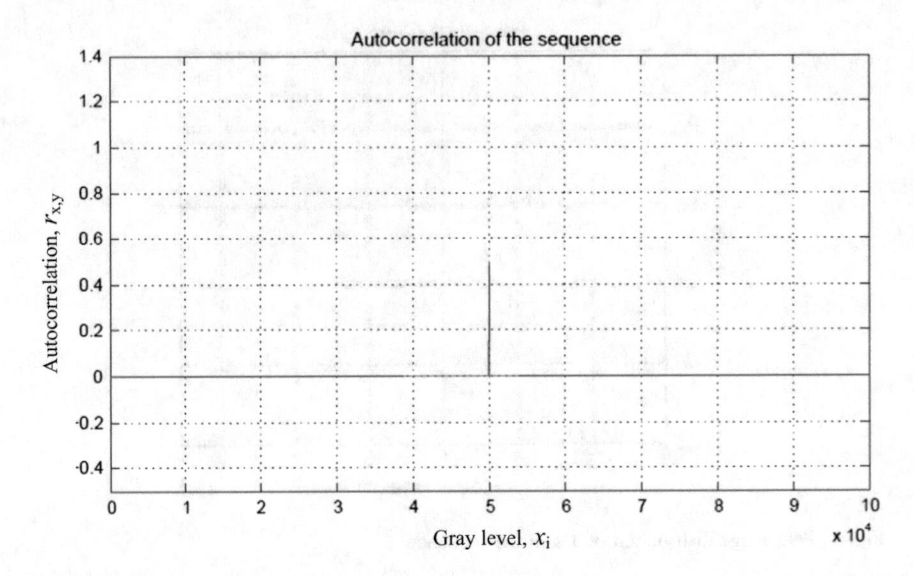


**Fig. 6** Correlation function of the sequence

properties to those of the white noise. To investigate it further, standard deviations were also calculated and were found out to be 14.315 and 14.168 respectively, which is lower than reported in [46].

*Correlation coefficient*: Generally, correlation coefficient is considered as a statistical parameter to measure quality of an encryption process. Theoretically, the autocorrelation function from the generated sequence should be a noise like impulse at the origin. The correlation coefficient was calculated using the following equation [12]:

$$r_{xy} = \frac{cov(x,y)}{\sqrt{d(x)}\sqrt{d(y)}}; d(x) = \frac{1}{N}\sum_{i=1}^{N}\left(x_i - \frac{1}{N}\sum_{i=1}^{N} xi\right)^2 \tag{15}$$

where *cov* (*x*, *y*) stands for covariance between two pixels *x* and *y*.

To verify experimentally for chaotic generator, this function was plotted using Eqs. (7–11) in Fig. 6, where *x* axis shows gray level $x_i$ while *y* axis shows autocorrelation function $r_{x,y}$ at that gray level. In this figure, good autocorrelation function can be seen clearly. The maximum value outside origin was observed to be 0.00215. In Table 3 is shown the correlation coefficient $r_{xy}$ of batch of ($x_i$, $y_i$, for i = 1, 2, 3 … N) pairs of gray values of two adjacent pixels in six encrypted images. It is clear that pixels have been completely decorrelated due to encryption.

**Table 3** Correlation coefficients of the original and encrypted images

| Image | Original | | | | | Encrypted | | | | |
|---|---|---|---|---|---|---|---|---|---|---|
| | Horizontal | Vertical | Diagonal | Averages | Standard deviation | Horizontal | Vertical | Diagonal | Averages | Standard deviation |
| Lena | 0.9855 | 0.9881 | 0.9667 | 0.9801 | 0.009534 | −0.00072 | 0.00053 | 0.00103 | 0.00028 | 0.00074 |
| Barbara | 0.9785 | 0.9574 | 0.9762 | 0.9707 | 0.009451 | 0.00082 | −0.00102 | 0.00063 | 0.000143 | 0.00083 |
| Yacht | 0.9592 | 0.9499 | 0.9487 | 0.9526 | 0.004693 | −0.000761 | 0.00052 | −0.00049 | −0.00024 | 0.00055 |
| Airport | 0.9572 | 0.9601 | 0.9318 | 0.9497 | 0.012712 | 0.000541 | −0.00061 | 0.000385 | 0.00024 | 0.000551 |
| Lake | 0.9752 | 0.95001 | 0.9621 | 0.962437 | 0.010287 | 0.000523 | 0.00052 | −0.00041 | 0.000105 | 0.00051 |
| Jet | 0.9821 | 0.94030 | 0.9531 | 0.9585 | 0.017487 | −0.000434 | 0.00061 | 0.00032 | 0.000151 | 0.000549 |

## 5 Discussion and Conclusions

A detailed overview of block cipher as applied to images has been presented in this chapter along with brief discussion of each typical step in image encryption. Furthermore, a JPEG2000 compatible block cipher is proposed that provides layered encryption using two independent chaotic behavior sequences. The first sequence is generated through a chaotic neural network, whereas second one is constructed through cellular neural network of size $5 \times 5$. The final key size shown was slightly higher than 256 bits. Since the proposed encryption process has a number of variables like number of wavelet decomposition levels, number of bit planes to encode and number of layers for encryption, the approach can be termed as flexible.

As the proposed approach is JPEG2000 format compliant, it can facilitate forensic investigators to decrypt wider range of encrypted files and folders to enable searching larger data volume from those folders for evidence. Likewise, the steps (like wavelet decomposition, key generation, etc.) in the proposed approach can also be implemented in hardware, and thus can save search time of investigators.

In the proposed approach, each of the sequences generated can also function independently providing single layer encryption. However, in order to increase complexity and thus key size to match that of two-layer encryption, either chaotic neural network architecture be expanded to include another hidden layer or cellular neural network size be increased. However this needs further research.

As stated before, since JPEG2000 process involves generating subbands of quarter size at each level, different neural network size(s) would be required to generate code sequences at each level if multiple level wavelet decomposition is used to start with encryption process. However, this direction of research requires further analysis.

For coding purposes, typically after wavelet decomposition the next step is quantization (if needed) followed by coding of the image. Considering our proposed approach, the encrypted image would now enter the next step of compression/coding. However, for optimization purposes, it needs further analysis whether encryption should follow compression or encryption can be done in the compressed domain.

The last but not the least is common use of authenticated watermark on images. In order to combine watermark with proposed encryption, it is easy to embed it either in the first layer before XOR operation or before second layer XOR operation during cellular neural network code generation. We intend to report on this development in our next research activity.

## References

1. Memon, Q., Khoja, S.: Academic program administration via semantic web—A case study. In: Proceedings of International Conference on Electrical, Computer, and Systems Science and Engineering, Dubai, vol. 37, pp. 695–698 (2009)

2. Memon, Q., Akhtar, S., Aly, A.: Role management in Adhoc networks. In: Proceedings of Spring Simulation Multi-conference, Virginia, USA, vol. 1, pp. 131–137 (2007)
3. Mallat, S.: A theory for multiresolution signal decomposition: the wavelet representation. IEEE Trans. Pattern Anal. Mach. Intell. **11**, 674–693 (1989)
4. Cohen, A., Daubechies, I., Feauveau, J.: Biorthogonal bases of compactly supported wavelets. Commun. Pure Appl. Math. **45**(5), 485–560 (1992)
5. Chua, L., Yang, L.: Cellular neural networks: theory. IEEE Trans. Circuits Syst. **35**(10), 1257–1272 (1988)
6. Xu, S., et al.: Novel global asymptotic stability criteria for delayed cellular neural networks. IEEE Trans. Circuits Syst. –II Express Briefs **52**(6), 349–353 (2005)
7. Yi, S., et al.: Two novel cellular neural networks based on Mem-elements. In: Proceedings of the 34th Chinese Control Conference, July 28–30, 2015, Hangzhou, China, pp. 3452–3456 (2015)
8. Wang, L., et al.: Cellular neural networks with transient chaos. IEEE Trans. Circuits Syst.-II: Express Briefs **54**(5), 440–444 (2007)
9. Ayhan, T., Yalcin, M.: Randomly reconfigurable cellular neural network. In: Proceedings of 20th European Conference on Circuit Theory and Design, pp. 604–607 (2011)
10. Shitong, W., Min, W.: A new detection algorithm based on fuzzy cellular neural networks for white blood cell detection. IEEE Trans. Inf. Technol. Biomed. **10**(1), 5–10 (2006)
11. Peng, J., Zhang, D., Liao, X.: A digital image encryption algorithm based on hyper-chaotic cellular neural network. Fundamenta Informaticae **90**, 269–282 (2009)
12. El Assad, S., Noura, H., Taralova, I.: Design and analyses of efficient chaotic generators for crypto systems. In: Advances in Electrical and Electronics Engineering, IAENG Special Edition of the World Congress on Engineering and Computer Science, pp. 3–12 (2008)
13. Chirikov, B.: A universal instability of many-dimensional oscillator systems. Phys. Rep. **52**(5), 263–379 (1979)
14. Chen, G., Mao, Y., Chui, C.K.: A symmetric image encryption scheme based on 3D chaotic cat maps. Chaos, Solitons Fractals **21**(3), 749–761 (2004)
15. Kuperin, Y., Pyatkin, D.: Two-dimensional chaos: the baker map under control. J. Math. Sci. **128**(2), 2798–2802 (2005)
16. Djellit, I., Kara, A.: One-dimensional and two-dimensional dynamics of cubic maps. Discrete Dyn. Nat. Soc. Article ID: 15840 (2006). doi:10.1155/DDNS/2006/15840
17. Gao, T., Chen, Z.: Image encryption based on a new total shuffling algorithm. Chaos, Solitan, and Fractals **38**(1), 213–220 (2008)
18. Dworkin, M.: Recommendation for block cipher modes of operation: the XTS-AES mode for confidentiality on storage devices, pp. 800–38E. NIST Special, Publication (2010)
19. IEEE Std 1619–2007: The XTS-AES Tweakable Block Cipher. Institute of Electrical and Electronics Engineers, Inc., 18 April 2008
20. Rogaway, P.: Efficient instantiations of tweakable blockciphers and refinements to modes OCB and PMAC. In: Advances in Cryptology—Asiacrypt 2004, Lecture Notes in Computer Science, vol. 3329, pp. 16–31. Springer (2004)
21. ENSIA Report: Algorithms, key sizes and parameters report. 2013 recommendations, version 1.0, October 2013. http://ensia.europa.eu
22. Daemen, J., Rijmen, V.: The Design of Rijndael: AES. The Advanced Encryption Standard. Springer (2002)
23. Fluhrer, S., Lucks, S.: Analysis of the E0 encryption system. In: Vaudenay and Youssef [338], pp. 38–48
24. Matsui, M., Nakajima, J., Moriai, S.: A description of the Camellia encryption algorithm. In: RFC 3713 (Informational) (2004)
25. NIST Special Publication 800-67-Rev1: Recommendation for the Triple Data Encryption Standard Algorithm (tdea) block cipher. National Institute of Standards and Technology (2012)
26. ETSI/SAGE Specification: Specification of the 3GPP Confidentiality and Integrity Algorithms. Document 2: Kasumi Algorithm Specification. ETSI/SAGE (2011)

27. Schneier, B.: Description of a new variable-length key, 64-bit block cipher (Blowfish). In: Anderson, R.J. (ed.) FSE. Lecture Notes in Computer Science, vol. 809, pp. 191–204. Springer (1993)
28. Matsui, M.: Linear crypto-analysis method for DES cipher. In: Tor Helleseth, editor, EUROCRYPT. Lecture Notes in Computer Science, vol. 765, pp. 386–397. Springer (1993)
29. Biham, E., Shamir, A.: Differential cryptanalysis of DES-like cryptosystems. J. Cryptol. **4**(1), 3–72 (1991)
30. Kada, A., Youlal, H.: E-financial secure solution using neural network techniques. Can. J. Electr. Comput. Eng. **30**(3), 145–148 (2005)
31. Malki, S., et al.: Vein feature extraction using DT-CNNs. In: 10th International Workshop on Cellular Neural Networks and Their Applications, Istanbul, Turkey, 28–30 Aug 2006
32. Shortt, A., et al.: Compression of optically encrypted digital holograms using artificial neural networks. J. Display Technol. **2**(4), 401–410 (2006)
33. Saeed, K., Nammous, M.: A speech-and speaker identification system: Feature extraction, description, and classification of speech-signal image. IEEE Trans. Ind. Electron. **54**(2), 887–897 (2007)
34. Moreno, F., et al.: Reconfigurable hardware architecture of a shape recognition system based on specialized tiny neural networks with online training. IEEE Trans. Ind. Electron. **56**(8), 3253–3263 (2009)
35. Eidinger, E., et al.: Age and gender estimation of unfiltered faces. IEEE Trans. Inf. Forensics Secur. **9**(12), 2170–2179 (2014)
36. Luo, H., et al.: Color Transfer Using Visual cryptography. Measurement Elsevier **51**, 81–90 (2014)
37. Roy, M., et al.: A novel approach for change detection of remotely sensed images using semi-supervised multiple classifier system. Inf. Sci. Elsevier **269**, 35–47 (2014)
38. Omer, G., et al.: Performance of SVM and artificial neural network for mapping endangered tree species using worldview-2 data in Dukuduku forest, South Africa. IEEE J. Sel. Topics Appl. Earth Obs. Remote Sens. **8**(10), 4825–4840 (2015)
39. Al-Jarrah, O., et al.: Data randomization and cluster-based partitioning for Botnet intrusion detection. IEEE Trans. Cybern. (2016)
40. Lian, S.: Image Authentication Based on Neural Network. Cornell University, CoRR, 2007 abs/0707.4524
41. Munukur, R., Gnanam, V.: Neural network based decryption for random encryption algorithms. In: 3rd International Conference on Anti-counterfeiting, Security, and Identification in Communication, pp. 603–605 (2009)
42. Joshi, S., Udupi, V., Joshi, D.: A novel neural network approach for digital image data encryption/decryption. In: Proceedings of IEEE International Conference on Power, Signals, Controls and Computation, pp. 1–4 (2012)
43. Lian, S.: A block chiper based on chaotic neural networks. Neurocomputing **72**, 1296–1301 (2009)
44. Lian, S., Chen, X.: Traceable content protection based on chaos and neural networks. Appl. Soft Comput. **11**, 4293–4301 (2011)
45. Bigdeli, N., Farid, Y., Afshar, K.: A novel image encryption/decryption scheme based on chaotic neural network. Eng. Appl. Artif. Intell. **25**, 753–765 (2012)
46. Bigdeli, N., Farid, Y., Afshar, K.: A robust hybrid method for image encryption based on hopfield neural network. Comput. Electr. Eng. **38**, 356–369 (2012)
47. Su, Z., et al.: Multimedia security: A survey of chaos based encryption technology. In: Karydis, I. (ed.) Book Chapter in Multimedia—A multidisciplinary approach to complex issues, InTech (2012)
48. Upmanyu, M., et al.: Blind authentication: A secure crypto-biometric verification protocol. IEEE Trans. Inf. Forensics Secur. **5**(2), 255–268 (2010)
49. Hassan, H., et al.: New chaotic image encryption technique. In: Symposium on Broadband Networks and Fast Internet, Baabda, pp. 103–108 (2012)

50. Xing-Yuan, W., Xue-Mei, B.: A novel image block cryptosystem based on a spatiotemporal chaotic system and a chaotic neural network. Chin. Phys. B **22**(5), 050508 (2013)
51. Chatzidakis, S., et al.: Chaotic neural networks for intelligent signal encryption. In: 5th International Conference on Information, Intelligence, Systems and Applications, Chania, pp. 100–105 (2014)
52. Mohamed, F.: A parallel block-based encryption schema for digital images using reversible cellular automata. Eng. Sci. Technol: Int J. Elsevier **17**, 85–94 (2014)
53. Ying-Qian, Z., Xing-Yuan, W.: A symmetric image encryption algorithm based on mixed linear–nonlinear coupled map lattice. Inf. Sci. Elsevier **273**, 329–351 (2014)
54. Baheti, A., et al.: Proposed method for multimedia data security using cyclic elliptic curve, chaotic system and authentication using neural network. In: 4th International Conference on Communication Systems and Network Technologies, pp. 664–668 (2014)
55. Zhou, S.: Image encryption technology research based on neural network. In: International Conference on Intelligent Transportation, Big Data & Smart City, pp. 462–465 (2015)
56. Shukla, P., et al.: Applied cryptography using chaos functions for fast digital logic-based systems in ubiquitous computing. Entropy **17**, 1387–1410 (2015)
57. Nguyen, T., Marpe, D.: Objective performance evaluation of the HEVC main still picture profile. IEEE Trans. Circuits Syst. Video Technol., 1–8 (2014). doi:10.1109/TCSVT.2014.2358000
58. Memon, Q.: A new approach to video security over networks. Int. J. Comput. Appl. Technol. **25**(1), 72–83 (2006)
59. Memon, Q.: On integration of error concealment and authentication in JPEG2000 coded images. Adv. Image Video Process. **2**(3), 26–42 (2014)
60. Yayık, A., Kutlu, Y.: Neural network based cryptography. Neural Netw. World **2**(14), 177–192 (2014)
61. Kumar, A., et al.: Application of Runge-Kutta method for the solution of non-linear partial differential equations. Appl. Math. Model. **1**(4), 199–204 (1977)
62. Rukhin, A., et al.: A statistical test suite for random and pseudorandom number generators for cryptographic application. NIST Special Publication 800-22, Revision 1a (Revised: April 2010), Bassham, L., III, 2010. http://csrc.nist.gov/rng/
63. Marsaglia, G.: DIEHARD: a battery of tests of randomness, http://www.fsu.edu/pub/diehard/
64. Walker, J.: ENT: a pseudorandom number sequence test program. http://www.fourmilab.ch/random/
65. Golomb, S.: Shift Register Sequences, Revised edn. Aegean Park, Laguna Hills, CA (1982)

# Part III
# Digital Forensic Applications

# Data Streams Processing Techniques

**Fatma Mohamed, Rasha M. Ismail, Nagwa L. Badr and Mohamed Fahmy Tolba**

**Abstract** Many modern applications in several domains such as sensor networks, financial applications, web logs and click-streams operate on continuous, unbounded, rapid, time-varying streams of data elements. These applications present new challenges that are not addressed by traditional data management techniques. For the query processing of continuous data streams, we consider in particular continuous queries which are evaluated continuously as data streams continue to arrive. The answer to a continuous query is produced over time, always reflecting the stream data seen so far. One of the most critical requirements of stream processing is fast processing. So, parallel and distributed processing would be good solutions. This paper gives (i) Analysis to the different continuous query processing techniques. (ii) A comparative study for the data streams execution environments. (iii) Finally, we propose an integrated system for processing data streams based on cloud computing which apply continuous query optimization technique on cloud environment.

**Keywords** Continuous queries optimization · Multiple plans · Query mesh · Continuous bounded queries · Continuous nearest neighbor queries · Pattern mining · Continuous Top-k queries · Continuous skyline queries · Multiple continuous queries · Map reduce · Elastic processing · Load balancing · Recourses provisioning

## 1 Introduction

Recently a new class of data-intensive applications has become widely recognized: applications in which the data are modeled best not as persistent relations but as transient data streams. However, their continuous arrival in multiple, rapid, time-varying, possibly unpredictable and unbounded streams appear to yield some

F. Mohamed (✉) · R.M. Ismail · N.L. Badr · M.F. Tolba
Faculty of Computer and Information Sciences, Ain Shams University, Cairo, Egypt
e-mail: fatma_najib1991@yahoo.com



A.E. Hassanien et al. (eds.), *Multimedia Forensics and Security*,
Intelligent Systems Reference Library 115, DOI 10.1007/978-3-319-44270-9_12

fundamentally new research problems. These applications also have inherent real-time requirements, and queries on the streaming data should be finished within their respective deadlines [1, 2]. In this context, researchers have proposed a new computing paradigm based on Stream Processing Engines (SPEs). SPEs are computing systems designed to process continuous streams of data with minimal delay. Data streams are not stored, but are processed on-the-fly using continuous queries. The latter differs from queries in traditional database systems because a continuous query is constantly "standing" over the streaming tuples and results are continuously output. In the last few years, there have been substantial advancements in the field of data stream processing. From centralized SPEs, to Distributed Stream Processing Engines (DSPEs), which distribute different queries among a cluster of nodes (interquery parallelism) or even distributing different operators of a query across different nodes (interoperator parallelism). However, some applications have reached the limits of current distributed data streaming infrastructures [3].

Because of the continuous changes in input rates, DSPSs need techniques for adjusting resources dynamically with workload changes. Making decisions when to update resource allocation in response to workload changes and how, is an important issue. Effective algorithms for elastic resource management and load balancing were proposed, where resizes the number of VMs in a DSPS deployment in response to workload demands by taking throughput measurements of each involved VM [3, 4, 5]. Thus, cloud computing has emerged as a flexible for facilitating resource management for elastic application deployments at unprecedented scale. Cloud providers offer a shared set of machines to cloud tenants, often following an Infrastructure-as-a-Service (IaaS) model. Tenants create their own virtual infrastructures on top of physical resources through virtualization. Virtual machines (VMs) then act as execution environments for applications [5].

Thus, we categorize research challenges in data streams to: (1) Continuous queries processing which focus on continuous queries optimization, how to provide real time answering of continuous queries, how to process different typed of continuous queries, and how efficiently process multiple continuous queries. (2) Data streams execution environments where different environments were proposed to execute data streams such as parallel, distributed, and cloud environments. Where, exploiting parallelism and distribution techniques to fast data streams processing, also exploiting virtualization strategies in cloud to provide elastic processing environment in response to workload demands.

The rest of the chapter is organized as follows: related background is introduced first, and then efficient processing techniques for continuous queries, including different algorithms for effective continuous query optimization are presented. Then, we present different execution environments for data streams, which include parallel, distributed, and cloud environments. Then, we present the related research issues. And then our proposed system for data streams processing over cloud computing is presented. Then future research directions are presented. Finally, we present the conclusion.

# 2 Background

## 2.1 *Data Streams*

Data streams are the data which generated from most of the recent applications such as sensor networks, real-time internet traffic analysis, and on-line financial trading. This data has a continuous, unbounded, rapid and time-varying nature rather than finite stored data sets which generated from the traditional applications. Thus the traditional database management systems (DBMSs) are not suitable with this data. And to match the data streams nature, the data stream management system (DSMS) was presented. DSMS are dealing with transient (time-evolving) rather than static data and answering persistent rather than transient queries.

## 2.2 *Data Steams Processing*

Because of the continuous and time-varying nature of the data streams, it needs real time and continuous processing. For query processing in the presence of the data streams, the continuous queries were considered. It evaluated continuously over time with the continuous arrive of the data streams. It makes sense for users to ask a query once and receive updated answers over time.

## 2.3 *Sliding Windows*

A challenging issue in processing data streams is that they are of unbounded length. Thus, storing the entire stream is impossible. To efficiently process and deal with the unbounded data streams, the sliding window model was used. In sliding window model, only the most recent N elements are used when answering the continuous queries and old data must be removed as time goes on, where N is the window size.

## 2.4 *Continuous Query Optimization*

Continuous query optimization is the concept which based on two factors, the time factor and the accuracy factor. The optimization based on the time factor focuses on how to minimize the processing time of the continuous and rapid data streams. And the optimization based on the accuracy factor focuses on providing accurate results for data streams queries.

Most of traditional database systems use single query optimizers, which select the best query plan to process the whole data based on the overall average statistics of data. However, this isn't efficient with data streams with the changeable statistics over time. Because of the disadvantages of using single plan for data streams processing, some database systems determine pre-computed query plans and provide the best execution plans for the tuples on-the-fly at runtime, as its optimization strategy.

## 3 Continuous Queries Processing

This section presents different algorithms for effective continuous query optimization, which provide real time answering with high accuracy. Also, it presents the processing for different types of continuous queries. In addition, it presents how to process multiple continuous queries using efficient methods.

### *3.1 Continuous Queries Processing Based on Multiple Query Plans*

In order to improve query performance, the query plan optimization and migration strategy based on multi-factors is generated. It cannot only chooses the optimized query plan according variable data stream character parameters, but also insures smooth migration between query plans [6].

The Query Mesh (QM) framework was proposed in [7], which compute multiple query plans, each designed for a particular subset of the data with distinct statistical properties. Traditional query optimizers select one query plan for processing all data based on overall statistics of all data. Because of data streams nature and their non-uniform distributions, selecting a single query plan provides ineffective query results. Thus, QM is a good alternative to data stream processing approaches in the literature. QM improves execution time by up to 44 % better than the state-of-art approaches, but this approach increases the memory usage for its offline work.

The semantic query optimization (SQO) approach was proposed in [8]. SQO uses dynamic substream metadata at runtime to find the best query plan for processing the incoming streams. The most important advantage of SQO approach is it depends on identifying four herald enabled semantic query optimization opportunities to reduce query execution cost. Thus, SQO doesn't need to use a cost model for applying these optimizations. Also, SQO optimization technique reduces query execution times, up to 60 %, compared to the approaches in the literature. Despite all these advantages of SQO, it is not applicable to the semantics of sensor observations especially in the case of mobile sensors.

Lim and Babu in [9] proposed Cyclops to efficiently process continuous queries. Cyclops is a continuous query processing platform that manages windowed aggregation queries in an ecosystem, which composed of multiple continuous query execution engines. Cyclops employs a cost-based approach for picking the most suitable engine and plan for executing a given query. The most important advantage of Cyclops is; it executes continuous queries using a combination selected from various execution plan and execution engine choices. Cyclops shows the cost spectrum of query execution plans across three different execution engines—Esper, Storm, and Hadoop. One of the important advantages of Cyclops is it presents an interactive visualization of the rich execution plan space of windowed aggregation queries.

Adaptive multi-route query processing (AMR) was proposed in [10], for processing stream queries in highly fluctuating environments. AMR dynamically routes the incoming tuples to operators based on up-to-date system statistics instead of process all incoming tuples with the same fixed plan. Adaptive Multi-Route Index (AMRI) was proposed, which employs a bitmap time-partitioned design. The main advantage of AMRI is it provides a balance between efficient processing of continuous data streams and reducing the index overhead. AMRI improves the overall throughput by 68 % better than the state-of-the-art approach.

Table 1 Presents comparison between four different techniques for optimizing continuous queries (QM in [7]), SQO in [8], Cyclops in [9] and AMR in [10]), where QM, Cyclops and AMR depend on cost model in their work but SQO don't need cost model because it depends on four herald enabled semantic query

**Table 1** Comparison between QM, SQO, Cyclops and AMR

| Query optimization technique | Objective | Offline work | Cost model | Assigning plan | Classifier |
|---|---|---|---|---|---|
| QM | Finding multiple query plans each for a substream | Select best query plans for training data | Use cost model to estimate cost for each plan | Online | Use classifier to assign plan to each incoming tuple |
| SQO | Finding multiple query plans each for a substream | Identify the semantic query optimization opportunities | No need to use cost model | Online | Isn't use classifier |
| Cyclops | Picking the most suitable engine and plan for the query execution | Select best plan engine pair | Use cost model to select the best pair | Online | Isn't use classifier |
| AMR | Continuous routing of tuples based on up-to-date statistics | Determine up-to-date statistics | Use cost model to find the best index configuration | Online | Isn't use classifier |

optimization opportunities; which ensure reducing query execution cost. QM needs a classifier to test the incoming tuples in runtime. However, SQO, Cyclops and AMR don't use a classifier in their work.

### 3.2 Continuous Queries with Probabilistic Guarantees

Two different techniques were proposed in [11, 12], which provide probabilistic guarantees for continuous queries answers. Sketching technique (termed ECM-sketch) was proposed in [11]. It considers the problem of complex query answering over distributed, high-dimensional data streams in the sliding-window model. It allows effective summarization of streaming data over both time-based and count-based sliding windows with probabilistic accuracy guarantees. Although the advantages of ECM-sketches, it doesn't support uncertain data streams.

Zhang and Cheng proposed the probabilistic filters protocol for probabilistic queries in [12], which consider data impreciseness and provide statistical guarantees in answers. The main advantage of this protocol that; it adapts with pervasive applications, which deploy a lot of sensor devices. It reduces the communication and energy costs of the sensor devices. Based on probabilistic tolerance, the probabilistic filters protocol; provides accurate answers for continuous queries and reduces the utilization of resources. All these advantages of probabilistic filters protocol lead to improve the overall performance.

The ECM-sketch which proposed in [11], provides answers for complex queries and the complex queries weren't handled in [12]. With regard to high-dimensional data streams processing, it handled only by ECM-sketch. ECM-sketch not suitable for sensor networks such as probabilistic filters protocol. Where, the probabilistic filters protocol; reduces the communication and energy costs of the sensor devices in sensor networks. The probabilistic filters protocol considers multiple user query requests which not considered by ECM-sketch.

### 3.3 Continuous Bounded Queries Processing

Different algorithms were proposed in [13, 14] for answering continuous bounded queries. Huang and Lin proposed in [13] the continuous Within Query-based (CWQ-based) algorithm and the continuous min–max distance bounded query (CM2DBQ) algorithm for efficiently processing of CM2DBQ. "Given a moving query object q, a minimal distance dm, and a maximal distance dM, a CM2DBQ retrieves the bounded objects whose road distances to q are within the range (dm, dM) at every time instant". Also, bounded objects updating mechanism was proposed for real time evaluation of continuous query results in the case of object updates.

Bhide and Ramamritham in [14] proposed Caicos system for efficiently processing continuous bounded queries over a data stream. The main advantage of Caicos is it tracks continuously the evolving information of a highly dynamic data streams. Also, it efficiently provides real time and up-to-date continuous queries answers. An iterative algorithm and efficient dynamic programming-based algorithm were proposed for identifying the accurate metadata that needs to be updated for providing efficient continuous results.

Caicos system which proposed in [14] considers scheduling multiple jobs on multiple parallel machines, which not considered in [13] because Caicos system runs in a multiprocessor environment. Concerning the accuracy, Caicos system depends on accurate metadata which provides accurate results, but algorithms in [13] didn't address it. Providing real time processing of distributed data streams in large road networks was handled only in [13].

### *3.4 Continuous Queries Processing Over Sliding Window Join*

Two approaches were proposed in [15, 16] for improving the processing of data streams over sliding window joins. Qian et al. in [15] proposed a hardware co-processor called Uncertain Data Window Join Special co-Processor (UWJSP) for accelerating join processing of data streams. It process window join operation over multiple uncertain data streams. The main advantage of UWJSP is it greatly promotes Uncertain Data Stream Management System (UDSMS) processing speed. Also, UWJSP provides low cost and high performance.

Kim proposed a structure for sliding window equijoins in [16] to efficiently process of data streams. They proposed an alternative hash table organization for sliding window equijoins. The most advantage of this proposed organization is it easily finds the expired tuples and discarded them; because these tuples are always grouped in the oldest hash tables. Their results ensure that the proposed method improves the overall performance of data streams processing.

UWJSP which proposed in [15], handles multiple uncertain data streams processing which not handled in [16]. Also UWJSP uses instruction sets to efficiently track the changing queries which not used in [16]. Thus, UWJSP provides high scalability and flexibility than the proposed structure in [16]. However, in [16] the proposed sliding window equijoins structure allocates a hash table for each set of arriving tuples for a sliding window interval not for each stream source, which improves the performance of the sliding window than in [16].

## 3.5 Continuous Pattern Mining

Different algorithms were proposed in [17–19] for continuous pattern mining. Dou et al. in [17] proposed indexing algorithms for efficiently providing real time answers of historical online queries (constrained aggregate queries, historical online sampling queries and pattern matching queries) in sensor networks. The answer of these queries based on flash storage of sensor devices. Index construction algorithm was proposed to create multi-resolution indexes. Also, two techniques called checkpointing and reconstruction were proposed for failure recovery in sensor devices. Their results prove the efficiency of the proposed algorithms.

Yang et al. proposed methods in [18] for the incremental detection of neighbor-based patterns, in particular, density-based clusters and distance-based outliers over sliding stream windows; because incremental computation for pattern detection continuous queries is an important challenge. Thus, the main advantage of these methods is providing real-time detecting of patterns within streams sliding window. Also they developed cost model to measure the performance of the proposed strategies.

A data structure called sliding window top-k pattern tree (SWTP-tree) was proposed in [19] for efficiently mining the frequent patterns in a sliding window. SWTP-tree scans the data streams in sliding window continuously. Also an improved FP-growth mining algorithm is proposed for generating the top-k frequent patterns based on the proposed SWTP-tree. It gets the top-k patterns in their descending frequency order.

Concerning constrained aggregate queries and historical online sampling queries; those are handled only in [17]. Only the algorithms in [18] are adaptable with traffic monitoring. The continuous pattern mining was introduced in-depth in [18, 19] than in [17].

## 3.6 Continuous Top-K Queries Processing

Li et al. proposed in [6] an approach called Voronoi Diagram based Algorithm (VDA) for efficiently answering top-n bi-chromatic reverse k-nearest neighbor (BRkNN) query. Also, they propose a method for finding the candidate region for a BRkNN query. In addition, they propose filter-refinement based algorithm to answer BRkNN query efficiently. Their results ensure that the proposed algorithms improve the overall performance for answering top-n BRkNN queries.

Sandhya and Devi describe how to efficiently answer Top-K dominating query in [20]. Top-K dominating query selects k data objects that influence the highest number of objects in a data set. Improved Event Based Algorithm (IEVA) was proposed to efficiently answer Top-K dominating query. IEVA uses event scheduling and rescheduling towards avoiding the examination of points for inclusion in TOPK.

A distributed quantile filter-based algorithm was proposed in [21] to answer top-k query in a wireless sensor network. It evaluate top-k query in wireless sensor networks to maximize the network lifetime. Also, an online algorithm was proposed for answering time-dependent top-k queries with different values of k.

Different approaches were proposed in [6, 20, 21] for answering top-k queries. The safe interval which used for getting the top-k data points computed only in [20]. This safe interval speeds up the query response time because it used for getting the top-k data points without the examination of all points for inclusion in TOPK. Also, there is no need to build the Voronoi Diagram (VD) as in [6] or to build SWTP-tree as [21]. In addition, updating the safe intervals is easier than updating nodes in [21]. In [6] it is important to build the Voronoi Diagram (VD) to answer the bi-chromatic reverse k-nearest neighbor (BRkNN) query without the need to compute the BRkNN answer set for every data point. Li et al. in [6] address the k nearest neighbors' answers which not addressed in the rest. Thus, algorithms which proposed in [6] outperform the rest in location-based applications. Quantile filter-based algorithm in [21] is very suitable for the wireless sensor network; it is the only one that addresses maximizing the wireless sensor network lifetime.

## *3.7 Continuous Nearest Neighbor Queries Processing*

Wang et al. in [22] proposed an effective filtering-and-refinement framework for efficiently answering visible k nearest neighbor (Vk NN) continuous query. "Vk NN query retrieves k objects that are visible and nearest to the query object". Two pruning algorithms called safe region based pruning and invisible time period based pruning was proposed to reduce the search space for query processing. Their results ensure that the proposed pruning techniques provide high efficiency.

Two-dimensional index structure called bit-vector R-tree (bR-tree) was proposed in [23] for efficiently processing generalized k-nearest neighbor (GkNN) queries not only for spatial data objects but also for non-spatial data objects in wireless networks. A search algorithm was proposed for efficiently pruning the search space of bR-tree. Their results proved the efficiency of bR-tree in terms of energy consumption, latency, and memory usage.

In [24] a method was proposed for efficiently answering of continuous view field nearest neighbor query processing. "Given the view field and location of a user, the view field nearest neighbor query retrieves a data object that is nearest to the user's location and falls within the user's view field". The processing of these queries occured on two phases: initial phase and update phase. Two algorithms called Naive Exploration Algorithm and Fan-shaped Exploration Algorithm were proposed for the initial phase. Also, Fan- shaped Monitoring Algorithm was proposed for of the update phase.

An approach was proposed in [25] for efficiently answering continuous range k-nearest neighbor (CRNN) query in vehicular ad hoc networks. Three algorithms were proposed called decide current valid interval, decide next split point, and find next split point; which efficiently decide the current valid interval of the results. The main advantages of this approach are minimizing network bandwidth usage, reducing the computational cost, and decreasing local memory usage.

Elmongui et al. proposed in [26] algorithms to answer continuous aggregate nearest neighbor (CANN) queries for moving objects in spatio-temporal systems. To compute CANN queries results; a holistic algorithm (H-CANN) was proposed. Also, a progressive algorithm for CANN (P-CANN) was proposed; it improves the performance of CANN query answer.

Aggregate nearest neighbor query in uncertain graphs (UG-ANN) was proposed in [27] for answering ANN queries through uncertain graphs to be suitable for such applications as social network analysis and road network monitoring. Two pruning algorithms were proposed to improve the performance of UG-ANN; called structural pruning and instance pruning.

In [28] a server-side spatial mashup framework called SMashQ was proposed to answer k-NN queries in time-dependent road networks. This framework addresses the problem of collecting real-time traffic data from vehicles or roadside sensors in road networks. The main advantage of this framework is enabling a database server to access the route and travel time information from external Web mapping services such as Google Maps. Also, greedy object grouping algorithm was proposed; to reduce the number of external web mapping requests. One of the most advantages of SMashQ is it can scale up to a large number of objects.

A two-layer index structure called virtual grid quadtree with Voronoi diagram (VGQ-Vor) to answer mobile k nearest neighbor (MkNN) queries was proposed in [29]. This structure depends on Voronoi diagram for efficient processing of MkNN queries. Also, a moving k nearest neighbor (kNN) query algorithm based on VGQ-Vor was proposed. It checks mobile objects in the neighboring regions the mobile object in the query.

Fangzhou, Guohui, Li, Xiaosong and Cong proposed methods based on Voronoi Diagrams in [30] for Uncertain Data (UV-Diagram) to answer probabilistic nearest neighbor queries of uncertain data objects via wireless data broadcast (BPNN). UV-Hilbert-Partition was proposed to partition the UV-Diagram into several grid cells. Also, organizing method was proposed; to organize UV-Diagram's cells. To address the BPNN query processing, a distributed index, named UVHilbert-DI was proposed.

Concerning data streams processing in emerging applications, only the proposed framework in [22] is suitable. Only the algorithms proposed [6, 29, 30] depend on Voronoi Diagram for answering nearest neighbor queries. BR-tree which was proposed in [23], handles spatial and non-spatial specifications of data objects which are not handled in the rest. Also, algorithms in [23, 30] are the only algorithms which address data streams processing in wireless broadcasting systems. Only the algorithms in [27, 30] consider uncertain data streams.

Only in [24] data streams processing in range based applications as augmented reality systems, tour guide systems, and CCTV-based surveillance systems were considered. The proposed algorithms in [25] outperform the rest techniques in processing data streams in vehicular ad hoc networks because the other techniques don't consider the specifications which needed for these networks. With regard to the aggregate nearest neighbor queries, it addressed only in [26, 27]. Concerning road network, it addressed only in [28].

## 3.8 Continuous Queries Processing Based Tree Structure

A query indexing structure called Query Region tree (QR-tree) was proposed in [31], for processing continuous range queries (CRQs) over moving objects. "CRQ is the query which continuously retrieves the moving objects that are currently within a given query region of interest". One of the advantages of QR-tree is it reduces the frequency when any moving object contacts the server to receive new resident domains or update the query results.

Different algorithms were proposed in [32] for efficiently answering continuous region queries of sensor networks. ECH-tree generation algorithm was proposed to reduce the energy consumption in sensor networks. Also, a time-correlated region query method was proposed to answer continuous efficiently. Their result ensures the efficiency of the proposed methods.

In [33] an overlay network topology called Virtual Tree (VT) was proposed for efficiently answering interval valid aggregation queries in large scale dynamic networks. Also, distributed query protocol called overlay management protocol (OMP) was proposed to guarantee connectivity and path stability in VT.

Merkle Skyline R-tree and Partial S4-tree were proposed in [34] for efficiently answering location-based arbitrary-subspace skyline queries (LASQs). One of the main advantages of LASQs is it considers spatial and non-spatial attributes. Prefetching-based approach was proposed to provide efficient one-shot authentication to LASQs in outsourced databases and to reduce the server processing time.

In [21, 23, 31–34] different tree based solutions were proposed. In [23, 31, 34] non-spatial specifications of data objects were handled; but not handled in the rest. Also, only the techniques proposed in [23, 31, 32, 34] are suitable for location-based services. BR-tree in [23] is the only adaptable one with wireless broadcasting systems. And, data streams processing in peer-to-peer systems were addressed only in [33]. Techniques in [33] are the best of them to answer aggregation queries. Concerning query authentication, it addressed only by [34]. Only the proposed structure in [21] considers patterns mining.

## 3.9 Continuous Skyline Queries Processing

Nagendra and Candan show in [35] how skyline-window-join (SWJ) queries over pairs of data streams are computed. Given a set of data objects, the skyline query returns the objects that are not dominated by others. A Layered Skyline-window-Join (LSJ) was proposed, where LSJ operator partitions the overall process into processing layers. Also, LSJ outperforms the existing approaches which are not designed to eliminate redundant work across multiple processing layers.

A two-phase approach for continuous skyline monitoring in two-tier streaming settings was proposed in [36]. In the initialization phase, the initial query result is obtained by correctly merging local skyline from all data sites. Then, all data tuples are categorized with respect to their membership in local skyline and global skyline. One of the most advantages of this approach is it minimizes the bandwidth consumption between the server and the data sites.

The k-Skyband algorithm was proposed in [37] for efficiently answering k-skyband (kSB) query on incomplete data. kSB query differ from the traditional skyline query; where some dimensional values are missing. This method based on three concepts called expired skyline, shadow skyline, and thickness warehouse to improve its results. Also, constrained skyline (CS) and group-by skyline (GBS) queries over incomplete data were handled.

Huang, Chang and Lee proposed algorithms in [38] for efficiently answering two distance-based skyline queries called Continuous de-skyline query (Cde-SQ) and continuous k nearest neighbor-skyline query (Cknn-SQ) in road networks. For efficiently answering Cde-SQ; Cde-SQ and Cde-SQ + algorithms were proposed. And, for efficiently answering Cknn-SQ; Cknn -SQ and Cknn -SQ + algorithms were proposed.

In [39] index-based (I-SKY) and non-index-based (N-SKY) algorithms were proposed for efficiently answering range-based skyline queries. Also, incremental versions of the two algorithms were proposed to avoid re-computing the query results. The main advantage of them is saving computation cost for highly dynamic data streams. In addition, efficient methods were proposed for computing the valid scope of each skyline object.

A sliding window skyline model was proposed in [40] for efficiently answering probabilistic skyline query over uncertain data streams. Also, candidate list approach was proposed to efficiently determine the candidate skylines; which may be skylines in the future. In addition, enhanced refinement strategy was proposed to reduce the computation cost and provide more accurate results.

A filtering algorithm called FSKY was proposed in [41] to efficiently answer skyline queries for anti-correlated and clustered databases in sensor networks. Also, they proposed a scheme for data cluster representation. In addition, a sampling method was proposed for reducing the communication cost and save energy. Although the advantages of the proposed algorithms, approximate skyline and subspace skyline queries not handled.

In [35] SWJ handles continuous skyline monitoring over multiple data streams, which not handled in all of the previously mentioned skyline queries processing techniques. Also, layering the sliding windows which eliminate redundant work across consecutive windows was considered only in [35].

Only the proposed algorithms in [36, 37, 41] address skyline monitoring in distributed environment such as a wide-area sensor network. Also, the two-tier streaming processing was handled in [36, 41] but not handled in the rest. With regard to skylines computation over incomplete data, constrained skyline and group-by skyline; those were handled only in [37]. Answering skyline queries for anti-correlated and clustered databases was handled only in [41].

Only algorithms proposed in [38] are suitable for road network. The uncertain data processing was treated only in [40]. Also, the probabilistic skyline computing was considered only in [39, 40]. Algorithms in [39] are the suitable algorithms for location-based services. Also, both spatial and non-spatial attributes of data were addressed only in [39].

## 3.10 Multi Continuous Queries Processing

To provide practical solutions for matching highly dynamic data streams with multiple and dynamically-updated continuous queries, a stream processing system should support incremental evaluation over new data and support query optimization for continuous queries including computation sharing among multiple queries. Multi-query optimization has been shown to be a powerful technique for improving the efficiency of query processing. By, exploiting the overlap between queries in terms of shared operators and streams [42].

Ray et al. in [43] showed how multiple queries can be represented in the form of an operator tree, such that their commonalities can be easily exploited for multi-query plan generation. Operator tree shows how the plans from different queries can be merged. Multiple queries on same or related data streams may share their processing to alleviate the processing, storage, and communication costs.

Continuous queries may be complex, involving various operators, not all these operators are supported at every node in operator tree. Consider the blocking operators, like join or count. These operators require a set of stream tuples to be stored before the operations can be applied. Some nodes may not have enough storage capability for processing these operations; this is an important problem that faces operator tree.

The field Programmable Gate Arrays (FPGAs) based real-time data analytics platform was proposed [44]. An FPGA is especially powerful in exploiting parallelism because any form of parallel execution can be directly mapped to logic circuits in hardware. In addition, FPGAs can meet the required elasticity in scaling out to meet increasing throughput demands. FPGA exploits the overlapping components among given query plans to further improve the resource utilization and generate global Select-Project-Join (SPJ) query plan. Thus, FPGA outperforms operator tree [43] when dealing with join operators.

Chen et al. show how to provide real-time response for multiple top-k queries in [45]. Sharing the results of queries is a key factor in saving the computation cost and providing real-time response. They based on the frequency, which specifies the upper bounds on the re-execution intervals of queries; in sharing results of multiple top-k queries.

In [46] the adaptive Sharing-based Extended Greedy Optimization Approach (A-SEGO) was proposed. A-SEGO can be used to produce an optimized global

**Table 2** Comparison between operator tree, FPGA, A-SEGO, and frequency based Top-K optimization

| Optimization Method | Query type | Sharing strategy | Output | Cost based |
|---|---|---|---|---|
| Operator tree | Select-project queries | Share queries processing | Single global query plan | Non-based |
| FPGA | Select-project-join queries | Share queries processing | Single global query plan | Non-based |
| A-SEGO | Multi-way join queries | Share common join operations results | Single global query plan | Based |
| Frequency based | Top-K queries | Share queries intermediate results | Single global query plan | Non-based |

execution plan for multiple continuous queries. A-SEGO uses a cost model to determine if the current optimized global plan becomes not efficient. When the current plan is no longer efficient, A-SEGO generate a newly optimized plan in a timely manner.

In [43–46] four techniques for multiple query optimization based sharing (operator tree, FPGA, A-SEGO and Frequency based Top-K). Table 2 presents comparison between the four techniques. Operator tree, FPGA and A-SEGO share the processing of multiple queries, but FPGA and A-SEGO outperforms operator tree when dealing with join operators. Frequency based Share intermediate results of Top-K queries which have the same frequency. A-SEGO depends on cost model but the rest of them not depend on cost model.

FAst Skyline compuTation for multiple queries (FAST) was proposed in [47], for processing multiple continuous skyline queries over a data stream. FAST uses a filtering technique that can early discard an object that will not be a member of any future skyline of continuous queries, and uses a discriminant that can efficiently determine which objects in memory are skyline objects for which queries.

# 4 Data Streams Execution Environments

This section presents different execution environments for data streams, which achieve high performance and low-latency. Also, presents elastic stream processing algorithms and dynamic load balancing techniques.

## *4.1 Parallel and Distributed Stream Processing*

One of the most critical requirements of query processing in data stream systems is fast processing. So, parallel processing of queries using multiple processing units would be a good solution. Achieving data parallelism through multiple processing nodes requires partitioning the input stream of an operator into distinct and independent sub-streams for nodes to process concurrently [48, 49].

Also distributed stream processing systems must function efficiently for data streams that fluctuate in their arrival rates and data distributions. In a Distributed Data Stream Management System (DDSMS), the task of data stream processing is distributed on several processing nodes, and then results are assembled by cooperating of these nodes. Thus, DDSMS achieve higher performance, in terms of supported data stream rates, and better scalability, and the number of concurrent queries [42, 50].

### 4.1.1 Data Streams Processing Based on Map Reduce Framework

The MapReduce framework has been introduced as a scalable and fault-tolerant data processing framework that enables the processing of a massive volume of data in parallel on clusters of horizontally scalable commodity machines. However, it is not adequate for supporting real-time stream processing tasks [51, 52].

Main-Memory MapReduce (M3) was introduced in [51] as a framework for parallel computing in which, continuous queries over streams of data can be efficiently answered with high scalability, performance and short response time. M3 extends Hadoop where, M3 supports continuous execution of the Map and Reduce phases where individual Mappers and Reducers never terminate.

MapUpdate framework was proposed in [53] for parallel and distributed processing of data streams. MapUpdate is an extension of MapReduce framework. MapUpdate enables the developer to write a few functions and these functions are automatically executed over a cluster of machines. Also, it achieves low latency and high scalability. The main drawback of MapUpdate is it doesn't efficiently provide dynamic partitioning of the incoming data streams.

### 4.1.2 Data Streams Processing Based on the Query Mega Graph (QMG)

Safaei et al. in [48] introduced Dynamic Tuple Routing (DTR) algorithm for parallel processing of continuous queries in a multiprocessing environment. In DTR, each input data stream tuple, is processed through the shortest path in Query Mega Graph (QMG) [54].

Table 3 presents comparison between M3, MapUpdate and QMG frameworks in [48, 51, 53]. The three frameworks provide parallel processing for data streams and efficiently answered queries with high scalability, performance and short response time. M3 and MapUpdate extend MapReduce framework to process continuous streams. But QMG process data streams over the Query Mega Graph.

**Table 3** Comparison between M3 and QMG

| Framework | Tuples execution | Dividing tuples decisions over execution units | Number of processing units for each tuple | Routing through processing |
|---|---|---|---|---|
| M3 | Through group of mappers and reducers | Based on rate split | One mapper and one reducer | Through RMI |
| Map Update | Through group of mappers and updaters | Using deterministic tie-breaking procedure | One mapper and one updater | Through Deterministic tie-breaking procedure |
| QMG | Through K logical machines | Based on cost model | One for each step | Through processing node itself |

### 4.1.3 Data Streams Processing Based on Task Graph Structure

Two replication methodologies called Data Parallel Replication Mechanism (DPRM) and Task Copy Replication Mechanism (TCRM) were proposed in [55] to efficiently improve the throughput of parallel and distributed data streams processing. The main advantage of DPRM is applying concurrence processing on different pieces of the same, which leading to reducing processing computation time and decreasing the latency.

Ajwani et al. in [56] proposed a graph based framework for generating the task graphs for processing data streams. Graph generation techniques were proposed to generate directed graphs with a specified degree distribution. One of the most important advantages of these techniques is the correct generation of the undirected graphs.

In [55] data and task parallelism were considered to improve the throughput; but not considered in [56]. Also, task replication was handled only by [55]. But, the proposed techniques in [56] are more suitable for generating large streaming task-graphs than those in [55] because they generate it in a reasonable time.

### 4.1.4 Data Streams Processing in Traffic Distributed Networks

Anceaume and Busnel in [57] proposed AnKLe distributed algorithm to efficiently process data streams in distributed systems. It does online estimation of the similarity between observed data streams and expected ones, to detect in real time the presence of intrusions in network traffic. One of the most advantages of AnKLe distributed algorithm is estimation results with guaranteed error bounds and with little capacities in terms of storage and processing.

A framework is built on Discretized Streams was proposed in [58] to provide scalable traffic estimation for large scale data streams. Thus, an online Expectation Maximization (EM) algorithm was proposed. The main advantage of EM algorithm is; it is efficiently computes travel time distributions of traffic by incremental online updates. Also, it validated with a large dataset of GPS traces. In addition, this algorithm can scale to very large road networks and can update traffic state in a few seconds.

In [57, 58] two different frameworks were proposed for data streams processing in distributed systems. Both of them are suitable for traffic networks, but EM which proposed in [58] is suitable for estimating traffic on a very large city network. AnKLe in [57] provides more accurate results because its approximations with guaranteed error bounds. Also, it decreases the memory usage.

### 4.1.5 Controlling Streams Processing on Overload Conditions

Shan et al. in [50] proposed the improved Pair-Wise Algorithm to adjust the load of processing nodes in DDSMS and balance the load among all the computing nodes. The Pair-Wise Algorithm is composed of two main steps: (1) Initial distribution (2) Improved load balancing. Based on the initial distribution algorithm, an improved Pair-Wise algorithm is used to adjust the load of processing nodes.

In [59] Robust Load Distribution (RLD) was proposed to provide robust query processing performance in distributed processing environments. RLD provides ϙ-optimal query performance under load fluctuations without suffering from the performance penalty caused by load migration.

A co-scheduling framework based on fuzzy logic was proposed in [60] for processing data streams. A dynamic control method is proposed, the main advantage of it is avoiding resource shortage as well as overprovision. In this control method, fuzzy logic control is applied, where CPU can be co-scheduled and co-allocated for data streams processing. Also, iterative algorithm is adopted to iteratively allocate bandwidth by closely watching processing and storage status.

A streaming warehouse model was proposed in [61] to update scheduling in real-time data streams warehouses. This model combines the main features of traditional data warehouses, and real time and continuous processing of data streams systems. To solve the streaming warehouse update problem, scheduling algorithm was proposed.

In [62] autopipelining solution for Data Stream Processing was proposed. It takes advantage of multicore processors to improve throughput of streaming applications, in an effective and transparent way. This solution is effective in the sense that it provides good utilization of resources by dynamically finding and exploiting sources of pipeline parallelism in streaming applications. Autopipelining used a base optimization algorithm to provide good utilization of resources.

Cho et al. proposed a scheduling algorithm named power and deadline-aware multicore scheduling (PDAMS) algorithm in [63] for real-time processing in multicore systems. One of the advantages of PDAMS is workload balancing and energy saving of multicore systems. In addition, each processor core In PDAMS, can manage itself. Although all of these advantages, PDAMS didn't consider both of static power consumption, the overhead of scaling voltage and frequency.

In [50, 59–63] different algorithms were proposed to avoid the system failure and the decrease performance occurred as a result of overload condition. All of them except RLD depend on dynamic updating and scheduling for real time workload balancing. But, RLD depends on providing robust physical solution as its control methodology. Also, RLD is only one which process data streams over multiple logical plans. With regard to the efficient parallel processing for good resources utilization, it was handled only in [62, 63].

## 4.2 Data Streams Processing Based on Cloud Environment

Cloud computing offers an elastic infrastructure that distributed stream processing systems (DSPSs) can use to obtain resources on-demand, but an open problem is to provide a scalable and elastic stream processing engine for processing large data stream volumes. In [64, 65] a transparent query parallelism technique was mentioned "StreamCloud", which its elastic protocols exhibit low intrusiveness, and enabling effective adjustment of resources to the incoming load. In StreamCloud, users express regular continuous queries that are automatically parallelized.

Cervino et al. propose an adaptive algorithm in [4] that resizes the number of VMs in a DSPS deployment in response to input streams rates, called "Adaptive Cloud Stream Processing". It maintains low latency with a given throughput while keeping VMs operating to their maximum processing capacity. The Adaptive Cloud Stream Processing algorithm is invoked periodically and calculates the new number of VMs that are needed to support the current workload demand.

Fernandez et al. in [5] describe an integrated approach for stateful operator scale out and recovery called "Fault-Tolerant Scale Out". It scales out the number of cloud-hosted machines on demand, parallelizing operators when the workload increases and efficiently recover resource failure. In Fault-Tolerant Scale Out algorithm, to support dynamic scale out, they add a bottleneck detector that based on system statistics to identify the bottleneck operators in the query. To support dynamic Fault-Tolerant, they add a failure detector to recover a failed operator.

In [66] they develop a Complex Event Processing (CEP)-based resource monitoring framework for data stream processing on cloud, which continuously monitor resource utilization, manage, and adjust these resources in real-time to meet the SLAs while not overprovisioning resources. It controls hosts' status automatically according to the user-specified rules and metrics such as increasing or decreasing the amount of memory, number of cores, and disk resources for VMs (scaling up and down).

The auto-scaling techniques were proposed in [67] for Elastic Data Stream Processing. Auto-scaling techniques solve the problem of finding the right time to scale in or to scale out resources, which is one of the major challenges for the elastic systems. The goal of the auto-scaling techniques is to maximize the system utilization and at the same time to guarantee a low end to end latency.

Hu et al. in [68] proposed multi-step-ahead load forecasting method to adjust number of resources on demand; in data streams processing on cloud computing. Because of the complex and dynamic characteristics of cloud computing, this method based on statistical learning technology and support vector regression (SVR) to provide efficient solution which adaptable with cloud computing.

In [69] an autonomic cloud bursting approach was proposed for optimizing ordered throughput for near real-time, data-intensive, independent computations. They proposed three scheduling heuristics as part of the autonomic cloud bursting approach. These scheduling heuristics optimize ordered throughput although the changes on workload characteristics, variation in bandwidth, and available resources. Based on workload characteristics the autonomic cloud bursting approach can provision appropriate number of resources in EC, because it determines the number of instances which can be optimally utilized in the external cloud. The main advantage of using cloud bursting is the efficient execution of online data streams workloads across multiple clouds.

An elastic auto-parallelization solution for data stream processing was proposed in [70]. It dynamically adjusts the number of channels used for processing, to achieve high throughput without unnecessarily wasting resources. The auto-parallelization solution uses control algorithm, which periodically re-evaluates the number of channels to be used based on local run-time metrics it maintains.

In [71] Cloud MapReduce (CMR) was proposed. CMR overcomes the limitations of traditional MapReduce where, it supports data stream processing. Also, CMR uses pipeline between Map and Reduce phases. This Pipelined MapReduce approach leads to increase parallelism between the Map and Reduce phases. In addition, CMR takes the advantages of cloud computing. CMR supports flexible pricing using Amazon Cloud's spot instances.

In [51, 53, 71] three extension frameworks of MapReduce framework were proposed. Table 4, presents comparison between M3, Map Update and CMR frameworks, where the three frameworks extend MapReduce framework to support streams processing. Furthermore, CMR takes the advantages of cloud computing.

An information flow control model was proposed in [72] for securely and efficiently processing data streams. Also, they share the processing of multiple queries on the cloud using the operator tree [43]. The main advantage of this model is it protects the information against unauthorized disclosure and modification and

**Table 4** Comparison between M3 and CMR

| Framework | Supporting streams processing | Basic environment | Cloud based | Connection between map and reduce phases |
|---|---|---|---|---|
| M3 | Yes | MapReduce framework | Non-Cloud based | Through RMI |
| Map Update | Yes | MapReduce framework | Non-Cloud based | Through deterministic tie-breaking procedure |
| CMR | Yes | MapReduce framework | Cloud based | Through pipelining model |

prevents the leakage of information across organizations which process their data based on cloud computing.

All of the previously mentioned cloud based frameworks except information flow control model which proposed in [72] consider cloud resources provisioning on demand. However, information flow control model outperforms the rest with regard to the security and processing multiple queries as it is the only one which handles these issues.

Only the frameworks in [17, 64, 65, 71] handled parallel processing of data streams on cloud computing. The proposed approach in [5] is the only one which addressed fault recovery based on resources provisioning. Method in [68] is the only one which based on SVR to be more adaptable with the cloud computing environment. Also, Complex Event Processing was considered only in [66].

## 5 Research Issues

Based on our survey on the different environments for executing data stream, we conclude that the best environment for executing data stream is the cloud environment, where it provides great elasticity and high performance than other environments. Cloud computing can scale the number of processing virtual machines (up/down) on demand based on the continuous change in the input rate of data streams. It provides continuous adaptation with the incoming workload where it efficiently detects the overload conditions and then scale up the number of processing resources (virtual machines). Also, it detects the reduction of the streams input rate and then scale down the number of processing resources (virtual machines). There are many techniques were proposed to manage the continuous and changeable streams workload and provision the processing virtual machines based on this changeable workload. The cloud computing provides real time load balancing over the processing resources. Also, there are many research on how to efficiently recover failure occurred during data streams processing on the cloud environment. Cloud computing has additional solution over other environments for failure recovery, where it can scale up the number of processing resources and route the workload from the damaged virtual machines to the new ones. Also, multiple continuous queries are efficiently processed on the cloud computing. In addition, there are many proposed parallelization techniques on the cloud environment, where it provides parallel and distributed processing for data streams. Also Cloud computing provides great security for streams processing. For all these reasons the cloud environment is the best environment for executing data stream.

In Table 5 We present a comparison between cloud based [4, 5, 64–71] and non-cloud based [48, 50, 51, 59] execution environments.

**Table 5** Comparison between cloud based and non-cloud based environments

| Environment type | Methodologies | Overload controlling | Elasticity | Overload effect on data | Overall Performance |
|---|---|---|---|---|---|
| Cloud based | (1) StreamCloud method (2) Fault-Tolerant Scale Out algorithm (3) Adaptive Provisioning algorithm (4) CMR (5) CEP-based resource monitoring (6) multi-step-ahead load forecasting (7) Auto-scaling | Scaling the number of VMs | Elastic environment | No affection (data is processed on provisioned VMs) | Higher performance and higher accuracy |
| Non-Cloud based | (1) M3 method (2) DTR algorithm (3) RLD algorithm (4) Improved Pair-Wise Algorithm | Load rebalancing | Non-Elastic environment | Data is lost | Lower performance and lower accuracy |

# 6 Solutions and Recommendations

From the previous sections it is clear that real time answering for data streams is a very important issue. And, all of the existing systems give solutions for data stream processing from one perspective. Some systems focus on the optimization techniques used regardless of the processing environment. And other systems only focus on improving the processing environment. Thus, we propose the optimized cloud query mesh system which based on the idea of the query mesh (QM) solution for data streams processing. The basic idea of the QM solution is data streams processing over multiple query plans. Our proposed system solves the problems of existing systems (improving streams processing from one perspective), where it combines the two improvements' viewpoints. It applies continuous query optimization technique on the cloud environment. It takes the advantage of cloud processing environment where it exploits the virtualization of cloud computing to process streams in elastic and scalable way. Also, it applies continuous query optimization technique which not applied before in the cloud, where it processes the data streams over multiple query plans each plan is suitable for subset of data with the same statistics instead of using a single query plan to execute the whole data based on the average statistics of the whole data streams.

We divide our system to two subsystems, each subsystem responsible for some functions. The first subsystem is the Continuous Query Optimization Sub-System, which is our offline phase. It consists of two main blocks continuous query optimizer and operator's distributer. The continuous query optimizer is responsible for getting set of query plans each plan for a subset of data with distinct statistical properties. And, operator's distributer is responsible for generating physical plan which suitable to all query plans generated by the query optimizer. The second subsystem is the Streams Cloud-Based Execution Sub-System, which is our online phase. It consists of four main blocks (the input manager, the execution machines, the observer and the global manager). The input manager is responsible for managing the incoming tuples and assigning the suitable plan to each tuple. And, the execution machines which execute each tuple based on the assigned plan. And the observer is responsible for detecting the overload conditions. And the global manager is responsible for provisioning the demanded VMs in the case of overload conditions (Fig. 1).

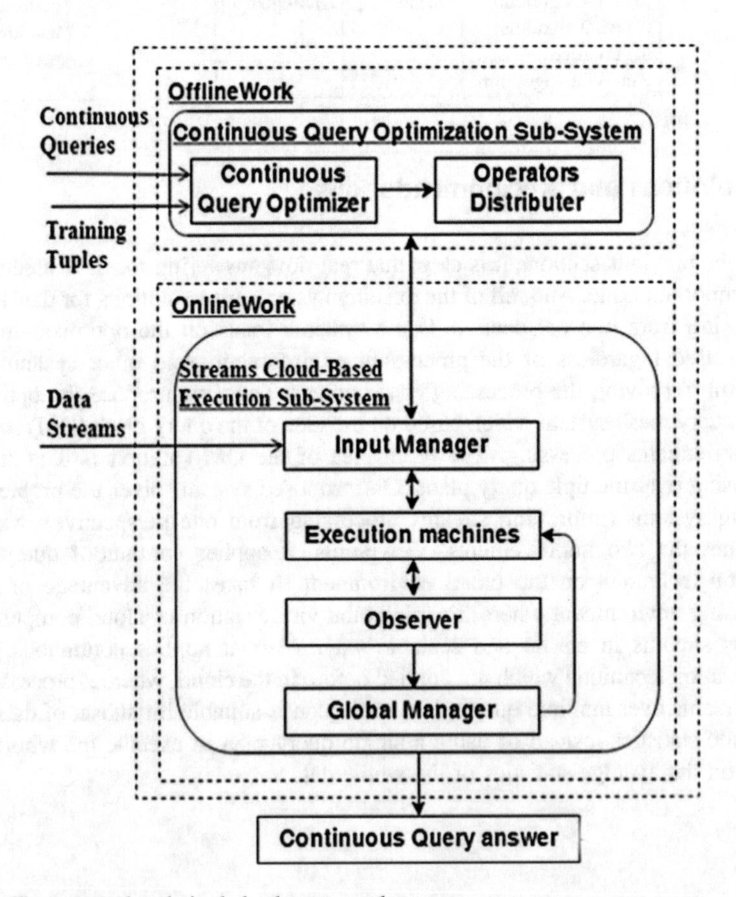


**Fig. 1** The proposed optimized cloud query mesh system

## 7 Future Research Directions

Most of researches on data stream processing in the literature focus on improving the performance either by proposing new query optimization techniques without any interest to the processing environment or by only improving the processing environment. However, there are many challenges on data stream processing such as the execution time, memory usage and online overheads. These challenges aren't handled efficiently by focusing on only one viewpoint of improvement.

Thus the future improvements should be in the two perspectives. And to provide real time processing of data streams, there is a need to future research which proposes hybrid techniques. These techniques should focus on the two viewpoints improvement. Also, data streams should be processed on elastic environment to adapt with the changeable nature of the data streams over time. Thus, we will get real time and more accurate answers for the continuous queries.

## 8 Conclusion

Many research challenging issues are appeared when processing data streams because real time answering needed for continuous queries. Where, the ability to provide real time answering is an important consideration in data stream processing because of its changeable nature. Also, accuracy of results and reducing data loss are important issues. Many recent researches focus on providing frameworks and proposing new techniques for data streams processing. Also, different environments were proposed for executing continuous queries. In this paper we present a survey on recent algorithms for data streams processing and Continuous queries optimization. Also, present techniques for efficiently process multiple continuous queries. In addition, we present different execution environments for data streams. We show the recent work for parallel and distributed processing data streams. Also, present novel techniques for cloud based stream processing. Based on the research challenges on data streams, we propose system for continuous query optimization based on the cloud computing as our future work. This system will provide real time answers for continuous queries. Also, it will provide elastic scaling of cloud resources based on streams' input rates.

## References

1. Kapitanova, K., Son, S.H., Kang, W., Kim, W.T.: Modeling and analyzing real-time data streams. In: Proceedings of International Symposium on Object/Component/Service-Oriented Real-Time Distributed Computing, pp. 91–98. IEEE (2011)
2. Lijie, Z., Yaxuan, W.: Query plan optimization and migration strategy over data stream. In: Proceedings of International forum on Information Technology and Applications, vol. 3, pp. 72–75. IEEE (2010)

3. Gulisano, V., Jimenez-Peris, R., Patino-Martinez, M., Soriente, C., Valduriez, P.: Streamcloud: an elastic and scalable data streaming system. IEEE Trans. Parallel Distrib. Syst. (TPDS) **23**(12), 2351–2365 (2012)
4. Cervino, J., Kalyvianaki, E., Salvachua, J., Pietzuch, P.: Adaptive provisioning of stream processing systems in the cloud. In: Proceedings of Data Engineering Workshops, pp. 295–301. IEEE (2012)
5. Castro Fernandez, R., Migliavacca, M., Kalyvianaki, E., Pietzuch, P.: Integrating scale out and fault tolerance in stream processing using operator state management. In: Proceedings of the 2013 International Conference on Management of data, pp. 725–736. ACM (2013)
6. Li, C.L., Wang, E.T., Huang, G.J., Chen, A.L.: Top-n query processing in spatial databases considering bi-chromatic reverse k-nearest neighbors. J. Inf. Syst. **42**, 123–138 (2014)
7. Nehme, R.V., Works, K., Lei, C., Rundensteiner, E.A., Bertino, E.: Multi-route query processing and optimization. J. Comput. Syst. Sci. **79**(3), 312–329 (2013)
8. Ding, L., Works, K., Rundensteiner, E.A.: Semantic stream query optimization exploiting dynamic metadata. In: Proceedings of IEEE Conference on Data Engineering (ICDE), pp. 111–122. IEEE (2011)
9. Lim, H., Babu, S.: Execution and optimization of continuous queries with cyclops. In: Proceedings of the 2013 international conference on Management of data, pp. 1069–1072. ACM (2013)
10. Works, K., Rundensteiner, E.A., Agu, E.: Optimizing adaptive multi-route query processing via time-partitioned indices. J. Comput. Syst. Sci. **79**(3), 330–348 (2013)
11. Papapetrou, O., Garofalakis, M., Deligiannakis, A.: Sketch-based querying of distributed sliding-window data streams. Proc. VLDB Endow. **5**(10), 992–1003 (2012)
12. Zhang, Y., Cheng, R.: Probabilistic filters: A stream protocol for continuous probabilistic queries. J. Inf. Syst. **38**(1), 132–154 (2013)
13. Huang, Y.K., Lin, L.F.: Efficient processing of continuous min–max distance bounded query with updates in road networks. J. Inform. Sci. **278**, 187–205 (2014)
14. Bhide, M., Ramamritham, K.: Category-based infidelity bounded queries over unstructured data streams. Knowl. Data Eng. IEEE Trans. **25**(11), 2448–2462 (2013)
15. Qian, J., Li, Y., Wang, Y., Chen, H., Dong, Y.: An embedded co-processor for accelerating window joins over uncertain data streams. J. Microprocess. Microsyst. **36**(6), 489–504 (2012)
16. Kim, H.G.: A structure for sliding window equijoins in data stream processing. In: 2013 IEEE 16th International Conference on Proceedings of the Computational Science and Engineering (CSE), pp. 100–103. IEEE (2013)
17. Dou, A., Lin, S., Kalogeraki, V., Gunopulos, D.: Supporting historic queries in sensor networks with flash storage. J. Inf. Syst. **39**, 217–232 (2014)
18. Yang, D., Rundensteiner, E.A., Ward, M.O.: Mining neighbor-based patterns in data streams. J. Inf. Syst. **38**(3), 331–350 (2013)
19. Chen, H.: Mining top-k frequent patterns over data streams sliding window. J. Intell. Inf. Syst. **42**(1), 111–131 (2014)
20. Sandhya, G., Devi, S.K.: An adaptive sliding window based continuous Top-K dominating queries. In: 2013 7th International Conference on Proceedings of the Intelligent Systems and Control (ISCO), pp. 349–353. IEEE (2013)
21. Chen, B., Liang, W., Yu, J.X.: Energy-efficient top-k query evaluation and maintenance in wireless sensor networks. J. Wireless Netw. **20**(4), 591–610 (2014)
22. Wang, Y., Zhang, R., Xu, C., Qi, J., YuGu, GeYu.: Continuous visible k nearest neighbor query on moving objects. J. Inf. Syst. **44**, 1–21 (2014)
23. Jung, H., Chung, Y.D., Liu, L.: Processing generalized k-nearest neighbor queries on a wireless broadcast stream. J. Inform. Sci. **188**, 64–79 (2012)
24. Yi, S., Ryu, H., Son, J., Chung, Y.D.: View field nearest neighbor: a novel type of spatial queries. J. Inform. Sci. **275**, 68–82 (2014)
25. Cho, H.J.: Continuous range k-nearest neighbor queries in vehicular ad hoc networks. J. Syst. Softw. **86**(5), 1323–1332 (2013)

26. Elmongui, H.G., Mokbel, M.F., Aref, W.G.: Continuous aggregate nearest neighbor queries. J. GeoInformatica **17**(1), 63–95 (2013)
27. Liu, Z., Wang, C., Wang, J.: Aggregate nearest neighbor queries in uncertain graphs. J. World Wide Web **17**(1), 161–188 (2014)
28. Zhang, D., Chow, C.Y., Li, Q., Zhang, X., Xu, Y.: SMashQ: spatial mashup framework for k-NN queries in time-dependent road networks. J. Distrib. Parallel Databases **31**(2), 259–287 (2013)
29. Wang, B., Qu, J., Wang, X., Wang, G., Kitsuregawa, M.: VGQ-Vor: extending virtual grid quadtree with Voronoi diagram for mobile k nearest neighbor queries over mobile objects. J. Frontiers Comput. Sci. **7**(1), 44–54 (2013)
30. Fangzhou, Z., Guohui, L., Li, L., Xiaosong, Z., Cong, Z.: Probabilistic nearest neighbor queries of uncertain data via wireless data broadcast. J. Peer-to-Peer Netw. Appl. **6**(4), 363–379 (2013)
31. Jung, H., Kim, Y.S., Chung, Y.D.: QR-tree: an efficient and scalable method for evaluation of continuous range queries. J. Inform. Sci. **274**, 156–176 (2014)
32. Tang, J., Zhou, Z., Niu, J., Wang, Q.: An energy efficient hierarchical clustering index tree for facilitating time-correlated region queries in the Internet of Things. J. Netw. Comput. Appl. **40**, 1–11 (2014)
33. Baldoni, R., Bonomi, S., Cerocchi, A., Querzoni, L.: Virtual Tree: a robust architecture for interval valid queries in dynamic distributed systems. J. Parallel Distrib. Comput. **73**(8), 1135–1145 (2013)
34. Lin, X., Xu, J., Hu, H., Lee, W.: Authenticating location-based skyline queries in arbitrary subspaces. IEEE Trans. Knowl. Data Eng. **26**(6) (2013)
35. Nagendra, M., Candan, K.S.: Layered processing of skyline-window-join (SWJ) queries using iteration-fabric. In: 2013 IEEE 29th International Conference on Proceedings of the Data Engineering (ICDE), pp. 985–996. IEEE (2013)
36. Lu, H., Zhou, Y., Haustad, J.: Efficient and scalable continuous skyline monitoring in two-tier streaming settings. J. Inf. Syst. **38**(1), 68–81 (2013)
37. Gao, Y., Miao, X., Cui, H., Chen, G., Li, Q.: Processing k-skyband, constrained skyline, and group-by skyline queries on incomplete data. J. Expert Syst. Appl. **41**(10), 4959–4974 (2014)
38. Huang, Y.K., Chang, C.H., Lee, C.: Continuous distance-based skyline queries in road networks. J. Inf. Syst. **37**(7), 611–633 (2012)
39. Lin, X., Xu, J., Hu, H.: Range-based skyline queries in mobile environments. IEEE Trans. Knowl. Data Eng. **25**(4), 835–849 (2013)
40. Ding, X., Lian, X., Chen, L., Jin, H.: Continuous monitoring of skylines over uncertain data streams. J. Inf. Sci. **184**(1), 196–214 (2012)
41. Yin, B., Lin, Y., Yu, J., Luo, Q.: Energy-efficient filtering for skyline queries in cluster-based sensor networks. J. Comput. Electr. Eng. **40**(2), 350–366 (2014)
42. Kalyvianaki, E., Wiesemann, W., Vu, Q.H., Kuhn, D., Pietzuch, P.: SQPR: stream query planning with reuse. In: 2011 IEEE 27th International Conference on Proceedings of the Data Engineering (ICDE), pp. 840–851,), Apr 2011. IEEE (2011)
43. Ray, I., Madria, S.K., Linderman, M.: Query plan execution in a heterogeneous stream management system for situational awareness. In: Proceedings of Reliable Distributed Systems (SRDS), pp. 424–429, Oct 2012
44. Sadoghi, M., Javed, R., Tarafdar, N., Singh, H., Palaniappan, R., Jacobsen, H.A.: Multi-query stream processing on fpgas. In: 2012 IEEE 28th International Conference on Proceedings of Data Engineering (ICDE), pp. 1229–1232, Apr 2012. IEEE (2012)
45. Chen, T., Chen, L., Ozsu, M.T., Xiao, N.: Optimizing multi-top-k queries over uncertain data streams. J. Knowl. Data Eng. IEEE Trans. **25**(8), 1814–1829 (2013)
46. Park, H.K., Lee, W.S.: Adaptive optimization for multiple continuous queries. J. Data Knowl. Eng. **71**(1), 29–46 (2012)
47. Lee, Y.W., Lee, K.Y., Kim, M.H.: Efficient processing of multiple continuous skyline queries over a data stream. J. Inform. Sci. **221**, 316–337 (2013)

48. Safaei, A.A., Sharifrazavian, A., Sharifi, M., Haghjoo, M.S.: Dynamic routing of data stream tuples among parallel query plan running on multi-core processors. J. Distrib. Parallel Databases **30**(2), 145–176 (2012)
49. Backman, N., Fonseca, R., Çetintemel, U.: Managing parallelism for stream processing in the cloud. In: Proceedings of the 1st International Workshop on Hot Topics in Cloud Data Processing, p. 1, Apr 2012. ACM (2012)
50. Shan, L., Xuejiao, H., Li, Y., Lizhen, X.: Research and improvement of load balancing algorithm in distributed sonar data stream management system. In: Proceedings of Web Information Systems and Applications (WISA), 2012 Ninth, pp. 163–169, Nov 2012. IEEE (2012)
51. Aly, A.M., Sallam, A., Gnanasekaran, B.M., Nguyen-Dinh, L., Aref, W.G., Ouzzani, M., Ghafoor, A.: M3: stream processing on main-memory mapreduce. In: 2012 IEEE 28th International Conference on Proceedings of Data Engineering (ICDE), pp. 1253–1256, Apr 2012. IEEE (2012)
52. Bedini, I., Sakr, S., Theeten, B., Sala, A., Cogan, P.: Modeling performance of a parallel streaming engine: bridging theory and costs. In: Proceedings of the 4th ACM/SPEC International Conference on Performance Engineering, pp. 173–184, Apr 2013. ACM (2013)
53. Lam, W., Liu, L., Prasad, S.T.S., Rajaraman, A., Vacheri, Z., Doan, A.: Muppet: MapReduce-style processing of fast data. Proc. VLDB Endow. **5**(12), 1814–1825 (2012)
54. Safaei, A.A., Haghjoo, M.S.: Parallel processing of continuous queries over data streams. J. Distrib. Parallel Databases **28**(2–3), 93–118 (2010)
55. Guirado, F., Roig, C., Ripoll, A.: Enhancing throughput for streaming applications running on cluster systems. J. Parallel Distrib. Comput. **73**(8), 1092–1105 (2013)
56. Ajwani, D., Ali, S., Katrinis, K., Li, C.H., Park, A.J., Morrison, J.P., Schenfeld, E.: Generating synthetic task graphs for simulating stream computing systems. Parallel Distrib. Comput. **73** (10), 1362–1374 (2013)
57. Anceaume, E., Busnel, Y.: A distributed information divergence estimation over data streams. J. Parallel Distrib. Syst. IEEE Trans. **25**(2), 478–487 (2014)
58. Hunter, T., Das, T., Zaharia, M., Abbeel, P., Bayen, A.M.: Large-scale estimation in cyberphysical systems using streaming data: a case study with arterial traffic estimation. J. Autom. Sci. Eng. IEEE Trans. **10**(4), 884–898 (2013)
59. Lei, C., Rundensteiner, E.A., Guttman, J.D.: Robust distributed stream processing. In: 2013 IEEE 29th International Conference on Proceedings of the Data Engineering (ICDE), pp. 817–828, Apr 2013. IEEE (2013)
60. Cao, J., Zhang, W., Tan, W.: Dynamic control of data streaming and processing in a virtualized environment. J. Autom. Sci. Eng. IEEE Trans. **9**(2), 365–376 (2012)
61. Golab, L., Johnson, T., Shkapenyuk, V.: Scalable scheduling of updates in streaming data warehouses. J. Knowl. Data Eng IEEE Trans. **24**(6), 1092–1105 (2012)
62. Tang, Y., Gedik, B.: Autopipelining for Data Stream Processing. J. Parallel Distrib. Syst. IEEE Trans. **24**(12), 2344–2354 (2013)
63. Cho, K.M., Tsai, C.W., Chiu, Y.S., Yang, C.S.: A high performance load balance strategy for real-time multicore systems. J. Scientific World Article ID 321231, 12p (2014)
64. Gulisano, V.M.: StreamCloud: an elastic parallel-distributed stream processing engine. Doctoral dissertation, Informatica. Ph.D. Thesis (2012)
65. Gulisano, V., Jimenez-Peris, R., Patino-Martinez, M., Valduriez, P.: Streamcloud: a large scale data streaming system. In: 2010 IEEE 30th International Conference on Proceedings of the Distributed Computing Systems (ICDCS), pp. 126–137, June 2010. IEEE (2010)
66. Saleh, O., Gropengießer, F., Betz, H., Mandarawi, W., Sattler, K.U.: Monitoring and autoscaling IaaS clouds: a case for complex event processing on data streams. In: Proceedings of the 2013 IEEE/ACM 6th International Conference on Utility and Cloud Computing, pp. 387–392. IEEE Computer Society (2013)
67. Heinze, T., Pappalardo, V., Jerzak, Z., Fetzer, C.: Auto-scaling techniques for elastic data stream processing. In: Proceedings of the 30th International Conference on Data Engineering Workshops (ICDEW), pp. 296–302, Mar 2014. IEEE (2014)

68. Hu, R., Jiang, J., Liu, G., Wang, L.: Efficient resources provisioning based on load forecasting in cloud. J. Scientific World, Article ID 10152, 14p (2014)
69. Kailasam, S., Gnanasambandam, N., Dharanipragada, J., Sharma, N.: Optimizing ordered throughput using autonomic cloud bursting schedulers. IEEE Trans. Software Eng. **39**(11), 1564–1581 (2013)
70. Gedik, B., Schneider, S., Hirzel, M., Wu, K.: Elastic scaling for data stream processing. IEEE Trans. Parallel Distrib. Syst. **25**(6), 1447–1463 (2013)
71. Dahiphale, D., Karve, R., Vasilakos, A.V., Liu, H., Yu, Z., Chhajer, A., Wang, C., et al.: An advanced MapReduce: Cloud MapReduce, enhancements and applications. J. Netw. Service Manage. IEEE Trans. **11**(1), 101–115 (2014)
72. Xie, X., Ray, I., Adaikkalavan, R., Gamble, R.: Information flow control for stream processing in clouds. In: Proceedings of the 18th ACM Symposium on Access Control Models and Technologies, pp. 89–100, June 2013. ACM (2013)

# Evidence Evaluation of Gait Biometrics for Forensic Investigation

**Imed Bouchrika**

**Abstract** Due to the unprecedented growth of security cameras and impossibility of manpower to supervise them, the integration of biometric technologies into surveillance systems would be a critical factor for the automation of security and forensic analysis. The use of biometrics for people identification is considered as a vital tool during forensic investigation. Forensic biometrics concerns the use of biometric technologies to primarily determine whether the identity of the perpetrator recorded during the crime scene can be identified or exonerated via a matching process against a list of suspects. The suitability of gait recognition for forensic analysis emerges from the fact that gait can be perceived at distance from the camera even with poor resolution. The strength of gait recognition is its non-invasiveness nature and hence does not require the subject to cooperate with the acquisition system. This makes gait identification ideal for situations where direct contact with the perpetrator is not possible.

## 1 Introduction

Security has become a major concern in modern society. This is due to the proliferating number of crimes and terror attacks as well as the vital need to provide safer environment. Because of the rapid growth of security cameras and impossibility of manpower to supervise them, the integration of biometric technologies into surveillance systems would be a critical factor for the automation of security and forensic analysis. More importantly, by early recognition of suspicious individuals who may pose security threats via the use of biometrics, the system would be able to deter future crimes as it is a significant requirement to identify the perpetrator of a crime as soon as possible in order to prevent further offenses and to allow justice to be administered. Biometrics is concerned with deriving a descriptive measurement based on either the human behavioural or physiological characteristics which should distinguish a person uniquely among other people. Examples of physiological-based

I. Bouchrika (✉)
Faculty of Science and Technology, University of Souk Ahras, Souk Ahras, Algeria
e-mail: imed@imed.ws



A.E. Hassanien et al. (eds.), *Multimedia Forensics and Security*,
Intelligent Systems Reference Library 115, DOI 10.1007/978-3-319-44270-9_13

biometrics include face, ear, fingerprint and DNA whilst the behavioural features include gait, voice and signature. Apart from being unique, the biometric description should be universal and permanent. The universality condition implies that it can be taken from all the population meanwhile the permanentness signifies that biometric signature should stay the same over time. As opposed to traditional identification or verification methods such as passports, passwords or pin numbers, biometrics cannot be transferred, forgotten or stolen and should ideally be obtained non-intrusively. Biometrics can work either in verification or identification mode. For verification, the system performs a one-to-one match for the newly acquired person signature against a pre-recorded signature in the database to verify the claimed identity. For identification mode, a one-to-many matching process is conducted against all subjects already enrolled in the database to infer the subject identity. Biometrics is now emerging in regular use being deployed in various applications such immigration border control, forensic systems and payment authentication.

The use of biometrics for people identification is considered as a vital tool during forensic investigation. Forensic science can be defined as the method of gathering, analysing and interpreting past information related to criminal, civil or administrative law. This includes the perpetrator identity and the modus operandi [31]. Forensics involves several processes including: investigation, evaluation, forensic intelligence, automated surveillance and forensic identity management [40]. Forensic analysis is performed in order to conclude further evidence to exonerate the innocent and corroborate the identity of the perpetrator through producing a well-supported evidence. The term evidence spans to include physical evidence, scientific statement or expert witness testimony. Scientific statements are usually supported by hypothesis and experiments driven by statistical-based evidence and biometrics. Forensic biometrics is the scientific discipline that concerns the use of biometric technologies to primarily determine whether the identity of the perpetrator recorded during the crime scene can be identified or excluded via a matching process against a list of suspects.

Many biometric features can be used in forensic analysis such as face, ear [2], speech and gait [8]. However, the availability of biometric features for identification is limited to forensic experts depending on the nature of the crime scene and perpetrators. Expert witness is usually based on a body of knowledge or experience provided by an individual who is formally qualified and broadly experienced in a particular domain. There are considerable factors contributing to establish the credibility of an individual acting as an expert including educational qualifications and relevant experience. However, qualitative and descriptive-based expert opinions are argued to be insufficient and less credible [6, 18, 37] in contrast to empirical-based statements which are gaining wider acceptance.

The admissibility of forensic evidence is administered by the court juries or the legal system depending on the juridical framework of the country. In the United States, the Daubert standard was conceived in 1993 when the Supreme Court handed down the case of *Daubert v. Merrel Dow Pharmaceurticals, Inc*. The court insisted on the need for reliability into rule 702 of the *Federal Rules of Evidence* which governs the admission of scientific evidence [17]. The Daubert standard requires to show that the proffered science has been tested based on a sound methodology and whether it is published and peer-reviewed taking also into account that the technique has received general acceptance in the relevant scientific community. The Daubert standard has replaced Frye as the practiced *gate-keeping* standard, causing a paradigm shift in the way expert witness testimony is rigorously analysed and as a result, lowering the threshold for cutting-edge science and raising it for preceding expertise lacking scientific foundation [39]. Within the United Kingdom, the *Forensic Science on Trial* report [13] highlights known concerns for the admissibility of scientific evidence. In response, the Law Commission of England and Wales provisionally proposed a new statutory test in 2009 which determines the admissibility of expert evidence via a set of guidelines and stronger legislation, preventing unreliable expert opinion to be admitted in a court of law [13]. The recommendations build on the US Daubert standard, providing the judge with support to apply scientific scrutiny and an obligation to the party tendering expert evidence to demonstrate its reliability. Unfortunately, as of 2015, none of these recommendations have yet been acted on.

For the history of biometrics, people have been identifying each other based on face, voice, appearance or gait for thousands of years. However, the first systematic and scientific basis for people identification dates back to 1858 when William Herschel recorded the handprint of each employee on the back of a contract whilst working for the civil service of India [5]. This was used as a way to distinguish employees from each other on payday. It was considered the first systematic capture of hand and fingerprints that was used primarily for identification purposes. For the use of forensic biometrics, Alphonse Bertillon who was a French police officer was the pioneer to employ the first biometric evidence into the judicial system presented as anthropometric measurements of the human body to counter against repeat offenders who could easily fake or change their names. He introduced the bertillonage system such that each person is identified through detailed records taken from their body anthropometric measurements and physical descriptions as they are impossible to spoof or change them. The system was adopted by the police authorities throughout the world. In 1903, the Bertillon system was challenged on the basis that anthropometric measurements are not adequate to differentiate reliably between people. This was fuelled by the case of identical twins who have almost the same measurements using the Bertillon system. In 1890s, Sir Francis Galton and Edward Henry described separately a classification system for people identification based on the person fingerprints taken from all ten fingers [2]. The characteristics that Galton described to recognize people are still used today. The Galton system was adopted by New York state prison where fingerprints were employed for the identification of apprehended inmates. Biometric systems are sold mainly for the following purposes: physical access control, logging attendance and personal identification [25].

The suitability of gait recognition for forensic analysis emerges from the fact that gait can be perceived at distance from the camera even with poor resolution as opposed to other biometric traits where their performance deteriorates severely.

Furthermore, the strength of gait recognition is its non-invasiveness nature and hence does not require the subject to cooperate with the acquisition system. This makes gait identification ideal for situations where direct contact with the perpetrator is not possible. Furthermore, as the purpose of any reliable biometric system is to be robust enough to reduce the possibility of signature forgery and spoofing attacks, the gait signature which is based on human motion is the only likely identification that can be taken in covert surveillance. Thus, gait analysis is more suited to forensic investigations as other biometric-based traits that could link to the presence in a crime scene can be obliterated and concealed as opposed to the gait pattern where the mobility of the perpetrator is a must as they have to walk or run to escape from the scene. In a recent empirical study conducted by Lucas and Henneberg [29], they argued that a combination of eight body measurements is sufficient to achieving a probability of finding a duplicate to the order of $10^{-20}$ comparing such findings to fingerprints analysis. Interestingly, in one of the high profile murder cases in the UK where a child was abducted and killed, the identity of the murderer could not be revealed directly from the surveillance video footage. The only inspiring solution that could be employed to determine the suspect's identity in this situation was gait recognition as proposed by researchers from the University of Southampton [33]. Gait analysis which is a well-established science in clinical and laboratory settings is becoming common as evidence in criminal trials with the advent of forensic podiatry which is concerned to examine foot-related evidence [16]. The notion that people can be recognized by the way they walk has gained an increasing popularity and produced impacts on public policy and forensic practice by its take up by researchers at the Serious Organized Crime Agency after its invention at Southampton in 1994 [33].

## 2 Gait Biometrics

Gait is defined as the manner of locomotion characterised by consecutive periods of loading and unloading the limbs. Gait includes running, walking and hopping. However the term gait is most frequently used to describe the walking pattern. The rhythmic pattern of human gait is performed in a repeatable and characteristic way [45] consisting for consecutive cycles. A gait cycle is defined as the time interval between successive instances of initial foot-to-floor contact for the same foot [14], and the way a human walks is marked by the movement of each leg such that each leg possesses two distinct phases. When the foot is in contact with the floor the leg is at the *stance phase*. The time when the foot is off the floor to the next step is defined as the *swing phase*. Each phase is marked by a start and an end; the stance phase begins with the *heel strike* of one foot when the leg strikes the ground. The locomotion process involves the interaction of many body systems working together to yield the most efficient walking pattern. The locomotion system consists of four main subtasks that are fulfilled at the same time to produce the walking pattern [46]. These four functions are: (i) initiation and termination of locomotor movements (ii) the generation of continuous movement to progress toward a destination (iii) adaptability to meet any changes in the environment or other concurrent tasks (iv) maintenance of the equilibrium during progression. Compared with quadrupeds, the maintenance of stability and balance for humans during walking is a particularly difficult task for the postural control system. This is mainly because for most of the gait cycle, the human body is supported by only a single leg with the centre of mass passing outside the base of support provided by the foot in contact with the floor.

Early medical investigations conducted by Murray [32] in 1964 produced a standard gait pattern for normal walking people aimed at studying the gait pattern for pathologically abnormal patients. The experiments were performed on sixty people aged between 20 and 65 years old. Each subject was instructed to walk for a repeated number of trials. For the collection of gait data, special markers were attached on every subject. Murray [32] suggested that human gait consists of 24 different components which render the gait pattern unique for every person if all gait movements are considered. It was reported that the motion patterns of the pelvic and thorax regions are highly variable from one subject to another. Furthermore, Murray observed that the ankle rotation, pelvic motion and spatial displacements of the trunk embed the subject individuality due to their consistency at different trials. In 1977, Cutting et al. [15] published a paper confirming the possibility of recognizing people by gait via observing moving lights mounted on the joints positions. Although, there is a wealth of gait studies in the literature aimed for medical use with a few referring to the discriminatory nature of the gait pattern, none is concerned with the automated use of gait for biometrics and recognizing people. The gait measurements and results introduced by Murray are to be of benefit for the development of automated gait biometric systems. However, the extraction of the gait pattern is proven complex using computer vision methods.

An automated vision-based system for people identification via the way they walk, is designed to extract gait features without the need to use markers or special sensors to aid the extraction process. In fact, all that is required is an ordinary video camera linked to a special vision-based software. Marker-less motion capture systems are suited for applications where mounting sensors or markers on the subject is not an option as the case of forensic analysis. Typically, gait biometric system consists of two main components: (i) a hardware platform dedicated for data acquisition. This can be a single CCTV camera or distributed network of cameras. (ii) a software platform for data processing and recognition. The architecture of the software side for gait biometric system is composed broadly of three main components: (i) detection and tracking of the subject, (ii) feature extraction and (iii) classification stage. Figure 1 shows the flow diagram for gait identification outlining the different subsystems involved in the process of an automated people recognition.

- *Subject Detection and Tracking*: a walking subject is initially detected within a sequence of frames using either simple background subtraction techniques or other methods such as the Histogram of Oriented Gradients. Subsequently, intra-camera tracking is performed to establish the correspondence of the same person across consecutive frames. Tracking methods are supported by simple low-level features

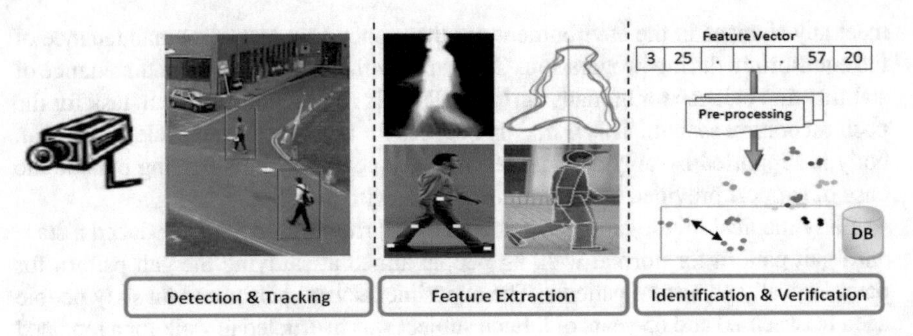


**Fig. 1** Overview of gait biometric system

such as blob size, speed and color in addition to the use of prediction algorithms to estimate the parameters of moving objects in the next frame. This is based on motion models which describe how parameters change over time. The most popular predictive method used for tracking is the Kalman filter [44], the Condensation algorithm [22], and the mean shift tracker [12].

- *Feature Extraction*: this is the most important stage for automated marker-less capture systems whether for gait recognition, activity classification or other imaging application. This is because the crucial data required for the classification phase are derived at this stage. Feature extraction is the process of estimating a set of measurements either related to the configuration of the whole body or the configuration of the different body parts in a given scene and tracking them over a sequence of frames. The features should bear certain degree of the individuality of the subject. High-level features estimated at this level for gait recognition can be categorized into two types: static and dynamic features. Examples of static features include the subject height and other anthropometric measurements meanwhile dynamic or kinematic features are such joints angular measurements and displacement of the body trunk.
- *Identification or Verification Phase*: it is mainly a classification process which involves matching a test sequence with an unknown label against a group of labelled references considered as the gallery dataset. At this stage, a high-level description is produced from the features extracted at previous phases to infer or confirm the subject identity. The classification process is normally preceded by pre-processing stages such as data normalisation, feature selection and dimensionality reduction of the feature space through the use of statistical methods. A variety of pattern recognition methods are employed in vision-based systems for gait recognition, including Neural Networks, Support Vector Machines (SVM) and K-Nearest Neighbour classifier (KNN). The latter is the most popular method for classification due to its simplicity and fast computation.

## 2.1 Gait Datasets

As gait biometrics has gained an increasing interest in surveillance and forensic cases, the establishment of gait databases becomes vital for the evaluation and assessment of research theories and systems proposed for gait analysis, automated markerless extraction and gait recognition. There were two early gait databases which are the UCSD and Southampton datasets. The first database was collected by the Visual Computing Group at the University of California San Diego containing 6 people in the database filmed in outdoor environment. The Southampton database is recorded at indoor laboratory and consists of 16 video sequences for 4 subjects wearing special trousers [33]. Subsequently, several gait databases were developed primarily for the HumanID at a Distance research program [35] funded by the Defence Advanced Research Projects Agency (DARPA). The program was aimed to improve technologies for facial and gait recognition as well as new technologies for people identification. The HumanID project included the following research institutions: University of Southampton, University of Maryland, Georgia Institute of Technology and Massachusetts Institute of Technology. Within this program, the University of Southampton released publicly the largest dataset for over 100 people containing over 20,000 video sequences accounting for different conditions as footwear, clothing and walking speed. Recently, the Chinese Academy of Sciences released the CASIA Gait Database for gait recognition and analysis which is itself composed of three datasets. The Institute of Scientific and Industrial Research (ISIR) at Osaka

**Table 1** Human Gait Databases

| Database Name | Subjects | Sequences | Environment | Description |
|---|---|---|---|---|
| Covariate SOTON | 12 | 12,730 | Indoor | Footwear, clothing, walking speed, viewpoint and carrying |
| CASIA Database A | 20 | 240 | Outdoor | 4 Viewpoints |
| MIT AI Database | 24 | 194 | Indoor | Viewpoint, Time |
| Georgia Tech | 20 | 188 | Outdoor | Viewpoint, Speed |
| CMU Mobo DB | 25 | 600 | Indoor, Treadmill | Walking speed, viewpoint, surface and carrying conditions |
| Large SOTON | 118 | 10,442 | Indoor, Outdoor, Treadmill | Viewpoint |
| Gait Challenge | 122 | 1,870 | Outdoor | Viewpoint, surface, footwear, time and carrying conditions |
| CASIA Database B | 124 | 13,640 | Indoor | 11 Viewpoints, clothing, carrying condition |
| CASIA Database C | 153 | 1,530 | Outdoor, Thermal | Speed, carrying condition |
| TUM-GAID DB | 305 | 3,370 | Indoor | Carrying, clothing, time |
| OU-ISIR Large Population | 4,016 | 7,860 | Indoor | Viewpoint |

University constructed the OU-ISIR Gait Database to aid research studies related to developing, testing and evaluating algorithms for gait-based human recognition. Table 1 surveys the different gait databases made publicly available to the research community.

## 2.2 Gait Recognition Methods

Much of the interest in the field of human gait analysis was limited to physical therapy, orthopedics and rehabilitation practitioners for the diagnosis and treatment of patients with walking abnormalities. As gait has recently emerged as an attractive biometric, gait analysis has become a challenging computer vision problem. Many research studies have aimed to develop a system capable of overcoming the difficulties imposed by the extraction and tracking of biometric gait features. Various methods were surveyed in [33, 47]. Based on the procedure for extracting gait features, gait recognition methods can be divided into two main categories which are model-based and appearance-based (model-free) approaches.

### 2.2.1 Model-Based Approaches

For the model-based approach, a prior model is established to match real images to this predefined model, and thereby extracting the corresponding gait features once the best match is obtained. Usually, each frame containing a walking subject is fitted to a prior temporal or spatial model to explicitly extract gait features such as stride distance, angular measurements, joints trajectories or anthropometric measurements. Although model-based approaches tend to be complex requiring high computational cost, these approaches are the most popular for human motion analysis due to their advantages [47]. The main strength of model-based techniques is the ability to extract detailed and accurate gait motion data with better handling of occlusion, self-occlusion and other appearance factors as scaling and rotation. The model can be either a 2 or 3-dimensional structural model, motion model or a combined model. The structural model describes the topology of the human body parts as head, torso, hip, knee and ankle by measurements such as the length, width and positions. This model can be made up of primitive shapes based on matching against low-level features as edges. The stick and volumetric models are the most commonly used structural-based methods.

Akita [1] proposed a model consisting of six segments comprising of two arms, two legs, the torso and the head. Guo et al. [19] represented the human body structure by a stick figure model which had ten articulated sticks connected with six joints. Rohr [38] proposed a volumetric model for the analysis of human motion using 14 elliptical cylinders to model the human body. Karaulova et al. [26] used the stick figure to build a hierarchical model of human dynamics represented using Hidden Markov Models. For the deployment of structural model-based methods for gait recognition, Niyogi and Adelson [34] was perhaps the pioneer in 1994 to use a model-based method for gait recognition. Gait signature is derived from the spatio-temporal pattern of a walking subject using a five stick model. Using a database of 26 sequences containing 5 different subjects, a promising classification rate of 80 % was achieved.

The motion model describes the kinematics or dynamics of the body or its different parts throughout time. Motion models employ a number of constraints that aid the extraction process as the maximum range of the swinging for the low limbs. Cunado et al. [14] was the first to introduce motion model using the Velocity Hough Transform to extract the hip angular motion via modelling human gait as a moving pendulum. The gait signature is derived as the phase-weighted magnitudes of the Fourier components. A recognition rate of 90 % was achieved using the derived signature on a database containing 10 subjects. Yam et al. [48] modeled the human gait as a dynamic coupled oscillator which was used to extract the hip and knee angular motion via evidence gathering. The method was evaluated on a database of 20 walking and running subjects, achieving a recognition rate of 91 % based on gait signature derived from the Fourier analysis of the angular motion. Wagg and Nixon [42] proposed a new model-based method for gait recognition based on the biomechanical analysis of walking subjects. Mean model templates are adopted to fit individual subjects. Both the anatomical knowledge of human body and hierarchy of shapes are used to reduce the computational costs. The gait feature vector is weighted using statistical analysis methods to measure the discriminatory potency of each feature. On the evaluation of this method, a correct classification rate of 95 % is achieved on a large database of 2,163 video sequences of 115 different subjects. Bouchrika et al. [7, 9] proposed a motion-based model for the extraction of the joints via the use a parametric representation of the elliptic Fourier descriptors describing the spatial displacements. As most of the model-based are exploiting 2D images, there are recent work aimed for introducing 3-dimensional model for the extraction of gait features including the work of Ariyanto and Nixon [3] (Fig. 2).

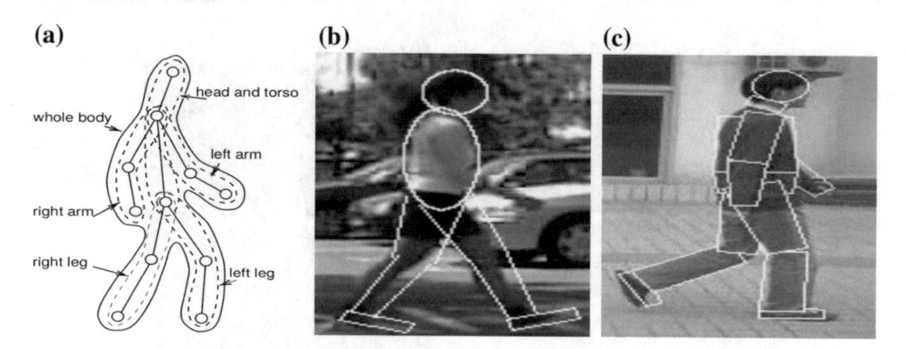


**Fig. 2** Model-based approaches for gait feature extraction: **a** Karaulova et al. [26]. **b** Wagg and Nixon [42]. **c** Wang et al. [43]

### 2.2.2 Appearance-Based Approaches

Appearance-based or model-free approaches for gait recognition do not need a prior knowledge about the gait model. Instead, features are extracted from the whole body without the need to explicitly extract the body parts. The majority of appearance approaches depends on data derived from silhouettes which are obtained via background subtraction. The simplest method is called the Gait Energy Image (GEI) introduced by Han and Bhanu [20] in which gait signature is constructed through taking the average of silhouettes for one complete gait cycle. Experimental results confirmed that higher recognition rates can be attained to reach 94.24 % for a dataset of 3,141 subjects [24]. However, such method performs poorly when changing the appearance. Gait Entropy Image (GenI) is a silhouette-based representation introduced by Bashir et al. [4] which is computed by calculating the Shannon entropy for each pixel achieving a correct classification rate of 99.1 % on dataset of 116 subjects. The Shannon entropy estimates the uncertainty value associated with a random variable. Other similar representations include Motion Energy Image, Motion Silhouette Image, Gait History Image and Chrono-Gait Image. Hayfron-Acquah et al. [21] introduced a method for constructing a gait signature based on analysing the symmetry

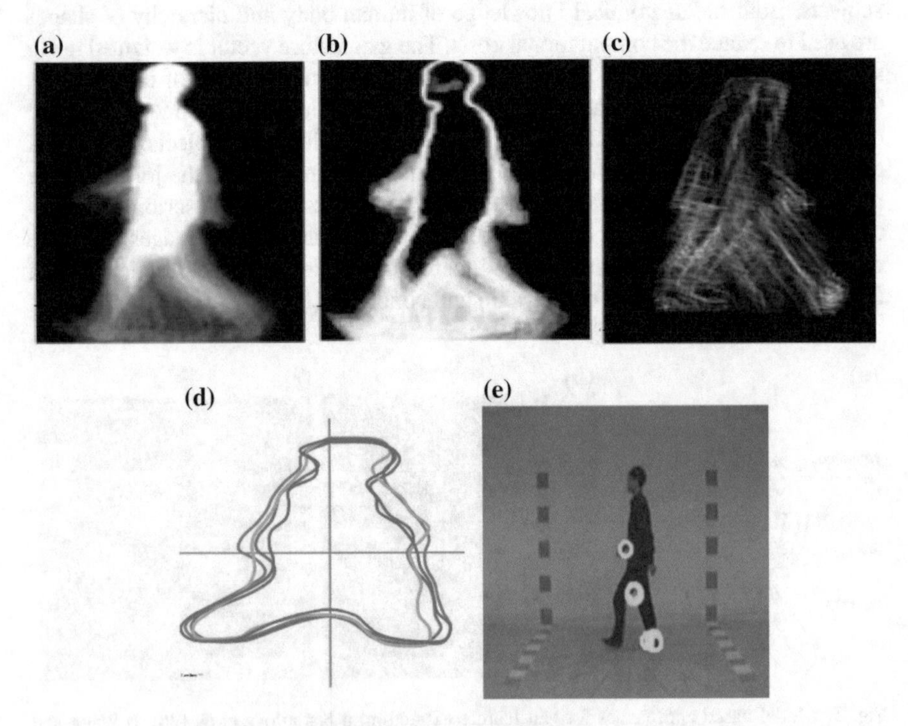


**Fig. 3** Appearance-based methods for gait recognition: **a** Use gait energy image [20]. **b** Gait entropy image [4]. **c** Symmerty map [21]. **d** Procruste shape analysis [11]. **e** STIP descriptors

of human motion. The symmetry map is produced via applying the Sobel operator on the gait silhouettes followed by the Generalised Symmetry Operator (Fig. 3).

As the accuracy of silhouette-based methods depends on the background segmentation algorithm which is not reliable for the case of real surveillance footage in addition to the sensitivity issue to varied appearances, a number of appearance-based methods have emerged recently that use instead interest-point descriptors. Kusakunniran [27] proposed a framework to construct gait signature without the need to extract silhouettes. Features are extracted in both spatial and temporal domains using Space-Time Interest Points (STIPs) by considering large variations along both spatial and temporal directions at a local level. Appearance-based Method relies pivotally on statistical methods to reduce or optimize the dimensionality of feature space using methods such as Principal Component Analysis. In addition, advanced machine learning methods are usually applied as multi-class support vector machine and neural networks. Contentiously, recent investigations by Veres et al. [41] reported that most of the discriminative features for appearance-based approaches are extracted from static components of the top part of the human body whilst the dynamic components generated from the swinging of the legs are ignored as the least important information.

## 3 Gait Analysis for Forensics

Forensic gait analysis has been recently applied in investigations at numerous criminal cases as law enforcement officers have no option to identify the perpetrator using well-established methods as facial recognition or fingerprints. This is partly due to the fact that key biometric features such as the perpetrator's face can be obscured or veiled and the CCTV footage is deemed unusable for direct recognition whilst the perpetrators are usually filmed at a distance walking or running away from the crime scene. Gait experienced specialists are consulted to assist with the identification process of an unknown person by their walking pattern through a qualitative or quantitative matching process. This would involve examining the unique and distinctive gait and posture features of an individual. Subsequently, a statement is written expressing an opinion or experimental results that can be used in a court of law. Meanwhile, the practice of forensic podiatry involves examining the human footprint, footwear and also the gait pattern using clinical podiatric knowledge [16]. However, gait analysis performed by a podiatrist involves the recognition and comparison of nominal and some ordinal data without quantitative analysis using numerical forms of data [16]. Because of the rising profile of gait biometrics and forensic podiatry, gait is used numerously as a form of evidence in criminal prosecutions with the inauguration of the American Society of Forensic Podiatry in 2003. Recently, Iwama et al. [23] developed a software with a graphical user interface in order to assist non-specialists to match video sequences based on gait analysis. For the methods used for forensic gait analysis, we can classify them into two major categories which are: *Descriptive-based* or *Metric-Based* approaches.

### 3.1 Descriptive-Based Methods

For descriptive-based methods, forensic gait specialists observe visually the gait pattern looking for abnormalities, irregularities or clues that can be exploited to confirm a match or mismatch. Abnormalities can include legs inversion, swinging range and flexion during stance. Larsen et al. [28] proposed a checklist for conducting gait analysis involving a number of semantic questions related to the gait pattern and posture of the individual that can be used to conclude the possibility of identification of the perpetrator by their walk. The checklist criteria include questions such as long or short steps, stiff or relaxed gait, outward or inverted feet rotation and forward or backward learning of the upper body. There were a number of concerns and critics for the admissibility of descriptive methods into criminal investigations as they are considered unreliable partly due to the fact that images are open to different interpretation by the forensic experts analysing them [6]. Biber [6], Porter [37] and Edmond et al. [18] argued that there are technical and contextual factors that can visually distort the relationship between the image and the actual object it represents. Furthermore, Vernon highlighted that cautions are required with descriptive-based evaluation [16] as subjectivity can be prone to errors.

An incident of a bank robbery in 2004 was handled by the Unit of Forensic Anthropology at the University of Copenhagen [30]. The police observed that the perpetrator has a special gait pattern with a need to consult gait practitioners to assist with the investigation. The police were instructed to have a covert recording of the suspect walking pattern within the same angle as the surveillance recordings for consistent comparison. The gait analysis revealed that there are several matches between the perpetrator and the suspect as an outward rotated feed and inverted left ankle during the stance phase. Further posture analysis using photogrammetry showed that there is a resemblance between the two recordings including a restless stance and anteriour head positioning. There were some incongruities observed during the analysis including wider stance and the trunk is slightly leaned forward with an elevated shoulders. This is suspected to be related by the anxiety when committing a crime [28]. Based on the conducted analysis, a statement was given to the police regarding the identity however such methods are argued that they do not constitute the same level of confidence as well-established methods such as fingerprints. The findings were subsequently presented in court and the suspect was convicted of robbery whilst the court stressed that gait analysis is a valuable tool [28]. In a similar case handled by the Unit of Forensic Anthropology, a bank robbery was committed by two masked people wearing white clothing. The bank was equipped with several cameras capturing most of the indoor area. One of the camera showed one of the perpetrator walking rapidly from the entrance. The frame rate was low which

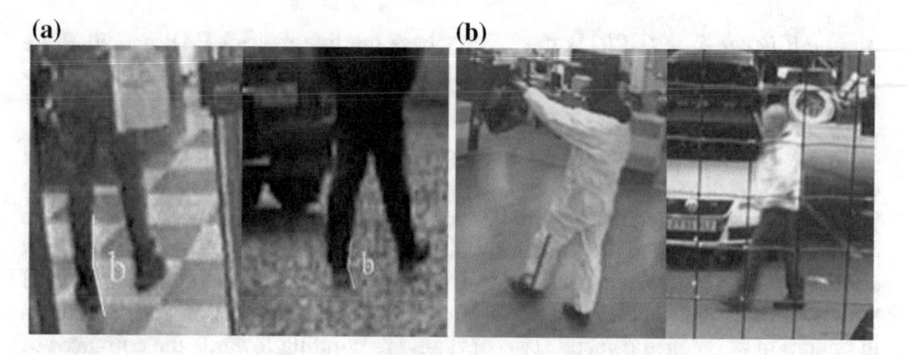

**Fig. 4** Forensic gait cases for bank robbery [28, 49]: **a** 2004. **b** 2012

left only few useful images showing the perpetrator gait. Based on experimental results showing the most discriminatory potency for the joints angles, Yang et al. [49] argued about the possibility of identification based on certain instances of the gait cycle using the observed angular swinging of the legs. Figure 4 shows the two discussed cases of the bank robberies handled by the forensic unit.

## 3.2 *Metric-Based Methods*

For this approach, a score value or a probability is produced based on a quantitative matching process of extracted measurements related to the human body parts or dynamics of the gait pattern. The extraction is done either manually or automatically through the use of vision-based marker-less methods. In forensics, it is desirable to use measurable characteristics for which matching probabilities can be estimated as it is more reliable for identification and evidence admissibility to be considered [29]. The Cumulative Match Score (CMS) is a useful measure for biometric systems which was introduced by Phillips et al. in the FERET protocol [36] for the evaluation of face recognition algorithm. The measure assesses the ranking capabilities of the recognition system by producing a list of scores that indicates the probabilities that the correct classification for a given test sample is within the top $n$ matched class labels. The Receiver Operating Characteristics (ROC) plot is more suited for forensic analysis which expresses the verification result based on a decision threshold to confirm the claimed identity of an individual against a probe sample. The threshold which is usually experimentally defined through matching all sample pairs in a larger gallery dataset refers to the separation of the genuine score distributions from the imposter values. The following three metrics are produced for ROC analysis:

1. *False Acceptance Rate (FAR)* which is the rate of samples erroneously accepted by the system as a true match.
2. *False Reject Rate (FRR)* refers to the percentage for cases when an individual is not matched to their genuine identity.

3. *Equal Error Rate (ERR)* is the value where the two metrics FAR and FRR are equal. Biometric systems with lower ERR values are considered to be more accurate and reliable.

In a recent case handled by the Metropolitan Police of London [8], a number of crimes include physical assaults and burglary against pedestrians walking on a short pathway near a subway in one of the London suburb. The same crime was reported to occur numerous times in the same fashion and at the same place. The police officers strongly suspected it was carried out by the same members of an organized gang of youngsters aged between 17 and 20 years old. There are a number of CCTV cameras in operation at the crime scene. Two of them are pointing towards the entrances of the subway as shown in Fig. 5. Two other cameras are set to record both views of the walking pass next the subway as shown in Fig. 5. The police provided a set of videos in order to deploy gait analysis to find further information that would assist them in their investigation. CCTV footage from all cameras for the crime scene at two different days was made available to the Image Processing Research group at the University of Southampton. The police provided another video of a suspect member of the gang being recorded whilst was being held at the police custody. The video was recorded at a frame rate of 2 frames per second and a resolution of $720 \times 576$ pixels. In one of the videos that was recorded on 4th April 2008, two members of the gang wore helmets to cover their faces and drove a scooter motorbike. A female pedestrian came walking through the subway where they followed her from behind on the walking path. When she entered the subway, one of them walked and snatched her bag violently using physical assault and even dragging her down on the ground. Afterwards they left away on a scooter. In a different CCTV footage recorded on the following day, the same crime was carried out with apparently the same looking perpetrators riding a scooter motorbike seen snatching a bag of another woman. The police managed to trace the suspects, partly using a helmet found near the crime scene. Facial recognition cannot be applied in such cases due the low-resolution of imagery data in addition to the fact that the perpetrators where hiding their faces

**Fig. 5** Sample frames from the crime scene CCTV cameras, 2008

completely. Such a challenging case is common for police authorities suggesting a need to explore innovative technologies in their investigation as gait biometrics.

The proposed method for gait analysis from video sequences acquired from CCTV cameras is based on Instantaneous Posture Matching (IPM). Medical and psychological studies confirmed that the task of natural walking is executed in a different way from every person [15, 32]. Therefore, the limbs position is unique in every instant of the movement and the kinematic properties of the human body can be efficiently used for identity matching between different videos. Further, recent investigation by Larsen et al. [28, 30, 49] confirmed the usefulness of using anatomical and biomechanical knowledge to recognize other individuals for different types of court cases. The Instantaneous Posture Matching approach aims to estimate the mean limbs distance between different video sequences wherein subjects are walking. The matching process is based on the anatomical proportion of the human body within a window of frames. We consider two different video sequences $v_1$ and $v_2$ recorded with the same frame rate. To compare the videos for identity matching purposes, a set of reference frames from the first video are matched progressively against a window of frames from the other video sequence. Given the joint coordinates $(x, y)$ for the hip $x_{h1}$, knee $x_{k1}$ and ankle $x_{a1}$ (two of each are extracted for the left and right legs; both sides of the hips are extracted since we consider front view video) of the human body of video $v$ at frames/time $t$. In order to define a position vector for the extracted joints for direct matching between subjects, we shift the extracted the joints to a new coordinate system whose origin point is set as the left ankle point. To alleviate the effects of different camera resolutions, the new shifted positions are normalised by the subject height. Therefore, a feature vector $P_v(t)$ of video $v$ at frame $t$ is defined as:

$$P_v(t) = \frac{\begin{bmatrix} x_{h1}(t) - x_{a1}(t) & x_{h2}(t) - x_{a1}(t) & x_{k1}(t) - x_{a1}(t) & x_{k2}(t) - x_{a1}(t) & x_{a2}(t) - x_{a1}(t) \\ y_{h1}(t) - y_{a1}(t) & y_{h2}(t) - y_{a1}(t) & y_{k1}(t) - y_{a1}(t) & y_{k2}(t) - y_{a1}(t) & y_{a2}(t) - y_{a1}(t) \end{bmatrix}}{L} \quad (1)$$

where $L$ is the subject's height in pixels. The joint coordinates are referred to the image reference system and it is assumed that the subjects in the $v$ video sequences have the same walking direction without any loss of generality. The walking direction, in fact, can be easily extracted as the angle of inclination of the straight line which approximates the heel-strike points [7]. The extraction of joint coordinates from the video sequences can be achieved with different approaches either manually or using the marker-less approach. After having extracted the normalised joints position vector, the two subjects of different video sequences $v_1$ and $v_2$ are considered to have the same identity if the joints distance $D$ defined in Eq. (2) (as the mean distance of the Euclidian distances between the poses of subjects in different videos starting from frames $t_1$ and $t_2$, over a window of $W$ consecutive frames) is less than a chosen factor:

$$D(v_1, v_2) = \min\{d(v_1, t_1, v_2, t_2) : 0 \le t_1 \le |v_1| - W, 0 \le t_2 \le |v_2| - W\} \le \tau \quad (2)$$

where $|v_n|$ is the number of frames for video $v_n$ and $d(v_1, t_1, v_2, t_2)$ is defined in Eq. (3) as:

$$d(v_1, t_1, v_2, t_2) = \left( \frac{\sum_{f=1}^{W} \| P_{v1}(t_1 + f) - P_{v2}(t_2 + f) \|}{W} \right) \quad (3)$$

The threshold value $\tau$ in Eq. (2) is chosen by analysis of intra- and inter-subject differences on a large gait database. $f$ refers to the frame number. A statement was written to the police based on the achieved value of $D$ confirming that the perpetrators on the processed videos are the same person. Other evidence gathered by the police was consistent with gait-based reported results. In fact, we believe that the use of vertex location is more favorable in forensic procedure because this can be more readily communicated to those without a technical background, but there are other approaches that might derive a better performance [33]. Moreover, it is well known that the perception of a subject's gait varies with change in direction of camera relative to the subjects path. There are now techniques that provide for viewpoint-invariant gait recognition and which have been used to track subjects across nonintersecting camera views.

## 4 Evidence Evaluation and Challenges

Evaluation of biometric-based evidence in forensic investigation plays a pivotal role for its admissibility in court. This is because legal cases would involve serious punishment or recurrent further crimes on the basis of the produced evidence. The state of biometric-based forensics is still considered nascent rather than established. Saks et al. [39] argued that descriptive-based or observational methods for identification are being increasingly challenged in court as they can be subjective in addition to the recent development of metric-based evidence governed by statistical and empirical methods. Conversely, Saks [39] reported that 63 % of 86 DNA exoneration cases were due to testing errors recommending that the lack of reported error rates must be addressed through blind testing and external proficiency analysis. Champod et al. [10] described a uniform framework based on the Bayesian theorem that attempts to quantify the evidence with a likelihood-Ratio (LR). Within the *LR* approach, biometric technologies are used to assess statistically the evidential value for a biometric signature associated to a reference sample. The *LR* is defined as given in Eq. (4):

$$LR = \frac{Pr(E|S, I)}{Pr(E|\overline{S}, I)} \quad (4)$$

Such that $E$ is the evidence value measured as the similarity score. $S$ is the hypothesis supported by the prosecution asserting the perpetrator identity as the suspect. $\overline{S}$ is the defense proposition that biometric features correspond to another individual. $I$ indicates relevant background information about the case [10].

In forensic biometrics, one of the key issues is what are the chances that another individual has the same biometric measurements. In other words, evidence can be challenged around the certainty that there exist no other people having the same signatures as the perpetrator at any one given time. As we are limited to screen the entire population, the certainty of finding a possible duplicate is supported using statistical probabilities based on research performed on relatively smaller datasets. For gait forensic analysis, a dataset of 101 subjects are taken from the CASIA-B dataset with an average of 35 video sequences for every subject. Automated marker-less extraction is applied to obtain the joints positions including the hip, knees and ankles. In the performance test, we defined a dataset of incremental size $n \in \{2, 3, 4...N = 101\}$ subjects. We compute the similarity scores $S_n^{Intra}$ and $S_n^{Inter}$ for all the match combinations of video sequences of the same subjects and different subjects. The $S_n^{Intra}$ and $S_n^{Inter}$ are computed as the mean values for the intra- and inter-match scores computed using the Instantaneous Posture Matching approach defined earlier as expressed in Eqs. (5) and (6):

$$S_n^{Intra} = \sum_{a=1}^{n} \frac{\sum_{i=1}^{L_a} \sum_{j=i+1}^{L_a} D(v_i^a, v_j^a)}{\frac{L_a(L_a-1)}{2}} \Big/ n \tag{5}$$

and

$$S_n^{Inter} = \sum_{a=1}^{n} \sum_{b=a+1}^{n} \frac{\sum_{i=1}^{L_a} \sum_{j=i+1}^{L_a} D(v_i^a, v_j^a)}{L_a \times L_b} \Big/ n(n-1) \tag{6}$$

where $v_i^a$ is the $i$th video sequence of subject $a$. $L_a$ is the number of video sequences belonging to subject $a$. $D$ is the distance computed as defined in Eq. (2). The general framework for performance analysis is outlined by starting with an initial dataset of size $n = 2$ and then the database size is progressively increased by including more (different) subjects in the experimental test. The selection of new subjects into the dataset is done at random. To avoid bias when selecting subjects, the similarity scores $S_n^{Intra}$ and $S_n^{Inter}$ are computed as the average for up to 100 different initial datasets selected at random. The experimental results are shown in Fig. 6 which illustrates the observed relationship between the database size and the similarity match scores of the intra- and inter-classes computed using the proposed Instantaneous Posture Matching algorithm for the different 100 subsets taken at random. The results show that when increasing the database size, the similarity scores tend to converge to fixed values that are well separated. This suggests that for larger population, gait analysis can be still deployed and the size of the database should not be a factor impacting on the analysis.

Compared to other well-established and widely used biometrics in forensic science as fingerprints and DNA, gait analysis is reported to be influenced by a number of external covariate factors that can affect the gait pattern and therefore undermine evidence credibility. Factors can be related to the appearance of the subject as clothing, footwear or psychological factors as anxiety state and medical conditions in

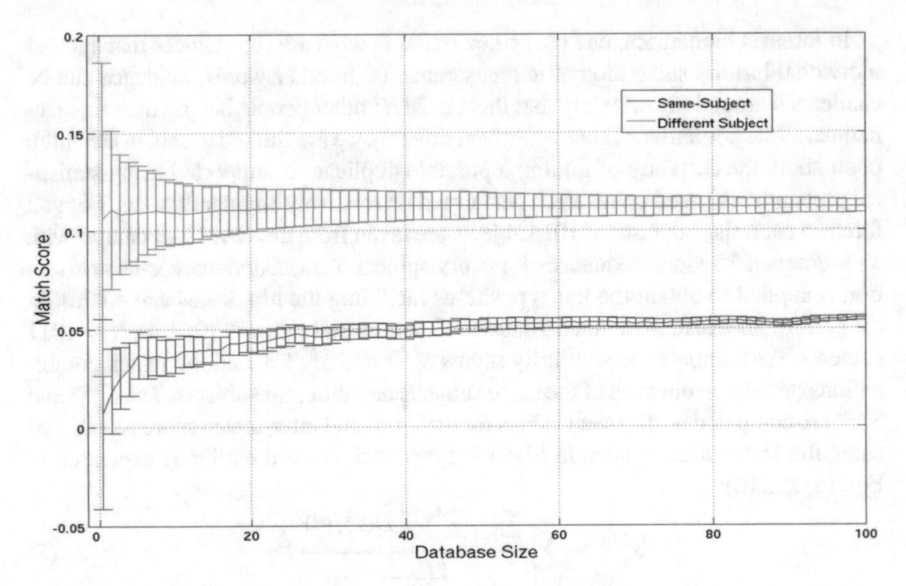


**Fig. 6** Sample frames from the crime scene CCTV cameras, 2008

addition to the acquisition environment as viewpoint and illumination. A number of studies emerged recently addressing such issues with promising identification rates of gait recognition under several covariate factors.

# 5 Conclusions

The notion that people can be recognized by the way they walk has gained an increasing popularity and produced impacts on public policy and forensic practice by its take up by researchers at the Serious Organized Crime Agency with numerous forensic cases where gait is used as a form of evidence in successful criminal prosecutions. This chapter outlines the different methods used for the automated extraction of gait features and recognition. We discuss the deployment of gait analysis into forensic investigation with detailed study of both descriptive and metric-based approaches. The chapter finally examines the evaluation of biometric-based evidence in forensics due to its pivotal role for the admissibility in criminal proceedings.

# References

1. Akita, K.: Image sequence analysis of real world human motion. Pattern Recogn. **17**(1), 73–83 (1984)
2. Arbab-Zavar, B., Xingjie, W., Bustard, J., Nixon, M.S., Li, C.T.: On forensic use of biometrics. In: Handbook of Digital Forensics of Multimedia Data and Devices (2015)

3. Ariyanto, G., Nixon, M.S.: Marionette mass-spring model for 3D gait biometrics. In: 5th International Conference on Biometrics, IEEE, pp 354–359 (2012)
4. Bashir, K., Xiang, T., Gong, S.: Gait recognition using gait entropy image. In: 3rd International Conference on Crime Detection and Prevention, pp 1–6 (2009)
5. Berry, J., Stoney, D.A.: The history and development of fingerprinting. Adv. Fingerpr. Technol. **2**, 13–52 (2001)
6. Biber, K.: Visual jurisprudence: the dangers of photographic identification evidence: katherine biber argues that caution is required when using photographic evidence in court. Crim. Justice Matters **78**(1), 35–37 (2009)
7. Bouchrika, I., Nixon, M.S.: Markerless feature extraction for gait analysis. In: Proceedings of IEEE SMC Chapter Conference on Advanced in Cybernetic Systems, pp. 55–60 (2006)
8. Bouchrika, I., Goffredo, M., Carter, J., Nixon, M.: On using gait in forensic biometrics. J. Forensic Sci. **56**(4), 882–889 (2011)
9. Bouchrika, I., Carter, J.N., Nixon, M.S.: Towards automated visual surveillance using gait for identity recognition and tracking across multiple non-intersecting cameras. Multimedia Tools Appl. **75**(2), 1201–1221 (2016)
10. Champod, C., Evett, I.W.: A probabilistic approach to fingerprint evidence. J. Forensic Identif. **51**(2), 101–122 (2001)
11. Choudhury, S.D., Tjahjadi, T.: Silhouette-based gait recognition using procrustes shape analysis and elliptic fourier descriptors. Pattern Recogn. **45**(9), 3414–3426 (2012)
12. Comaniciu, D., Ramesh, V., Meer, P.: Real-time tracking of non-rigid objects using mean shift. In: Proceedings IEEE Conference on Computer Vision and Pattern Recognition, vol. 2 (2000)
13. Commission, L.: Expert Evidence in Criminal Proceeedings in England and Wales, vol. 829. The Stationery Office (2011)
14. Cunado, D., Nixon, M.S., Carter, J.N.: Automatic extraction and description of human gait models for recognition purposes. Comput. Vis. Image Underst. **90**(1), 1–41 (2003)
15. Cutting, J.E., Kozlowski, L.T.: Recognizing friends by their walk: gait perception without familiarity cues. Bull. Psychonomic Soc. **9**(5), 353–356 (1977)
16. DiMaggio, J.A., Vernon, W.: Forensic podiatry principles and human identification. In: Forensic Podiatry, pp 13–24. Springer (2011)
17. Edmond, G.: The admissibility of incriminating expert opinion evidence in the US, England and Canada. Judic. Off. Bull. **23**(8), 67 (2011)
18. Edmond, G., Thompson, M.B., Tangen, J.M.: A guide to interpreting forensic testimony: scientific approaches to fingerprint evidence. Law, Probab. Risk **13**(1), 1–25 (2014)
19. Guo, Y., Xu, G., Tsuji, S.: Understanding human motion patterns. In: Proceedings of the 12th IAPR International Conference on Pattern Recognition, Conference B: Computer Vision & Image Processing, vol. 2 (1994)
20. Han, J., Bhanu, B.: Individual recognition using gait energy image. IEEE Trans. Pattern Anal. Mach. Intell. **28**(2), 316–322 (2006)
21. Hayfron-Acquah, J.B., Nixon, M.S., Carter, J.N.: Automatic gait recognition by symmetry analysis. Pattern Recogn. Lett. **24**(13), 2175–2183 (2003)
22. Isard, M.C., Blake, A.C.: Condensation: conditional density propagation for visual tracking. Int. J. Comput. Vis. **29**(1), 5–28 (1998)
23. Iwama, H., Muramatsu, D., Makihara, Y., Yagi, Y.: Gait-based person-verification system for forensics. In: IEEE Fifth International Conference on Biometrics: Theory, Applications and Systems (BTAS), pp. 113–120 (2012)
24. Iwama, H., Okumura, M., Makihara, Y., Yagi, Y.: The ou-isir gait database comprising the large population dataset and performance evaluation of gait recognition. IEEE Trans. Inf. Forensics Secur. **7**(5), 1511–1521 (2012)
25. Jain, A.K., Kumar, A.: Biometric recognition: an overview. In: Second Generation Biometrics: The Ethical, Legal and Social Context, pp 49–79 (2012)
26. Karaulova, I.A., Hall, P.M., Marshall, A.D.: A hierarchical model of dynamics for tracking people with a single video camera. In: Proceedings of the 11th British Machine Vision Conference, vol. 1, pp. 352–361 (2000)

27. Kusakunniran, W.: Recognizing gaits on spatio-temporal feature domain. IEEE Trans. Inf. Forensics Secur. **9**(9), 1416–1423 (2014)
28. Larsen, P.K., Simonsen, E.B., Lynnerup, N.: Gait analysis in forensic medicine. J. Forensic Sci. **53**(5), 1149–1153 (2008)
29. Lucas, T., Henneberg, M.: Comparing the face to the body, which is better for identification? Int. J. Legal Med. 1–8 (2015)
30. Lynnerup, N., Vedel, J.: Person identification by gait analysis and photogrammetry. J. Forensic Sci. **50**(1), 112–118 (2005)
31. Meuwly, D., Veldhuis, R.: Forensic biometrics: from two communities to one discipline. In: Conference of the Biometrics Special Interest Group, pp. 1–12 (2012)
32. Murray, M.P.: Gait as a total pattern of movement. Am. J. Phys. Med. **46**(1), 290–333 (1967)
33. Nixon, M.S., Tan, T.N., Chellappa, R.: Human Identification Based on Gait. Springer, Secaucus, NJ, USA (2005)
34. Niyogi, S.A., Adelson, E.H.: Analyzing and recognizing walking figures in XYT. In: Proceedings of the IEEE Computer Society Conference on Computer Vision and Pattern Recognition, pp. 469–474 (1994)
35. Phillips, P.J.: Human identification technical challenges. In: Proceedings of the International Conference on Image Processing, vol. 1, pp. 49–52 (2002)
36. Phillips, P.J., Moon, H., Rizvi, S.A., Rauss, P.J.: The FERET evaluation methodology for face recognition algorithms. IEEE Trans. Pattern Anal. Mach. Intell. **22**(10), 1090–1104 (2000)
37. Porter, G.: CCTV images as evidence. Aust. J. Forensic Sci. **41**(1), 11–25 (2009)
38. Rohr, K.: Towards model-based recognition of human movements in image sequences. CVGIP Image Underst. **59**(1), 94–115 (1994)
39. Saks, M.J., Koehler, J.J.: The coming paradigm shift in forensic identification science. Science **309**(5736), 892–895 (2005)
40. Tistarelli, M., Grosso, E., Meuwly, D.: Biometrics in forensic science: challenges, lessons and new technologies. In: Biometric Authentication. Springer, pp. 153–164 (2014)
41. Veres, G.V., Gordon, L., Carter, J.N., Nixon, M.S.: What image information is important in silhouette-based gait recognition? In: Proceedings of the 2004 IEEE Computer Society Conference on Computer Vision and Pattern Recognition, CVPR 2004, vol. 2, pp. II–776. IEEE (2004)
42. Wagg, D.K., Nixon, M.S.: On automated model-based extraction and analysis of gait. In: Proceedings of the Sixth IEEE International Conference on Automatic Face and Gesture Recognition, pp. 11–16 (2004)
43. Wang, L., Ning, H., Tan, T., Hu, W.: Fusion of static and dynamic body biometrics for gait recognition. IEEE Trans. Circ. Syst. Video Technol. **14**(2), 149–158 (2004)
44. Welch, G., Bishop, G.: An introduction to the Kalman Filter. In: ACM SIGGRAPH 2001 Course Notes (2001)
45. Winter, D.A.: The Biomechanics and Motor Control of Human Movement, 2nd edn. Wiley (1990)
46. Woollacott, M.H., Tang, P.F.: Balance control during walking in the older adult: research and its implications. Phys. Therapy **77**(6), 646 (1997)
47. Yam, C.Y., Nixon, M.: Gait recognition, model-based. In: Li, S., Jain, A. (eds.) Encyclopedia of Biometrics, pp. 633–639. Springer, US (2009)
48. Yam, C.Y., Nixon, M.S., Carter, J.N.: Automated person recognition by walking and running via model-based approaches. Pattern Recogn. **37**(5), 1057–1072 (2004)
49. Yang, S.X., Larsen, P.K., Alkjær, T., Simonsen, E.B., Lynnerup, N.: Variability and similarity of gait as evaluated by joint angles: implications for forensic gait analysis. J. Forensic Sci. 1556–4029 (2013)

# Formal Acceptability of Digital Evidence

**Jasmin Cosic**

**Abstract** In this chapter author will try to explain the concept of acceptability of digital evidence, and presents the research results on the subject of acceptability of digital evidence in courts in Bosnia and Herzegovina. The purpose is to gain insight into the manner in which judges resonate during making decision on the (not)acceptability of digital evidence, and explore current situation when acceptability and maintaining the chain of custody comes in question. Within the chapter results of preliminary research conducted at the courts in Bosnia and Herzegovina will be presented, on the subject of a digital evidence acceptability in criminal procedure. At the end of chapter will be proposed a model which can help and support forensic investigator, court experts and finally the judges to decide of admissibility of digital evidence more clearly and systematically, using scientific methods and tools.

## 1 Introduction

This book chapter is divide into three main section. First section clarifies a concept of acceptability and admissibility of digital evidence and present a research goal, methodology and tools. In second section, a term "formal acceptability" of digital evidence is explained and formalized in ontology language. In the third section DEMF framework and Protégé tools are used for the realization and concretization of this idea. The results of the research and conclusion are presented at the end of this chapter.

Admissibility or acceptability of digital evidence are the terms which determines whether the potential digital evidence will eventually become a true evidence, that usually ends in court epilogue in case of official, government investigations. The judge, or judicial council decide which evidence will be presented or not in their courtroom. Likewise, the judges decide on the involvement of the court expert

J. Cosic (✉)
Ministry of Interior, University of Bihac, Bihac, Bosnia and Herzegovina
e-mail: jasmincosic@mupusk.gov.ba



A.E. Hassanien et al. (eds.), *Multimedia Forensics and Security*,
Intelligent Systems Reference Library 115, DOI 10.1007/978-3-319-44270-9_14

witness for ICT[1] who testify on scientific aspects of potential digital evidence. Today there are a very few published papers describing how judges make this decision. There weren't publicly published results of such or similar scientific research in the region (Balkan Country's, The South East Europa and wider).

Essential for understanding the acceptability of digital evidence is to understand two basic principles which are mentioned within the acceptability of digital evidence in courts. These are Daubert principle and Frye test. Daubert principle, as mentioned, replaced longtime used "Frye test", and according to it, science and scientific methods are increasingly introduced as mandatory in expertise and presentation of digital evidence. American Rule 702, as part of the Federal Rules of Evidence, provides guidance on expert qualification and reduces the possibility of bias in expertise. Rule 702 of the American legislation is used as a preventive for possible speculations by experts, and which judge could use. In order to be properly informed and to be able to make decisions on the acceptability of digital evidence in court, and to understand expert witness testimony, the judges and the jury must own certain knowledge about information and communication technologies (ICTs) [1–3].

In most case that knowledge is not based on formal training and education, rather than on personal experience and knowledge which is gained through computer and Internet use [4, 5].

In his research Mason [6] presented several complex situations with which the judges encounter and have to cope with at courtroom in which they preside:

- The judge was presented—so called "*log file*"[2] from a network server that indicates that the attack on the server came from specific IP address. Internet Service Provider (ISP) records show that IP address belongs to a computer with a particular residence at the time of the incident. This information should be used to "improperly" identify a person as offender.
- The judge was presented a "*call log*" history and phone service provider records that indicate that one mobile phone was used to make a call towards the other mobile phone. The judge and the jury may mistakenly believe that this evidence proves that the owners of these phones had a conversation.
- Microsoft Word document metadata include name of the person who registered the software. Unless this information is intentionally deleted, it stays in each document that is generated with Office application. The judge leading the case usually concludes that the metadata included in document prove that the name of that person identifies the creator of the document.
- Digital signature on an electronic document can be accepted by the court as evidence if the signature owner is also creator of the electronic document. If the electronic signature is compromised function of electronic signature can be

[1]Court expert witness for Information and Communication Technology (or in some country called Informatics expert witness).

[2]The file located on a network server on which all actions are recorded, as on the system so on the network and application part of the server.

misused. If the judge or the jury are not aware of that, counterfeit document can be accepted as legitimate in the process [6].

One of the situations that judges encounter during the criminal proceedings in which they need to know how to make the right decision is the following:

- The following data were obtained from administrator of web portal and from Internet Service Provider (ISP)—a complete data on the person who made DDoS attack (Distributed Denial of Service) on a web portal (IP address, MAC address, date and time of attack, etc.). Computer seizing and examination revealed that computer is used by a person who has no knowledge or any motive to do such attacks, and that his computer was so called "*zombie*" and a part of *bots* network, remotely controlled from a completely different location.

Another situation in which judges are encountered with the complexity of the problem of the acceptability of digital evidence is the following:

- Suspect for committing a criminal offence was recorded on bank video surveillance, which is located close to a place of committing an offence. The police confiscated a video with footage and extracted a part on which offence committing was recorded. The integrity of digital record is secured with SHA-256 function. Whereas, it was dark and bad lighting in the vicinity of the camera, face could not be recognized, therefore the expert, with help of a special methods, singled out a photograph of a face and "retouched" it, in order to obtain picture with more quality. This method completely changed original digital evidence and the following taking of SHA-256 hash gave a completely different result! [7].

Authors the judge's role is a key and he plays a "gatekeeper's" role in deciding which evidence will be accepted in his courtroom [4, 8, 9]. Rule 702 of Federal Rules of Evidence[3] provides a guidance for courts on court expertise qualification and for assurance of scientifically based testimony. "*Daubert versus Merrell" Dow Pharmaceuticals Inc* in 1993 describes a test with 4 steps that determine which evidence may be admissible in courts in the USA [1, 2, 10]:

- Testing: Whether the theory or technique, scientific procedures can be (has been) independently tested?
- Publishing: Are the theory or technique, scientific procedures published and whether have they passed a scientific review?
- Frequency of error: Is the frequency of error known, and can it be potentially used in application of theory or technique, scientific procedures?
- Acceptance: Are the theory or technique, scientific procedures generally accepted by the relevant scientific community?

[3]FRE—Federal Rule of Evidence is a Federal Rule of Evidence in USA.

Before Daubert principle of admissibility, the judges decided according to "*Fry v. United States*", which demanded that scientific evidence presented in court are product of generally accepted methods by the scientific community, but it allowed the judge to make a decision on his own regarding the general acceptance. Daubert principle or test provided judges with guide for evidence admissibility.

As already asserted, there is a very little literature that describe how judges accept digital evidence [2, 5, 11], all published paper are from 2002-2007, and there is no published results on this subject in region of SE Europa and/or Balkan countries. According research published in paper in 2007 [12] the principles that affect electronic evidence are basically the respect for data protection standards and the respect for the secrecy of communications and for the right of freedom of expression. According this author, European jurists say that the principles of legitimacy, the pertinence of the relevance, and the use of such evidence have greater influence. Technical experts stress the fact that they act accounting for respect for individual rights. For example experts in computer forensics from France, Luxembourg, and Ireland highlight the preservation of confidentiality, while in Germany and Greece mention the respect for data protection standards. The specialists in Italy and the United Kingdom prefer to develop their functions through encrypted material as basic principles. Furthermore, other experts say that they can count on the legal support coming from a notary or a witness, as in Spain and Romania [12]. Similar situation is in Balkan country and South East Europa.

## 2 Research Goals

Research goal was to give an insight into the current state of, primarily, the way that courts and judges formally accept digital evidence, whether the chain of custody is supervised, and what is the general situation at courts in B&H[4] when digital evidence and acceptability of digital evidence from the aspect of chain of custody are in question. Additional goal to be achieved is to draw attention to this problem and offer a model for future research and education required in this domain.

### *2.1 Research methodology*

The survey method proved to be the most appropriate method for this study. Questionnaire was compiled, whereby when composing the questions particular attention was paid "for every question to relate on indicator, indicator on variable, variable on hypothesis and hypothesis on problem" [13]. Target group were different level of courts (from municipality level, to canton and state level). The goal was to collect about 25–30

[4]Bosnia and Herzegovina

completed survey forms. It should be noted that courts are organized in a way that judges are divided on judges who lead “civil case” and judges who lead “criminal proceedings”. Digital evidence in Bosnia and Herzegovina (similar situation is in Croatia, Serbia and other West Balkan Country[5]) are not used in civil proceedings, or they are used in a very small ratio in comparison with criminal proceedings. Due to this fact the survey and survey forms sending was only based on contact with judges who work in criminal proceedings, and who come into contact with digital evidence every day.

## 3 Data Collection and Sample Characteristic

The study was conducted on a sample of 80 judges working in the criminal courts at different levels in different parts of the country B&H. The questionnaire was developed like on-line survey, it was sent on May 2013 to the targeted judge’s addresses (80 judges).

E-mail addresses of judges were found on the High Judicial and Prosecutorial Council[6] website. After one month, a questionnaire (survey) which was designed as a web form, was filled only by four judges, which meant that it was necessary to animate them further in order for percentage to be higher. Above was done over the phone, through the court presidents and court secretaries who took over an obligation to initiate filling in the survey by the targeted judges. Only one president of the court insisted on performing a survey by sending in printed forms, which would be subsequently filled and returned. After four months, the result was 30 survey forms filled which according to was a return rate of 37.5 %, which is considered acceptable in this form of study [14]. It should be noted that in all survey forms all fields were not filled because several judges did not answer the questions which could be directly connected with their identity. Although in the letter that was sent to courts was clearly indicated and the statement of data confidentiality and personal data confidentiality was given, request for anonymity was high, and it was insisted for the names of the courts to be coded.

Gender ratio of respondents is 60 % male versus 40 % female (Table 1). The largest number of judges respondents is aged 41–50 years (36.6 %), followed by 31–41 years 23 %, followed by judges between ages of 51 and 60 years (20 %), and over 60 years (20 % or 6 judges), Table 1.

The most judges have up to 10 years of work experience (31 %), followed by the judges with over 30 years of work experience (24 %), 10–15 years (17 %) and 16–20 years (17 %). Work experience of 21–30 years has 10 % of judges (3 judges), while one judge did not respond to this question.

[5]Confirmation of this statement is in the presentations from the conference on digital evidence organized by MOI HR and INSIG2 Company (http://www.mup.hr/main.aspx?id=121965).

[6]An independent and autonomous body which is tasked to provide independent, impartial and professional judiciary in Bosnia and Herzegovina (http://www.hjpc.ba).

**Table 1** Table of modalities frequency of respondent's demographic characteristics

| | Number of respondents | Percentage amount (%) |
|---|---|---|
| *Respondents gender* | | |
| Male | 18 | 60 |
| Female | 12 | 40 |
| Total: | 30 | 100 |
| *Respondents age* | | |
| 31–40 | 7 | 23 |
| 41–50 | 11 | 36.6 |
| 51–60 | 6 | 20 |
| Over 60 years | 6 | 20 |
| Total: | 30 | 100[a] |
| *Experience as a judge* | | |
| Up to 10 years | 9 | 31 |
| 10–15 years | 5 | 17 |
| 16–20 years | 5 | 17 |
| 21–30 years | 3 | 10 |
| Over 30 years | 7 | 24 |
| Total: | 29 | 100[a] |

[a]Sum of the percentages of certain modalities is not 100 % due to rounding

**Table 2** Table of modalities frequency of respondent's demographic characteristics No. 2

| | Number of respondents | Percentage amount (%) |
|---|---|---|
| *Level of court* | | |
| Municipal | 17 | 59 |
| Cantonal | 5 | 17 |
| State | 7 | 24 |
| Total | 29 | 100 |
| *Population covered by the court* | | |
| Up to 20.000 | 1 | 3.3 |
| 20.001–50.000 | 1 | 3.3 |
| 50.001–100.000 | 10 | 33.3 |
| Over 100.000 | 18 | 60 |
| Total | 30 | 100[a] |

[a]Sum of the percentages of certain modalities is not 100 % due to rounding

As for the level of the court most judges respondents belong to a municipal level that solve criminal offences based on territorial jurisdiction in the computer crime field (criminal offences that carry a lower punishment or smaller damage was made), 17 of them, accounting for 59 %, followed by the State Court with 7 judges (24 %) and the Cantonal Court with 5 respondents (17 %). One judge did not fill this field in the survey. Most courts cover areas with over 100.000 inhabitants, 18 of them, accounting for 60 % of the sample. Results are shown in Table 2.

## 4 Research Results

In the following section, will be presented a research results, apropos answers to the following questions:

- At what level judges ICT knowledge is,
- At what level the knowledge about digital investigation process and digital evidence is,
- What factors were key to knowledge of digital evidence,
- Are certain rules and standards followed during the acceptance of digital evidence,
- Is the court expert engaged to assist in the clarification of digital evidence,
- Are judges familiar with the concept of "chain of evidence preservation" and methods for preserving the integrity of digital evidence,
- Do judges in cases, in which they are Chair of the Council, insist on proving the integrity of chain of digital evidence, and
- What factors mostly contribute to the effective presentation of digital evidence as key to a certain case?

As mentioned earlier, the link to survey was sent by e-mail at 80 different addresses of judges in different parts of the Bosnia and Herzegovina, but it was returned only from 4 judges. That accounts only 0.05 %. Similar situation is with the doctoral dissertation [2]. This meant, that it was necessary to somehow initiate and motivate judges to fill out the survey. After further efforts were made, and data obtained from 30 returned survey, the result are summarized, anonymized and presented in Tables 3 and 4. Data that could potentially identify the court or a particular judge at that court are not shown since that is specifically requested by the judges when publishing research results.

*Explanatory note*: ID = ID of the survey; HS = Knowledge of hardware and software; IN = Internet and networks knowledge; WE = Web technologies and e-mail communication knowledge; DF = Knowledge of digital forensics, DD = Knowledge of digital evidence; XP: Experience and knowledge in relation to the other judges; DDD = Concurrence with digital evidence definition; Standard = Standard or rule followed during the digital evidence acceptance; The expert = Engaging an expert witness in the case; 5WS&1h = Knowledge of the chain of custody and proving 5WS&1h in court; Age = age of the respondents; Gender = Respondents gender; Pop = Population covered by court jurisdiction expressed in thousands (1 K = 1.000 residents).

Since it was necessary to determine whether there is a link between the two variables, the best method appeared to be statistical correlation test. The degree of connection between two different numeric indicators tried to be calculated by calculating the correlation factor.

**Table 3** Anonymized survey results

| ID | HS | IN | WE | DF | DD | XP | DDD | Standard |
|---|---|---|---|---|---|---|---|---|
| I1 | 5.0 | 5.0 | 5.0 | 5.0 | 5.0 | > | Yes | DAUBERT |
| I2 | 3.0 | 3.0 | 3.0 | 3.0 | 3.0 | < | Yes | FRY |
| I3 | 4.0 | 4.0 | 4.0 | 3.0 | 3.0 | < | Yes | FRY |
| I4 | 5.0 | 5.0 | 5.0 | 5.0 | 5.0 | < | Yes | DONOTKNOW |
| I5 | 3.0 | 4.0 | 4.0 | 2.0 | 2.0 | > | Yes | DONOTKNOW |
| I6 | 3.0 | 3.0 | 3.0 | 1.0 | 2.0 | = | Yes | DONOTKNOW |
| I7 | 5.0 | 5.0 | 5.0 | 4.0 | 4.0 | < | Yes | DONOTKNOW |
| I8 | 3.0 | 2.0 | 2.0 | 2.0 | 3.0 | = | Yes | DONOTKNOW |
| I9 | 2.0 | 3.0 | 2.0 | 3.0 | 2.0 | > | Yes | DONOTKNOW |
| I10 | 4.0 | 4.0 | 4.0 | 3.0 | 3.0 | < | Yes | FRY |
| I11 | 2.0 | 2.0 | 2.0 | 1.0 | 1.0 | = | Yes | DONOTKNOW |
| I12 | 4.0 | 4.0 | 4.0 | 1.0 | 2.0 | = | Yes | DONOTKNOW |
| I13 | 1.0 | 3.0 | 4.0 | 1.0 | 1.0 | = | Yes | DONOTKNOW |
| I14 | 1.0 | 2.0 | 2.0 | 2.0 | 3.0 | = | Yes | DONOTKNOW |
| I15 | 2.0 | 3.0 | 2.0 | 2.0 | 3.0 | < | Yes | DONOTKNOW |
| I16 | 2.0 | 2.0 | 2.0 | 2.0 | 1.0 | > | Yes | DONOTKNOW |
| I17 | 2.0 | 2.0 | 2.0 | 2.0 | 2.0 | = | Yes | DONOTKNOW |
| I18 | 5.0 | 4.0 | 3.0 | 3.0 | 3.0 | > | Yes | DONOTKNOW |
| I19 | 1.0 | 3.0 | 4.0 | 2.0 | 2.0 | = | Yes | DONOTKNOW |
| I20 | 5.0 | 5.0 | 4.0 | 4.0 | 4.0 | < | Yes | FRY |
| I21 | 3.0 | 2.0 | 3.0 | 2.0 | 2.0 | = | Yes | FRY |
| I22 | 1.0 | 1.0 | 1.0 | 1.0 | 1.0 | = | Yes | DONOTKNOW |
| I23 | 4.0 | 4.0 | 4.0 | 1.0 | 3.0 | = | Yes | FRY |
| I24 | 3.0 | 3.0 | 3.0 | 4.0 | 3.0 | = | Yes | DAUBERT |
| I25 | 4.0 | 5.0 | 5.0 | 4.0 | 4.0 | < | Yes | DONOTKNOW |
| I26 | 4.0 | 3.0 | 3.0 | 3.0 | 3.0 | = | Yes | DONOTKNOW |
| I27 | 3.0 | 2.0 | 1.0 | 1.0 | 1.0 | = | Yes | FRY |
| I28 | 1.0 | 1.0 | 1.0 | 1.0 | 1.0 | = | Yes | DONOTKNOW |
| I29 | 2.0 | 4.0 | 2.0 | 3.0 | 2.0 | = | Yes | DONOTKNOW |
| I30 | 2.0 | 3.0 | 2.0 | 2.0 | 2.0 | = | Yes | DONOTKNOW |

Based on the results in Table 3, 4 necessary relevant correlations were calculated.

- For 30 respondents (data) df = 28, alpha = 0.5 (two tailed), and critical Pearson's value = 0.361.
- Correlation of Age and DD values (age of judges and knowledge of digital evidence) will provide correlation factor r = –0.6779,
- Correlation of Age and DF values (age of judges and digital forensics process knowledge) will provide correlation factor value r = –0.6223,

**Table 4** Anonymized survey results

| ID | Expert | 5ws&1h | Age | Gender | Pop |
|---|---|---|---|---|---|
| I1 | Yes | Yes | 45 | M | >100 K |
| I2 | Yes | Yes | 37 | F | >20 K |
| I3 | Yes | No | 42 | M | >100 K |
| I4 | Yes | No | 43 | M | >100 K |
| I5 | Yes | No | 50 | F | >100 K |
| I6 | Yes | Yes | 55 | F | >100 K |
| I7 | Yes | Yes | 32 | M | >100 K |
| I8 | Yes | No | 62 | M | >100 K |
| I9 | Yes | No | 55 | F | <100 K |
| I10 | Yes | Yes | 56 | M | <100 K |
| I11 | Yes | No | 63 | M | <300 K |
| I12 | Yes | No | 49 | F | <100 K |
| I13 | Yes | No | 64 | M | <100 K |
| I14 | Yes | No | 47 | M | <100 K |
| I15 | Yes | No | 48 | M | <100 K |
| I16 | Yes | No | 59 | F | <100 K |
| I17 | Yes | No | 64 | M | <100 K |
| I18 | Yes | Yes | 37 | F | >50 K |
| I19 | Yes | No | 31 | F | >100 K |
| I20 | Yes | Yes | 33 | F | >100 K |
| I21 | Yes | Yes | 51 | F | <100 K |
| I22 | Yes | Yes | 59 | M | <100 K |
| I23 | Yes | No | 44 | M | <100 K |
| I24 | Yes | No | 43 | F | <100 K |
| I25 | Yes | Yes | 37 | M | <100 K |
| I26 | Yes | No | 39 | M | <100 K |
| I27 | Yes | Yes | 61 | M | <100 K |
| I28 | Yes | Yes | 65 | M | <100 K |
| I29 | Yes | Yes | 48 | F | <100 K |
| I30 | Yes | No | 50 | M | <300 K |

- Correlation of Age and WE values (age of judges and technologies and e-mail communication knowledge) will provide value of correlation factor $r = -0.6202$,
- Correlation of Age and HS values (age of judges and knowledge of hardware and software) will provide value of correlation factor $r = -0.4482$.

Considering that r is negative this indicates that younger judges have higher level of knowledge about these technologies.

When Age and 5ws&1h values are put in correlation (age of judges and knowledge of the chain of custody and proving 5WS&1h in court) it will provide $r = -0.1358$, where r converges toward 0, which means there is a weak connection, or no connection between these two variables. This means that knowledge of these concepts has no connection to age of judges.

When asked which standard do they follow during the acceptance of digital evidence in court chaired by:

- 7 judges responded that they follow the "FRY" principle or standard, 2 follow "DAUBERT" principle, while most judges, even 21 (accounting for 70 % of respondents) responded that they do not know (they did not know the answer to this question).
- All responders (100 %) agreed with proposed definition of digital evidence and had nothing to add.
- All respondents (100 %) also declared of using expert witness services in court cases which involved digital evidence:
  - 63 % considered that the expert must be listed on Ministry of Justice official list, while
  - 37 % considered that the expert does not need to be listed on the official Ministry of Justice expert list
- 12 judges (40 %) are familiar with the chain of custody concept and insist on proving it in court which they preside, while 18 (60 %) of judges are not familiar with the term and do not insist on proving the consistency of digital evidence.

In response to the question: "What factors were crucial to yours knowledge of digital evidence in question No. 1 (personal training, education through High Judicial and Prosecutorial Council, personal experience, training)":

- 18 judges stated it was "personal training",
- 10 it was education through High Judicial and Prosecutorial Council,
- 2 it was education through court experts work.
- 17 judges (57 %) considered to have same experience of knowledge of digital evidence in relation to other fellow judges, while 5 (17 %) considered to have less experience.
- 8 judges (26 %) considered to have more experience of knowledge of digital evidence in relation to other fellow judges.

Very interested answers were to the question: "Is there a standard of technical competencies which prosecutors (general attorney) or attorneys must have? In what way (not) understanding of digital evidence at hearings affect your decision?"

- 6 judges considered that court expert competencies are essential, and that is expected of him/her to "*know the case*",
- 3 considered "*da mihi factum dabo tibi ius*" (lat. give me the facts and I'll give you the law),

- 3 considered their knowledge of digital evidence as most important,
- 3 considered that there is no any standard.

Remaining 15 judges (50 %) did not answer this question!

To the question: "Looking at in terms of presenting digital evidence in courtroom in which you preside, which factors are decisive and contribute the most to the efficient presentation of digital evidence as key factors related to a specific case":

- 10 judges (33 %) answered that IT profession expert expertise and what he/she elucidates is the most important,
- 5 (17 %) declared "do not know",
- While remaining 15 (50 %) did not answer this question.

# 5 Defining the Formal Acceptability of Digital Evidence

In order for process of digital investigation to be complete, and digital evidence to be formally acceptable, certain conditions must be fulfilled:

- Hardware that is exempt from a person (suspect, victim) can contain original digital evidence.
- The person who is the owner of the hardware can be a suspect.
- A person who have court order for expertise are court experts in that case.
- Copy of digital evidence that contains a calculated summary function ("hash value") is the digital evidence best copy.
- Digital evidence best copy containing calculated summary function ("hash value") is the original digital evidence.
- For such copy it can be said that it has preserved integrity.
- Original evidence containing *person's fingerprint, so-called hash value, location coordinates—geo data, procedures, reason of access, time of access*, is considered formally acceptable evidence.

For formalizing these statement SWRL (Semantic Web Rule Language) [15] will be used. This is language based on a combination of the OWL DL and OWL Lite sublanguages of the OWL Web Ontology Language with the Unary/Binary Datalog RuleML sublanguages of the Rule Markup Language.

SWRL rules consist of two parts: *premise (antecedent)* and *conclusion (consequent).* The premise is a body of SWRL rule, while the conclusion is so-called head of SWRL rule [16]. Each rule in its essence consists of atoms, and thus the premise and the conclusion, as follows:

$$\text{atom} \wedge \text{atom} \wedge \text{atom} \rightarrow \text{atom} \wedge \text{atom}$$

$$\text{premise} \rightarrow \text{conclusion}$$

Whereby conclusion is filled when all atoms of premise are satisfied. Every atom has following form:

$$p(arg_1, arg_2, \ldots arg_n)$$

where p is a symbol of predicate from OWL, while $(arg_1, arg_2, \ldots arg_n)$ are arguments of that predicate. Thereby, predicate symbol can represent OWL class, attribute or data type. Arguments can be OWL individuals, data values or variables related to them [16].

Now, let us formalize our few most commonly used definition [7]:

**Definition 1** Hardware that is seized from the suspect, can contain original digital evidence.

$$\text{ORIGINALNDE} \equiv \text{DIGITALEVIDENCE} \sqcap \exists \text{ SEIZED.SUSPECT}$$

```
Hardver(?x), isSeizedFrom(?x, ?susp),                [1]
Suspect(?susp) ->
OriginalnDigitalEvidence(?x)
```

In above mentioned SWRL statement is defined: Individual?x that belongs to *hardware* class, and that is seized from individual?*susp* and that belongs to *Suspect* class, is original digital evidence. It means that evidence will be original only if two conditions are fulfilled.

**Definition 2** The person who is the owner of that hardware "can be" a suspect.

$$\text{SUSPECT} \equiv \text{INDIVIDUAL} \sqcap \exists \text{ ISOWNER.HARDWARE}$$

```
Individuals(?x) ^ Device(?dev) ^ isOwner(?x,         [2]
?dev) → Suspect(?x)
```

SWRL statement No. 2 defines if individual?x belongs to *Individuals* class, and if it is an owner of some device?*dev* and that it belongs to *Device* class, than that person is the *Suspect*.

**Definition 3** Amici Curiae that have court order for expertise are court experts in that case.

$$\text{COURTEXPERTS} \equiv \text{AMICICURIAE} \sqcap \exists\ \text{HASCOURT.ORDER}$$

```
AmiciCuriae (?x) ^ hasCourtOrder(?x,                    [3]
?courtorder) → CourtExperts(?x)
```

SWRL statement No. 3 states that if individual?x that belongs to *AmiciCuriae* class and has a *court order*, belongs to *CourtExpert* class. This means that the person will also be formally involved in the case as a "key" person that will present digital evidence in the courtroom.

**Definition 4 and 5** Best copy of Digital evidence containing "*hash value*" is best copy of Digital evidence.

$$\text{ODE} \equiv \text{DE} \sqcap \exists\ \text{HASHASHVALUE.DECOPY}$$

```
DigitalEvidenceCopy(?x) ^                               [4],[5]
hasHashValues(?x, ?sha2) →
DigitalEvidenceBestCopy(?x)
```

SWRL rule No. 4 and 5 states that digital evidence to be the best copy of digital evidence,[7] must have secured integrity throughout unchanged "*hash value*".

**Definition 6** Digital evidence copy containing "*hash value*" is best copy digital evidence.

$$\text{BCDE} \equiv \text{KDD} \sqcap \exists\ \text{HASHASHVALUE.DECOPY}$$

```
OriginalDigitalEvidence(?x) ^                           [6]
hasHashValue(?x,?hash) ^
swrlb:Equal(?hash,
"cb8d50487ca35f064a16471a0ca0722e898def84c
b076c40903916a6223b3fdc") →
DigitalEvidenceIntegrityPreserved(?x)
```

SWRL rule No. 6 states that if original evidence has calculation of summary function (hash value), than integrity of digital evidence is preserved

[7]According to IOCE standards, original digital evidence are never handled. It must be made a two identical copies—copy and best copy.

**Definition 7** Original evidence containing fingerprint, hash value, location coordinates, procedures, access reason, access time, is considered formally acceptable evidence.

$$\text{DEACCEPTABLE} \equiv \text{DIGITALEVIDENCE} \sqcap \exists \text{HASFINGERPRINT.DE} \wedge \text{DIGITALEVIDENCE} \sqcap \exists \text{HASHASH.DE} \wedge \text{DIGITALEVIDENCE} \sqcap \exists \text{ HASCOORDINATESDE} \wedge \text{DIGITALEVIDENCE} \sqcap \exists \text{HASPROCEDURE.DE} \wedge \text{DIGITALEVIDENCE} \sqcap \exists \text{HASREASON.DE} \wedge \text{DIGITALEVIDENCE} \sqcap \exists \text{HASTIME.DE}$$

```
OriginalDigitalEvidence(?x) ^                                [7]
hasFingerprint(?x, ?fp) ^
hasHashValue(?x, ?sha2) ^
hasEvidenceLocationCoordinates (?x,
?gps) ^  hasEvidenceProcedure(?x, ?proc)
^ hasAccessReason(?x, ?razl) ^
hasAccessTime(?x, ?ts) →
DigitalEvidenceAcceptable(?x)
```

**Defining the rule for formal acceptability of digital evidence** SWRL statement above mentioned defines formal acceptability of digital evidence and states:

"Digital evidence is formally acceptable if" 5w's & 1h "is satisfied, or if it is known:

- Who handled digital evidence?
- What is digital evidence?
- When was digital evidence handled?
- How was digital evidence handled?
- Where was digital evidence handled?
- Why were digital evidence handled?"

This means that the court will formally accept digital evidence if it has: *fingerprint of a person* who handled it—it can be a digital signature, biometric feature (fingerprint, eye iris, some of bio characteristics etc.), *value of summary function* or "*hash value*" which will ensure the integrity of evidence, *time stamp* which will confirm when was digital evidence handled, SOPs (Eng. Standard Operating Procedures) *why* digital evidence were handled (*procedures*), *location coordinates (geo data)* that will confirm where digital evidence were handled, and the *reason why* it was all done and that is usually a Court or Prosecutor's order, etc. [17].

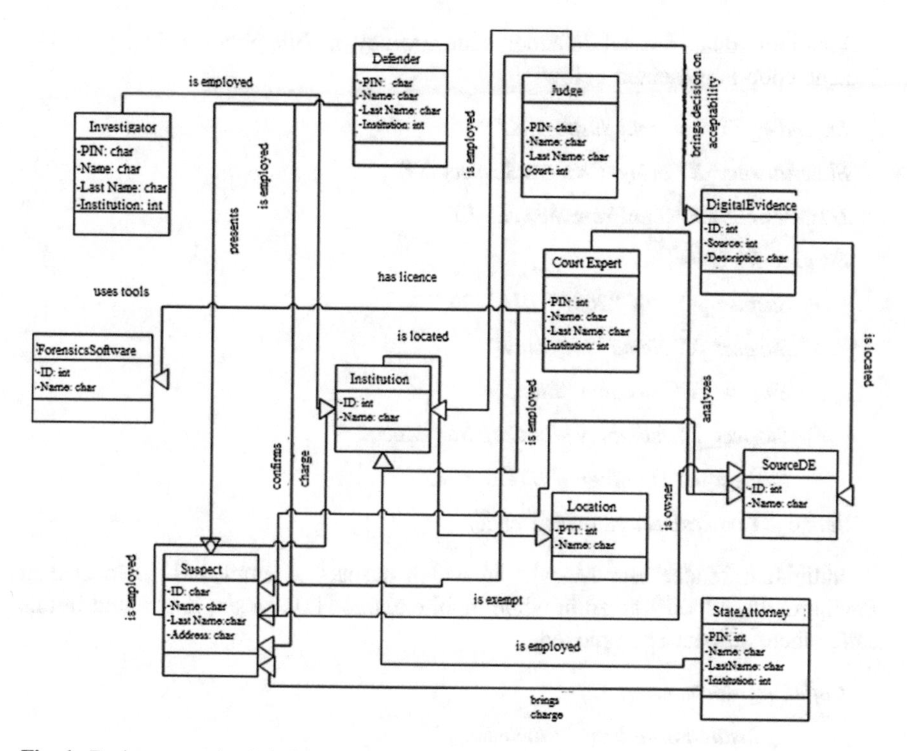


**Fig. 1** Basic process in digital investigation process

## 6 Implementing Rules Through DEMF

In the[8] Fig. 1 basic concepts in digital investigation process were described, their basic attributes, and relations between concepts that occur in this process at all stages. Now we can create a concrete individual to test this framework (Fig. 1). For this purpose we will use Protégé [18], an free, open-source ontology editor and framework for building intelligent systems. Protégé was developed at the Stanford University, and is supported by a strong community of academic, government, and corporate users, who use Protégé to build knowledge-based solutions in areas as diverse as biomedicine, e-commerce, and organizational modeling. In this case Protégé will be used in digital forensic domain to develop an intelligent system for decision support.

Let's implement an individuals with basic attributes and relation in Protégé program.

[8]Digital Evidence Management Framework (Conceptual framework by Cosic and Baca 2010).

First individuals present defender, state attorney, mobile phone and suspect. The Protégé code is presented below:

*Defender_XY represent Suspect_XY*

*StateAttorney_XY bringsCharges Suspect_XY*

*HTCDesireZ isExemptFrom Suspect_XY*

*Suspect_XY have*

*Suspect_XY PIN "2607970111120"*

*Suspect_XY Name "Miroslav"*

*Suspect_XY Surname "Bacic"*

*Suspect_XY isEmployedIn CaffeBar_Macak*

*Suspect_XY is OwnerOf HTCDesireZ*

*Judge_XY makesDecission Suspect_XY*

Individual "Cafee Bar Macak" in which suspect is employed, is located at Pavlinska 45, 42000 Varazdin, while mobile phone HTC Desire Z is found in this caffe where suspect is employed.

*CaffeBar_Macak have:*

*InstitutionAddress "Pavlinska 45"*

*HTCDesireZ isFound in CaffeBar_Macak*

*Suspect_XY isEmployed in CaffeBar_Macak*

A code for created individual CourtExpertWitnessICT_XY with basic attributes and relation is presented:

*CourtExpertWitnessICT_XY hasLicence EnCase*

*CourtExpertWitnessICT_XY Name "Jasmin"^^string*

*CourtExpertWitnessICT_XY Lastname "Cosic"^^string*

*CourtExpertWitnessICT_XY PIN "2102969112222"*

*CourtExpertWitnessICT_XY useForensicTools EnCase*

*CourtExpertWitnessICT_XY isEmployedat CZBPF_FOI_Varazdin*

*CourtExpertWitnessICT_XY analyse HTCDesireZ*

*CourtExpertWitnessICT_XY type CourtExperts*

Below are shown other relevant individuals and rules. Digital evidence 1 and 2 are located on mobile phone HTC. HTC phone have some basic attributes. A person who is suspect is owner of mobile phone, and Court expert witness for ICT analyze this phone.

*HTCDesireZ isFound in CaffeBar_Macak*

*DigitalEvidence_1 isLocatedOn HTCDesireZ*

*DigitalEvidence_2 isLocatedOn HTCDesireZ*

*HTCDesireZ have:*

*HTCDesireZ IMEI "35-209900-176148-23"^^string*

*HTCDesireZ hasOS "Android 4.0"^^string*

*HTCDesireZ serialNumber "321543SBK32"^^string*

*HTCDesireZ type MobilePhones*

*HTCDesireZ isExemptFrom Suspect_XY*

*Suspect_XY isOwner HTCDesireZ*

*CourtExpertWitnessICT_XY analyze HTCDesireZ*

Now let's formalize individual Judge with basic attributes and function in this process:

*Judge_XY PIN "3101949132113"^^string*

*Judge_XY Name "Neven"^^string*

*Judge_XY Lastname "Madicc"^^string*

*Judge_XY isEmployedIn CountryCourtZagreb*

*Judge_XY type Judges*

*Judge_XY makesDecission Suspect_XY*

*Judge_XY makesDecission DigitalEvidence*

*CountryCourtZagreb type Courts*

*CountryCourtZagreb InstitutionAddress "Trg Nikole Subica, Zagreb"^^string*

In the device (mobile phone HTC Desire Z), two digital evidence were found — *DigitalEvidence_1* and *DigitalEvidence_2*. The individual *DigitalEvidence_1* is complete, contains all attributes that are required and expected, while individual *DigitalnEvidence_2* is incomplete and don't have a key attributes—hash value and person's fingerprint:

*Individual: DigitalEvidence_1*

*DigitalEvidence_1 type OriginalDigitalEvidence*

*DigitalEvidence_1 haveCoordinates "44º53`N 16º09`E"^^string*

*DigitalEvidence_1 isLocatedOn HTCDesireZ*

*DigitalEvidence_1 haveHashValue "1fcaf179ba907fd1c2a75460f6596a81d682643ceed44b0626a6380b1a24008ccc19a69d212d21ff425c8f457a6a9ca4758fa76fe77832e068369704ea47f79b"^^string*

*DigitalEvidence_1 haveProcedure "Standard Operating Procedures"*^^string

*DigitalEvidence_1 haveReasonofAccess "Court Order No 321/13"^^string*

*DigitalEvidence_1 type CopyDigitalEvidence*

*DigitalEvidence_1 haveFingeprint "HMNpXRteitVd2tjSjzbEMma5i8OJgxV/jCWkGe1K2HfYQ9DFDaxB5eq8ncGTzlSdqQOpCkyKVHwzwlbcoaEC3Q=="^^string*

*DigitalEvidence_1 type BestCopyDigitalEvidence*

*DigitalEvidence_1 type haveTimeStamp "Thu, 25 Feb 2016 08:57:36 GMT"^^string*

*Judge_XY makesDecission DigitalEvidence*

*Individual: DigitalEvidence_2*

*DigitalEvidence_2 type OriginalDigitalEvidence*

*DigitalEvidence_2 haveCoordinates "44º53`N 16º09`E"^^string*

*DigitalEvidence_2 isLocatedOn HTCDesireZ*

*DigitalEvidence_2 haveProcedure "Standard Operating Procedures"*^^string

*DigitalEvidence_2 haveReasonofAccess "Court Order No 321/13"^^string*

*DigitalEvidence_2 type CopyDigitalEvidence*

*DigitalEvidence_2 type BestCopyDigitalEvidence*

*DigitalEvidence_2 type haveTimeStamp "Thu, 25 Feb 2016 08:57:36 GMT"^^string*

*Judge_XY makesDecission DigitalEvidence*

Now, when we have all necessary elements (individuals, attributes, relations and rules), we can, using Protégé, perform testing this framework. Key role here is "*reasoner*" or Protégé's deductive addition that provides reasoning over ontology. Pellet reasoner [19, 20] was used. Pellet is designed for applications where it is necessary to present information and allow reasoning over them by using OWL. It became an essential tool used with OWL, which has full support for OWL-DL. Implemented in Java, it is distributed under an open source license [20].

**Fig. 2** DL query that gives answer to question which digital evidence is acceptable

**Fig. 3** DL query that gives answer to question what digital evidence is unacceptable

Since *DigitalEvidence_1* individual is designed to have all attributes that define formal acceptability, when query in DL query TAB was launched, Pellet reasoner as a result gave *DigitalEvidence_1* Instance (Fig. 2).

*DigitalEvidence_2* individual do not have all necessary attributes that determine digital evidence. When we make query *DigitalEvidenceUnacceptable* through DL Query TAB, Pellet will answer with *DigitalEvidence_2* (Fig. 3). The same answer would be gained by using query:

```
NOT DigitalEvidenceAcceptable
```

Meaning the negation of the statement DigitalEvidenceAcceptable.

## 7 Conclusion

The book chapter attempts to draw attention on problem on acceptability—admissibility of digital evidence in the courtroom. Today this is a big problem in all world, and very small scientists dealing with this problem. Some studies in West Europa country indicates that only lawyers deals with this problem, while here should be focused on technical aspect, and knowing a real, basic concept of digital forensic and digital evidence. This is only the first step to this complex walking. Preliminary study on the subject of acceptability of digital evidence in courts in B&H was presented. Similar research have not been conducted in the region, while in the USA at the University of *New South Eastern* doctoral dissertation on similar subject was defended in 2010 [2].

Results of research conducted in B&H can be briefly summarized as follows:

- There is an inverse correlation and real significant relationship between age of judges and their knowledge of hardware and software, Internet and e-mail communication, digital evidence and digital forensics, which points to the fact that the ICT domain and digital forensics is more familiar to younger judges (r is from ±0.4 to ±0.7).
- There is little or almost no correlation between the age of judges and the knowledge of the term chain of custody and proving of 5WS&1h (r from ± 0:00 and ± 0.20), indicating the fact that the (no) knowledge of this process in general is independent of the age of the judges.
- The judges are the most reliant on expert witnesses and they believe in what they bring and clarify for them, and that is findings and opinion of the ICT profession expert witness.
- Lack of judges knowledge of "Fry" and "Daubert" principle is evident, and of a way that is eventually used during the digital evidence acceptance, and of which rule is followed. In this segment crucial role is played by court experts.

Research has shown that judges face with digital evidence at courtrooms in which they preside every day, in the form of electronic messages, text documents, photos, video clips (video surveillance, etc.), website, mobile phone short text messages, but that evidence are usually presented to them in material form, on paper. Consequently, they identify those evidence with the material evidence and they do not make a distinction. Most of them considered that those evidence should be treated like any other evidence, and that is the duty and obligation of an IT/ICT professional expert to clarify the evidence to them. The same troubling fact is that judges (nor prosecutors) do not know the boundary of expert's obligations, and commands that they give are usually superficial, vague, improperly written, and problems and disagreements occur from the start. Entire "*Rule 702 of the US* Federal Rules of Evidence" is completely dedicated to this type of situation, while neighboring countries (and many EU countries) do not own such law.

The survey results showed that judges are not aware of the fact that digital evidence is latent, and that they are easily changed and/or destroyed.

This shows that judges need serious and comprehensive training that will initially cover the basics of IT, Internet and network, hardware and software, and after that the process of digital forensic investigations, and ultimately digital evidence, their basic characteristics and principles of eligibility. Another problem is that, is some country in this kind of education are involved criminalist, lawyer and non-technical staff, which implies wrong and superficial training.

At the end of this book chapter it was presented a conceptual framework developed in author's dissertation, to help to develop an application for supporting judges to decide what evidence is formal acceptable and what is not. This framework is a ontology distributed under open-source license and can be downloaded from web site: https://ontohub.org/repositories/demf.

Today there is implemented a DEMF framework in java application [21] and is tested in few police Agency and prosecutor office [22].

## References


1. Casey, E.: Digital Evidence and Computer Crime, 3rd edn. Academic Press (2011)
2. Kessler, G.C.: Judges awareness, understanding, and application of digital evidence. Graduate School of Computer and Information Science, Nova Southeastern University (2010)
3. Frowen, A.: Computer forensics in the courtroom: is an IT Literate judge and jury necessary for a fair trial? (2009). http://www.intaforensics.com/Blog/Computer-Forensics-In-The-Courtroom-Is-An-IT-Literate-Judge-And-Jury-Necessary-For-A-Fair-Trial.aspx
4. Cohen, F.: Challenges to Digital Forensic Evidence. ASP Press (2008)
5. Losavio, M., Adams, J., Rogers, M.: Gap analysis: judicial experience and perception of electronic evidence. Digit. Forensic Pract. **1**(1), 13–17 (2006)
6. Mason, S.: Judges and technical evidence. In: The 2nd Internationa Conference on Cyberforensics Education and Training (2008)
7. Cosic, J.: Building an open framework for establishing and maintenance a chain of custody of digital evidence in forensic analysis of digital evidence. Doctoral Thesis, University of Zagreb (2014)
8. Zainudin, N.M., Merabti, M., Llewellyn-jones, D.: A digital forensic investigation model for online social networking (2010)
9. Kerr, O.S.: Computer Crime Law, 2d (American Casebook), 2 edn. West (2009)
10. Calhoun, M.C.: Scientific evidence in court: Daubert or Frye. 15 years later. Leg. Backgrounder (2008)
11. Frakes, K., Rogers, M., Martin, S., Scarborough, C.: Survey of law enforcement perceptions regarding digital evidence. In: Advanced in Digital Forensics III, International Federation for Information Processing, Digital Library, vol. 242, pp. 41–52 (2007)
12. Insa, F.: The admissibility of electronic evidence in court (AEEC): fighting against high-tech crime—results of a European study. J. Digit. Forensic Pract. **1**(4), 285–289 (2007)
13. Vujevic, M.: Uvođenje u znanstveni rad: u području društvenih znanosti. Informator, Zagreb (1988)
14. Rada, D.: Measure and control of non-response in a mail survey. Eur. J. Mark. **39**(1), 16–32 (2005)
15. W3C: SWRL: a semantic web rule language combining OWL and RuleML. https://www.w3.org/Submission/SWRL/. Accessed 01 Jan 2016
16. O'Connor, M.: SWRL language FAQ. ProtegeWiki. http://protege.cim3.net/cgi-bin/wiki.pl?SWRLLanguageFAQ (2013)
17. Cosic, J., Bača, M.: Do we have full control over integrity in digital evidence life cycle? In: 32nd International Conference on Information Technology Interfaces (ITI), 2010, pp. 429–434 (2010)
18. The Protégé Ontology Editor and Knowledge Acquisition System. http://protege.stanford.edu/

19. Clark&Parsia: Pellet: OWL 2 Reasoner for Java. http://clarkparsia.com/pellet/
20. Sirin, E., Parsia, B., Grau, B.C., Kalyanpur, A., Katz, Y.: Pellet: a practical OWL-DL reasoner. Web Semant. Sci. Serv. Agents World Wide Web **5**(2), 51–53 (2007)
21. Cosic, J., Baca, M.: Leveraging DEMF to ensure and represent 5ws&1h in digital forensic domain. Int. J. Comput. Sci. Inf. Secur. **13**(2) (2015)
22. Bača, M., Cosic, J., Cosic, Z.: Forensic analysis of social network. In: Proceeding of 35th Information Technology Interface Conference, IEEE, Cavtat, Croatia (2013)

# A Comprehensive Android Evidence Acquisition Framework

**Amir Sadeghian and Mazdak Zamani**

**Abstract** Android is the most popular operating system among all smart phones. This popularity increased the chances that, an Android phone be involved in a crime, either in possession of a criminal or in possession of a victim. There are many techniques exist which help the investigator to gather and extract evidence from the Android smart phones. Each of these techniques has some advantages, disadvantages, and limitations. Therefore the investigator should have knowledge of all available data acquisition techniques. The data that can be potential evidence presents in different part of an Android device. Therefore during the forensic acquisition process, the order of volatility should be considered. In this study we introduced a comprehensive framework for data acquisition from Android smart phones. Then we described the details of each step.

## 1 Introduction

Nowadays new generation of smart phones with their high performances, almost replaced the personal computers. The new series of smart phones are equipped with powerful processor, memory and graphic unit which allow them to do almost everything that a personal computer is able to do. Moreover the mobility, light weight, and access to high speed internet network over 4G are other benefits of new generation of smart phones. These advantages made most of people from different age and occupation groups to use smart phones to do their daily tasks such as checking email, writing a note, scheduling an event, write a document, and etc. Based on a

A. Sadeghian (✉)
Advanced Informatics School, Universiti Teknologi Malaysia,
Kuala Lumpur, Malaysia
e-mail: i@root25.com

M. Zamani
Department of Computer Science, Kean University, Union, NJ, USA
e-mail: mzamani@kean.edu





A.E. Hassanien et al. (eds.), *Multimedia Forensics and Security*,
Intelligent Systems Reference Library 115, DOI 10.1007/978-3-319-44270-9_15

study by Gartner, only in year 2013 near 968 million smart phones sold to the users which shows 42.3 % incline from 2012 [1, 2].

Currently smart phones are using different types of operating systems (OS). Based on a study by Gartner, in year 2013 the highest share of market was belong to Android OS with near 759 million out of 968 million devices sold [2]. The remaining part of the market share was belong to iOS, Microsoft, BlackBerry and other operating systems.

Android is a Linux based operating system which developed by Google for touch screen smart phones. Later the usage of this OS extended to other devices such as smart TVs, smart watches, digital camera, and etc. Google released Android source code by open source license. The open source feature of Android encouraged many developers to develop and release different types of applications for this OS.

The popularity of Android OS made it a good target for hackers [3, 4]. Moreover due to this popularity the chance of involvement of a smart phone with Android OS in a crime is high. An Android smart phone can be involved in crimes in three different ways. First, it can belong to a criminal. Second, it can belong to a victim. Third, it can be a repository of an offense.

- Android phone as possession of a criminal: The smart phone might use by a criminal during the process of committing the crime. For example the criminal might communicated with other criminal parties through his smart phone.
- Android phone as possession of a victim: The smart phone might use by the victim during the time of incident. For instance the smart phone might contain the communication of the victim and the criminal.
- Android phone as a repository of an offense: The smart phone might use to store and carry an offense. For example the smart phone might contain child pornography, stolen banking credentials or other sensitive information belong to the victims.

An Android smart phone can be an outstanding source of information that can be used to solve a crime case. This information can be evidence of a crime. Examining of Android smart phones always was a challenge for the investigators [5]. Some of these challenges are listed below:

- In some of Android smart phones access to external memory (SD card), without removing the battery is hard. If the SD card located under the battery, this requires to disconnect the battery which leads to losing of RAM information and in some cases activation of security mechanism.
- Android smart phones manufactures usually change the hardware and software structure of their products to remain competitive. These changes make the forensics procedure of phones harder and investigators need to have up-to-date information about the hardware and software structure of the new released models.
- The RAM and storage capacity of new smart phones are continuously increasing. Bigger size of memory and storage are equal to longer process of producing image from the smart phone.

- New generation of Android phones are equipped with encryption mechanism. Encryption mechanism encrypts the whole contents of the storage. Decrypting process of content of the phone can be very time consuming.

Mobile device forensics is the science of recovering digital evidence from a mobile device under forensically sound conditions using accepted methods [6]. Considering all the challenges about Android forensic, there should be a comprehensive framework for accessing, preserving and acquiring the evidences inside the Android smart phones. This framework should follow forensics guidelines and standards. In this research we introduce a forensic framework for collecting and preserving evidences from Android smart phones.

## 2 Problem Statement

When a smart phone with Android operating system found in a crime scene, there should be a defined framework for finding and preserving the evidences. The way an Android smart phone works is quite different from the way a personal computer does. These differences are due to both hardware and software structures of the Android phone. Therefore the investigator is not able to follow usual personal computer forensic methodologies.

## 3 Proposed Model

When an investigator decides to collect evidences, he should start from the most volatile to the less volatile [7, 8]. The order of volatility in Android phones is: memory, internal storage, externals (If exist), and SIM card. Moreover the service provider logs also can be very helpful in the forensics process. Even though they are not directly part of the Android smart phones, but they are playing an important role in the forensic process. For instance when the access to text messages and call log is restricted, the service provider can provide this information for the forensics purpose. However this is very dependent to the law of the country which the crime occurred in it [9]. More over the location of cell towers which the smart phone used them for communication also can provide an estimate location of the criminal/victim, in absence of access to Android phone GPS information [10].

The forensic process of smart phones is usually made of 3 main groups of procedures (Seizure, Acquisition, and Analysis). Each of main groups contains few sub procedures. In this paper we limit the scope of the study to evidence acquisition from the Android smart phones. Therefore the Seizure and analysis procedures are not part of the scope of this study. Acquisition is the process of retrieval of content from the smart phone memory. This process is similar to bit-by-bit copy imaging in personal computer forensics [6]. The data acquisition can be done either on-site

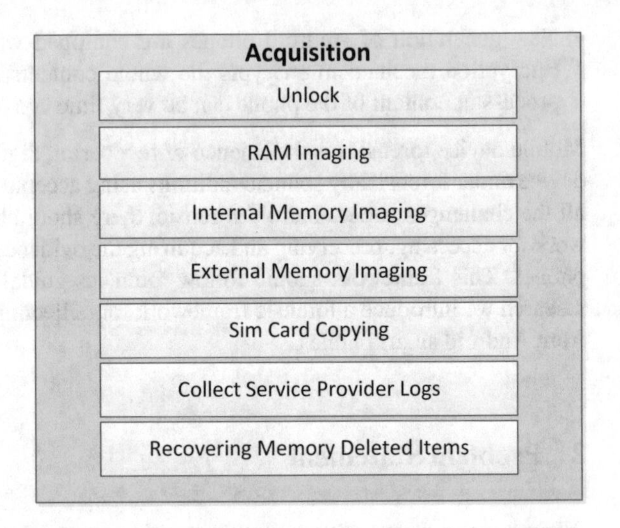


**Fig. 1** Proposed framework for android data acquisition

(live) or in laboratory. There are some considerations about battery drain of smart phone during transfer to the laboratory which is out of scope of this study [11]. In this study we assume that the smart phone found in the locked state, which is the most complicated scenario. However, if the smart phone found in unlock state, first the USB debugging should be enabled and then try to avoid the phone to go to locked mode. Then the investigator can skip the first process and go to Ram imaging step. Figure 1 illustrates the steps in our proposed framework for acquisition in Android smart phones.

## *3.1 Unlocking*

The main different of smart phones forensics with personal computers forensics is that, If the computer found in locked state, it is easy to take a bit-by-bit image of its memory. However in Android phones forensics, if the phone seize in locked state, the procedure of imaging and analysis become very complicated. The latest versions of Android OS are powered by encryption feature. Encryption feature encrypts the entire memory, and only decrypt the content after valid unlock procedure (Pattern, Pin, Password entry). Therefore in case the investigator cannot retrieve the authentication key, he will not be able to decrypt the content easily and need to do the brute force attack. In this section we review 12 techniques to retrieve the unlock key or bypass the authentication.

**Fig. 2** Sample of fingerprint traces remained on a smart phone screen (Ayers)

### 3.1.1 Smudge Attack

Smudge attack is the term which used for an attempt to understand the recent user sensitive inputs in the devices that use touch screen as their input method [12]. The constant remaining of the finger print smudges of the phone users on the touch screen can be a solution in finding the Android password pattern. Usually the password pattern smudges remain on the screen until the time that the user deliberately cleans the screen. In existence of proper lightening and high quality picture, by the help of a computer image editing software the investigator is able to make the pattern visible. However this method is not useful when the user heavily uses the touch screen after the unlocking the smart phone. Figure 2 shows a sample of finger print trace remained on a smart phone screen.

### 3.1.2 Brute-Force 4 Digit Pin

Older versions of Android smart phone used to suffer from lock screen vulnerability. Due to the short period of wait time between wrong tries, the brute force attack becomes feasible. It used to lock the user after each 5 unsuccessful attempts for 30 seconds. Therefore brute forcing the login pin from 0000 to 9999 only needs 16.6 hours. However this vulnerability is fixed in the latest version of Android, but still some of the smart phones might use the old version authentication mechanism.

### 3.1.3 Gmail Recovery

In the latest version of the Android after few unsuccessful attempts, the smart phone give a choice to the user to bypass the pattern by logging into his/her Gmail account. In case the username and password of the criminal found in the crime scene, it can be used to bypass the pattern. If the smart phone doesn't have access to the internet, this method becomes ineffective.

### 3.1.4 Unlock Using Apps

Certain types of Android applications are designed to unlock the screen. These applications can be divided into two groups.

First, those applications that are available on Google Play store. In this scenario there should be access to the owner of mobile phone Gmail account and a computer. Next the application will be install using the computer and his Gmail account through Google Play store. Then the application should be triggered to start (By a call, or a message). Finally the application unlocks the phone. This solution is only possible, if the Gmail credentials of the owner of the phone found somewhere in the crime scene.

Second group are applications that need to be transferred using the USB cable. By the help of Android SDK and Debug Bridge the unlock application should copied into the smart phone. Next using the command line the application should be installed and started to unlock the screen. This solution requires that the USB debugging to be enabled.

Both previously mentioned solutions will omit the integrity of memory data, which this is against forensics guidelines. However in the cases that the investigator doesn't have any other solution, he might use any of these procedures to access to the evidences in the smart phone.

### 3.1.5 Manipulate SQLite Database

The setting of screen lock is saved in the SQLite database. Changing the value of the setting in the database allows the investigator to bypass the lock. However this option is only available in rooted Android phones.

### 3.1.6 Cold Boot Attack

Cold boot attack allows the investigator to retrieve the password from the locked Android phones RAM by freezing the smart phone to 10 °C. Cold boot attack use remanence effect to preserve RAM content for a short period of time [13, 14]. The remanence effect says that RAM contents fade away gradually over time, as slower as colder the RAM chips are [15].

For the cold boot attack, investigator should place the phone in a freezer to reach to 5–10 °C. Then the battery should be disconnected and reconnected immediately. This method allows the investigator to save notable amount of the RAM. After rebooting the phone, the bootloader needs to become unlock which requires the erasing of the user partition. The drawback of this technique is that, during the unlocking of the bootloader the user partition will be wiped out and this can be equal to losing some amount of evidences. In a study by Muller and colleagues, they proved that by using cold boot, it is possible to recover noticeable amount of sensitive information form a locked Android phone RAM [15].

### 3.1.7 Password Cracking

Android stores the screen lock PIN as a hashed string in password.key file located at /data/system/password.key path. However the password is not in plain text. The password first salted using a random string and then hashed. Android keeps the salt for later authentication process in the /data/data/com.android.providers.settings/data bases/settings.db database. Therefore with existence of hash and the salt, it is feasible to brute force the PIN code. For doing the brute force, the sequential string first salted using the salt, then the salt hashed and finally compares against existing hash. If the hashes are equal then the string is the PIN. The limitation of this technique is in its need to file system which needs debugging to be enabled [16].

### 3.1.8 Gesture Key

Android stores the pattern in /data/system/gesture.key file. If the phone is already rooted, the investigator can simply remove the gesture.key file, which allows him to bypass the pattern. Since the possibilities of the pattern are limited, it is also possible to make a rainbow dictionary of all possible patterns and compare the hash inside the gesture.key against them [17].

### 3.1.9 JTAG

JTAG stands for Joint Test Action Group, an IEEE (1149.1) standard test access port. JTAG port usually exists on most of smart phones circuit board for debugging purposes. The JTAG can be used to access to memory and make a dump of it. The dump contains the hash and the salt, which allows the investigator to brute force and find the PIN [17, 18]. The process of producing JTAG dump can be slow. However when the developer mode is disabled, using JTAG can be one of the best solutions to bypass the authentication, while it is possible to preserve the integrity of the memory.

#### 3.1.10 Service Provider Codes

In some cases the phone might uses the SIM Card lock. SIM Card lock protects the SIM card consents from unauthorized access. However, the investigator with a letter from judiciary can ask the service provider for the PIN number.

#### 3.1.11 Check Confiscated Items

Usually the smart phone owners write down their password to avoid forgetting it. In some cases the password of smart phone and criminal email or computer is the same. Therefore the items that confiscated during the inspection should be checked carefully. There is a possibility that the password of the smart phone written somewhere on a note.

#### 3.1.12 Trial and Error

Some of the mobile phone users might chose weak pin numbers to avoid forgetting it (0-0-0-0 or 9-8-7-6). Therefore the investigator can try these possibilities to unlock the phone. However this technique is not recommended. Usually smart phones have limited number of unlock attempts, and after few try and error, the memory might be wiped or other security mechanism might be activated. If the investigator decide to use this technique, he should first check the available documents related to that specific model of smart phone to make sure how many unlock attempts are allowed.

### *3.2 RAM Imaging*

RAM is volatile type of memory which exists in Android smart phone [19]. The data on the RAM are volatile, which means they will be removed after the power source disconnected. Therefore the RAM Imaging process should take place while the Android phone is powered on and running (live). RAM usually contains information that is important for forensics investigation process. The RAM contains running processes, encryption keys, pieces of communications, open files, network structure and etc.

Forensic procedure for capturing memory should be in a way that minimizes the chance of alteration of the evidences [20]. However it is not possible to capture a

100 % exact duplicate of the volatile memory, because OS is continuously doing some changes in the RAM. The output of RAM imaging should not have noticeable different from the state before RAM acquisition.

In a study by Sylve and colleagues on volatile memory acquisition from Android devices, they come to the conclusion which, none of existing memory acquisition solutions are suitable for Android and they were not able to dump the full content of the RAM [21]. Then they introduced LiME for producing full RAM dump. In this section we will discuss the usage of JTAG and LiME for RAM imaging.

#### 3.2.1 JTAG

As we mentioned sooner in Sect. 3.1.9, Joint Test Action Group (JTAG) also known as boundary scan can be used to access directly to the memory using the processor. This technique can be used to produce an image of memory content. In this technique the risk of change in the content is minimized and memory chip doesn't need to be disordered [22].

#### 3.2.2 LiME

LiME is Linux Memory Extractor, formerly known as DMD [21]. Lime allows making a RAM dump. The investigator can choose either SD card or the network (TCP over Android Debug Bridge) as the destination of the dump. Since at this stage still the image of card (external memory) is not acquired, the only solution is to use network as the destination of the image. Using network for dumping the RAM image will minimize the overhead and footprints. Using this technique requires to have the Android Debug Bridge (ADB) feature enabled. Usually this feature is disabled by default. However there are few solutions which allow enabling ADB remotely [23].

### *3.3 Internal Memory Imaging*

The internal memory of Android smart phones should be one of the main focuses of forensics investigator. This memory contains variety types of data which can be potential evidence. Some of the examples of this information are GPS geographical locations, browser history, emails, chat histories, search histories, caches, logs, calendar events, notes, messages, voice messages, usernames, contact list, call log, social media activities, pictures, videos and etc.

In a study by Brothers, he divided the memory data acquisition to five main methods [24]. This grouping illustrated as a pyramid in Fig. 3. As we move from bottom of the pyramid to the top, each level becomes more sophisticated, costly and time consuming. In this section a description of each method is provided.

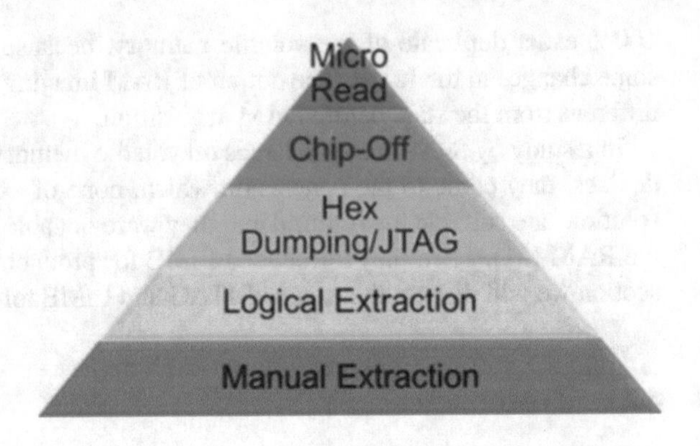


**Fig. 3** Memory acquisition techniques pyramid (Brothers)

### 3.3.1 Manual Acquisition

In this method there is a video camera which located over the smart phone and records any activity on screen of the phone. The investigator uses the touch screen and buttons to traverse through the content of the phone and in the same time the camera records all the content appears on the screen.

This technique should be used in cases which the integrity of information is not essential. Due to the fact that the integrity of the data might change during this process the data acquired from this method is not a reliable source to be used in the court [25].

This technique has some disadvantages. First disadvantage of this technique is that, it is not able to collect removed files and information which might be potential evidences. Moreover the process of traversing all sections manually can be very tedious. Second drawback of this method is that, using the touch screen and buttons can makes some unintentional changes which leads to data to modifies or overwritten. Lastly, if the smart phone found in state of broken screen or the touch screen and button are not working, this method will not be practical anymore.

### 3.3.2 Logical Acquisition

Logical acquisition refers to the process of copying of logical objects (files, folders, partitions) in the internal memory. First the connection between mobile and the computer should be established either from USB cable or other wireless connections. Then using proper tool the logical extraction of the internal memory will be produced. Logical extraction is not able to reproduce the deleted items [6].

### 3.3.3 HEX-Dumping and JTAG

Both Hex Dumping and JTAG (Explained in Sect. 3.1.9) provide more access to raw information in the internal memory. HEX-Dumping can be done either through connectivity with the phone or by removing the internal memory. The advantage of these methods is that the deleted information will be present in the produced image file. Using JTAG need high amount of training and should be done by a qualified person [26].

### 3.3.4 Chip-Off

In this method the internal memory chip physically will be removed from the mainboard and all the content of the memory will be extracted using another device [27]. For removing the memory, de-soldering process should be done. De-soldering process increases the chance of damage to the chip which may leads to lose of all data [26]. Chip-Off method is forensically-sound and assures the integrity of data. However doing the Chip-off procedure can be very costly and requires special training.

### 3.3.5 Micro Read

Micro Read is a technique which uses electron microscope to observe and record the state of gates on a NAND or NOR chip. This technique requires great amount of knowledge about electronic gates and the equipment are very costly. This technique only might be performed when the memory damaged and the phone was involved in a high profile case [6].

## *3.4 External Memory Imaging*

Usually Android smart phones are equipped with an external memory card. Memory cards designed to increase the capacity of smart phone storage. They usually used to archive less sensitive data such as music, picture, and videos. However this information can be a good source of evidence. These memory cards are removable and generally located underneath of battery. Often the content of external memory is not encrypted and it can be a rich source of evidence. Figure 4 shows different types of SD memory card.

The investigator should follow few steps for acquiring image from the external memory. First, the memory card should put on the write protected mode. Second, a hash of memory card produced for later integrity check. Then, a duplicate image of the memory should be produce using forensic tools. Finally, the hash of image should be obtained and compared against the memory card hash. Equality of both hashes shows the integrity of the produced image.

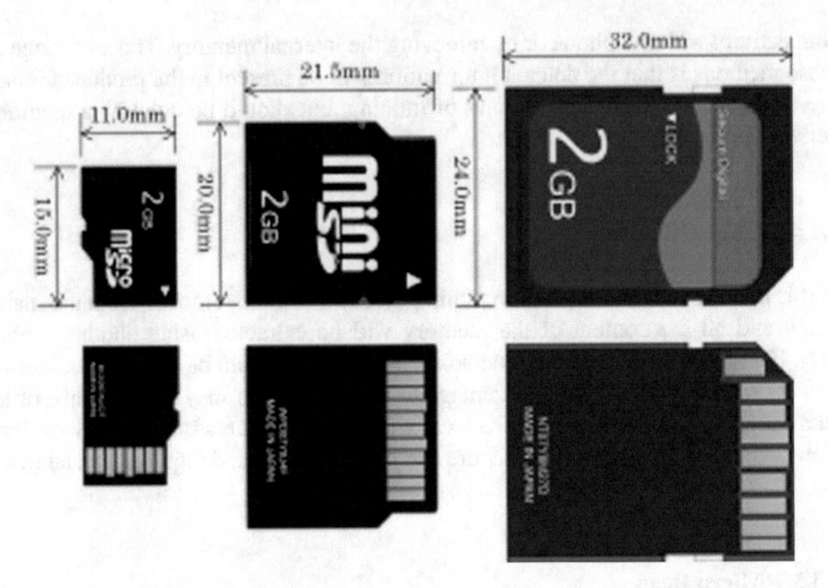


**Fig. 4** Different types of SD memory card (www.elinux.org)

## 3.5 *SIM Card Copying*

SIM stands for subscriber identity module. All mobile phones are powered by a SIM card for their communication with cellular network. The SIM card can be inserted and removed from a phone, which allows the user to move his information from one handset to another one. SIM cards have different size, but all of them used for identification and authentication purposes of the cell phone owners. SIM cards are similar to small computers, because they have their own processor, a read only memory, an operating system, a random access memory and a read only memory [6]. Figure 5 demonstrate different types of SIM card.

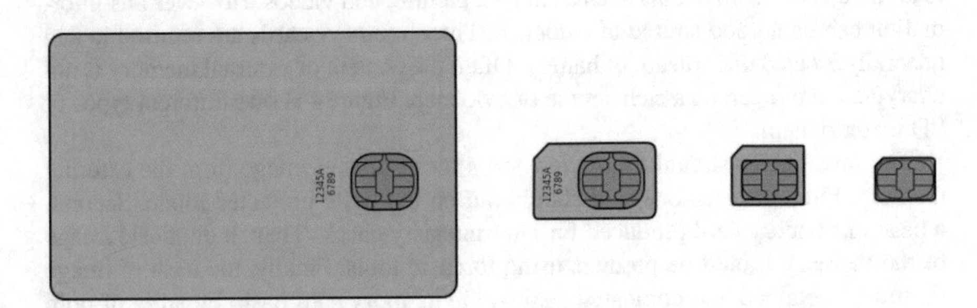


**Fig. 5** Different types of SIM card (www.JustinOrmont-Wikimedia.org)

**Table 1** SIM card access conditions

| Access condition | Description |
|---|---|
| Always (ALW) | The command always runs on the file without any limitation |
| Card Holder Verification 1 (CHV1) | The command runs only after PIN1 validation, or the time that PIN1 is not activated |
| Card Holder Verification 2 (CHV2) | The command runs only after PIN 2 validation, or the time that PIN2 is not activated |
| Administrative (ADM) | The service provider is the person who is in charge of allocation of this condition |
| Never (NEV) | The command never will run |

A SIM card contains several types of information that are valuable for a forensic investigator.

- Phone book
- Last dialed numbers
- Short message service (SMS)
- Enhanced message service (EMS)
- Incoming and Outgoing calls
- Location information for routing
- Provider information

SIM cards use different security measurement to protect their data. SIM card use access condition to assure that only user with proper authorization is allowed to run a command [28, 29]. The 5 SIM card access conditions are listed in Table 1.

For doing the forensic procedure on SIM card, first, it should be removed from the smart phone. Usually the SIM card is located underneath the battery. Then the SIM card should be cloned using a SIM card reader. The hash of image should be calculated to make sure from the integrity of image. The investigator should preserve the SIM card to avoid any changes in its data [29].

### *3.6 Collect Service Provider Logs*

Cellular network service providers record subscribers information to create usage bills. This information known as call detail records (CDR). Subscriber identity, dialed number, time, duration of the calls are few instances of this information. Moreover the location of telecommunication tower (BTS) which provided the service is also recorded. Changes between BTSs and their time are also recorded.

The period of keeping records by service provider is limited. Therefore the investigator should immediately proceed to collect this information before they get removed by service provider. By following the judicial procedures of the country, the content of SMS, MMS and voice mails may become available to the investigator.

In addition to communication information, the service provider is able to gives the following information about the SIM card owner:

- SIM card owner name and address
- SIM card billing name and address
- SIM card owner phone number
- Banking information used for billing and buying extra services
- Available services on the SIM card
- PIN1 and PIN2 numbers

This information can directly be used as an evidence or indirectly help the investigator in process of solving a case. For instance the record of cell tower location can give an estimate of the area which criminal was presented. The changes between cell towers and their time can help the investigator to understand an estimate path that the criminal traveled. However professional criminals are aware of these capabilities of service providers, and they might ask somebody to use their SIM card to produce false traces while they are in another location.

The banking information and address of the SIM card owner are also very useful evidences for proving the identity of the criminal/victim.

### 3.7 *Recovering Memory Deleted Items*

After the image from internal and external memory produced, the investigator should probe for deleted items. Usually criminals delete data related to their crime or they even might format the entire memory. When the user delete a file, the OS will not delete the actual file, but it will delete the pointer in the file table and mark the occupied space as free. Therefore the data is still in there and it is possible to recover it. However as soon as new data comes in, there is a chance that new data overwritten on the old deleted data. Different forensics tools are available which allows the investigator to recover the removed items.

## 4 Conclusion

The forensic procedure of smart phones is different from usual forensic procedure of personal computers. Therefore there should be a comprehensive framework for forensics of these devices. In this paper we focused on Android smart phones, and limit our scope to data acquisition. We proposed a complete forensic framework for acquisition of the data in android smart phones. First, we discussed different method to bypass or unlock the authentication mechanism of the Android phone. Next, we covered different acquisition methods to retrieve contents of RAM, internal memory, and external memory. Then we discussed how to make a copy of SIM card. Then we explain about the importance of service provider logs and what kind of useful information can be retrieved from them. Finally we discussed the procedure of removed data recovery. This framework will help forensic investigators to cautiously choose the proper method for acquiring data from the Android smart phone.

## References


1. Mylonas, A., Kastania, A., Gritzalis, D.: Delegate the smartphone user? Security awareness in smartphone platforms. Comput. Secur. **34**, 47–66 (2013)
2. Rivera, J., Van der Meulen, R.: Gartner says annual smartphone sales surpassed sales of feature phones for the first time in 2013. The Gartner, Egham (2014)
3. Eslahi, M., Var Naseri, M., Hashim, H., Tahir, N.M., Saad, E.H.M.: BYOD: Current state and security challenges. In: 2014 IEEE Symposium on Computer Applications and Industrial Electronics (ISCAIE), pp. 189–192. IEEE (2014)
4. Misra, A., Dubey, A.: Android Security: Attacks and Defenses. CRC Press (2013)
5. Murphy, C.A.: Developing process for mobile device forensics (2009)
6. Ayers, R., Brothers, S., Jansen, W.: Guidelines on Mobile Device Forensics, vol. 800, p. 101. NIST Special Publication (2013)
7. Brezinski, D., Killalea, T.: Guidelines for evidence collection and archiving. Request for Comments: 3227 (2002)
8. Farmer, D., Venema, W.: Forensic Discovery, vol. 6. Addison-Wesley Upper Saddle River (2005)
9. Taylor, M., Hughes, G., Haggerty, J., Gresty, D., Almond, P.: Digital evidence from mobile telephone applications. Comput. Law Secur. Rev. **28**(3), 335–339 (2012)
10. Smit, L., Stander, A., Ophoff, J.: An analysis of base station location accuracy within mobile-cellular networks. Int. J. Cyber-Secur. Digit. Forensics (IJCSDF) **1**(4), 272–279 (2012)
11. Al-Zarouni, M.: Mobile handset forensic evidence: a challenge for law enforcement (2006)
12. Aviv, A.J., Gibson, K., Mossop, E., Blaze, M., Smith, J.M.: Smudge attacks on smartphone touch screens. WOOT **10**, 1–7 (2010)
13. Gutmann, P.: Data remanence in semiconductor devices. In: Proceedings of the 10th Conference on USENIX Security Symposium, vol. 10, pp. 4–4. USENIX Association (2001)
14. Halderman, J.A., Schoen, S.D., Heninger, N., Clarkson, W., Paul, W., Calandrino, J.A., Feldman, A.J., Appelbaum, J., Felten, E.W.: Lest we remember: cold-boot attacks on encryption keys. Commun. ACM **52**(5), 91–98 (2009)
15. Müller, T., Spreitzenbarth, M.: Frost. In: Applied Cryptography and Network Security, pp. 373–388. Springer (2013)
16. Cannon, T., Bradford, S.: Into the droid: gaining access to android user data. In: DefCon Hacking Conference (DefCon12), Las Vegas, Nevada, USA (2012)
17. Munro, K.: Android scraping: accessing personal data on mobile devices. Netw. Secur. **2014**(11), 5–9 (2014)
18. Casey, E.: Handbook of Computer Crime Investigation: Forensic Tools and Technology. Academic Press (2001)
19. Berte, R., Dellutri, F., Grillo, A., Lentini, A., Me, G., Ottaviani, V.: A methodology for smartphones internal memory acquisition, decoding and analysis. In: Handbook of Electronic Security and Digital Forensics, p. 383 (2010)
20. Macht, H.: Live memory forensics on android with volatility. Friedrich-Alexander University Erlangen-Nuremberg (2013)
21. Sylve, J., Case, A., Marziale, L., Richard, G.G.: Acquisition and analysis of volatile memory from android devices. Digit. Invest. **8**(3), 175–184 (2012)

22. Breeuwsma, I., et al.: Forensic imaging of embedded systems using JTAG (boundary-scan). Digit. Invest. **3**(1), 32–42 (2006)
23. Sylve, J.: Android mind reading: memory acquisition and analysis with lime and volatility (2012)
24. Brothers, S.: How cell phone "forensic" tools actually work-cell phone tool leveling system. In: DoD Cybercrime Conference (2011)
25. Zhu, M.: Mobile cloud computing: implications to smartphone forensic procedures and methodologies. Ph.D. thesis. Auckland University of Technology (2011)
26. Hoog, A.: Android Forensics: Investigation, Analysis and Mobile Security for Google Android. Elsevier (2011)
27. Breeuwsma, M., De Jongh, M., Klaver, C., Van Der Knijff, R., Roeloffs, M.: Forensic data recovery from flash memory
28. Casadei, F., Savoldi, A., Gubian, P.: Forensics and sim cards: an overview. Int. J. Digit. Evid. **5**(1), 1–21 (2006)
29. Jansen, W.A., Delaitre, A.: Reference material for assessing forensic sim tools. In: 2007 41st Annual IEEE International Carnahan Conference on Security Technology, pp. 227–234. IEEE (2007)

# A New Hybrid Cryptosystem for Internet of Things Applications

**Ashraf Darwish, Maged M. El-Gendy and Aboul Ella Hassanien**

**Abstract** The Internet of Things or ''IoT'' defines a highly interconnected network of heterogeneous devices where all kinds of communications seem to be possible. As a result, the security requirement for such network becomes critical whilst these devices are connected. Today, all commercial applications will be performed via Internet; even the office environment is now extending to employ's home. This chapter presents a new proposed cyber security scheme for IoT to facilitate additional level of security through the involvement of a new level of key-hierarchy. In this chapter, we present the closed system environment, the proposed scheme, the services provided, the exchange of message format, and the employed four level key-hierarchies. We use application level security for selectively securing information to conserve power and increase computational speed which is useful for IoT and wireless applications. The analysis of the proposed scheme is discussed based on the strength of symmetric algorithms such as RSA and AES algorithms.

**Keywords** Cryptography · Cyber security · Hybrid cryptosystems · One-time pad · Internet of things (IoT)

## 1 Introduction

The main objective of cryptography is to allow users to communicate in a secure way over the Internet network. Cryptography is considered as two categories: symmetric key cryptography as described in [1–4] and public key-cryptography infrastructure (PKI) [5–7]. Both symmetric key and PKI possess the necessary

---

A. Darwish (✉) · M.M. El-Gendy
Faculty of Science, Computer Science Department, Helwan University, Cairo, Egypt
e-mail: ashraf.darwish.eg@ieee.org; mmostafa_fouad@yahoo.com

A.E. Hassanien
Faculty of Computers and Information, Cairo University, Cairo, Egypt
e-mail: aboitcairo@gmail.com


A.E. Hassanien et al. (eds.), *Multimedia Forensics and Security*,
Intelligent Systems Reference Library 115, DOI 10.1007/978-3-319-44270-9_16

characteristics to information security for a wide variety of applications such as electronic commerce, e-mail, e-voting systems via Internet, cloud computing, and IoT future applications.

There are four basic categories of attacks are described in [8]: (i) data disclosure, (ii) Fraud, (iii) data insertion, (iv) removal, and modification, and denial of service. Several security techniques and services are introduced in [9] to protect the mentioned threats which identified above as: authentication, access control, integrity, and privacy.

For some sensitive applications such as military applications, diplomatic communications, electronic commerce or electronic cash applications, the exchange of information through the Internet represents a sever vulnerability to such systems [10]. That is why we are in need to a secure cryptographic environment that supports a cryptographic technique that is proved to be very hard to break, through a closed system (business-to-; business connection).

Therefore, we demonstrate, in this chapter that represents an extended version [11], a new proposed unconditionally secure hybrid cryptosystem. This system is working in an environment that represents a closed system (business-to-business connection) by constructing an extranet connected over a VPN, through the tunneling technology and the e-mail exchange secure messages. The combination of the one-time pad with the RSA and AES cryptosystems, improves the one-time pad key management through the washing process. The proposed scheme provides four levels; of key-hierarchy. The first is the one-time pad key. It is used to encrypt the message after the washing process. The second is the AES session key that is used in the washing process. The third is the RSA public key, which is used to encrypt the AES session keys and the OTP pointer. The fourth is the pass phrase used to protect the encryption of the private keys. The proposed cryptosystem combines the one-time pad, known as Vernam cipher, which is theoretically unbreakable cipher, with the standard encryption algorithms, the RSA public-key algorithm, and the AES secret-key algorithm.

The one-time pad, under the assumption that it has been shared and exchanged between the users securely offers unbreakable cipher. The strength of the proposed cryptosystem will be formulated and presented. Hybrid systems are introduced in the literature based on the integration of the benefits of the symmetric and the asymmetric cryptosystems. There are public key algorithms have been presented in the literature that can provide authentication, and data integrity. Public key algorithms are computationally complex by their nature. These algorithms are known hash functions [12–15]. The Digital Signature Standard (DSS) [10, 16] can be applied in digital signature. Hybrid systems standards have been developed in the research for the exchange of encrypted data such as e-mails, including those for Privacy Enhanced Mail (PEM) [2], Pretty Good Privacy (PGP) [8], and Multipurpose Internet Mail Extensions (S/MIME).

This chapter establishes a secure cryptographic environment to sensitive applications such that their users can communicate securely, comfortably, and fast. Under this closed system environment, we are in need to a cryptosystem that achieve an ultimate security. With respect to matters related to the IoT management, it is not clear how this problem will be resolved in the context of this technology; however the distributed IoT technology can provide some solutions. Moreover, it will be difficult to retrieve all the relevant information throughout the network that might be needed for forensic analysis [17].

The rest of this paper is organized as follows. Section 2 introduces the related research work. Section 3 explores in details the proposed scheme. Section 4 describes the results and analysis of the proposed framework. A conclusion is presented in Sect. 5.

## 2 Related Work

A PKI algorithm allows client-server application to get trust in each other's authentication credentials in an efficient way. Hybrid cryptography systems can employ those credentials to improve authentication and make use of end-to-end confidentiality and integrity services. This public key functionality can enable a secure e-mail service, secure Web browsing, secure data storage, and secure networking.

A certificate is a statement issued by an authority of certification, according to a policy that binds an entity's public key to its name for a period of time. Entities can get private-public key pairs with associated public key certificates.

RSA algorithms are asymmetric cryptographic which encrypt symmetric keys in key exchange protocols and in hybrid cryptographic systems. In addition, entities X and Y can use end-to-end integrity services and confidentiality without the cooperation of any third entity.

Chatterjee [18] introduced a new method of combined cryptographic. In this paper, authors Vernam Cipher Method has been modified the standard for all characters (ASCII code 0-255) with randomized keypad, and introduced a feedback mechanism.

In [19], the Baek presented a hybrid identity-based encryption system which produce compact cipher texts while providing both efficiency and security. Baek provides a security analysis of Presented schemes against chosen cipher text attack under the well-known computational assumptions in the random oracle model.

There is computational soundness result for key exchange protocols with symmetric encryption has been presented in [20] by applying the lines of a chapter by Canetti and Herzog on protocols with public-key encryption.

Kusters [21], proposed a fully an encryption scheme. Kusters introduced a PKI scheme by using ideal lattices. The results of the proposed scheme in this paper are showed this scheme is simple and easy to be implemented for shadow images. In addition, this scheme can be applied in many electronic business areas.

Gentry [22] designed a hierarchical key assignment schemes which are secure and efficient. The Gentry applied two different constructions for time-bound hierarchical key assignment systems.

With the necessity need to store massive data in cyberspace and cross platforms, a certain security requirements must be met to efficiently protect confidentiality and privacy and this mechanism has been introduced in [23]. In this chapter, the authors review the state;-of-the-art of hybrid cryptosystems. Then, a novel scheme has been proposed for lightweight encryption of bulk data based on recursive cryptographic hashes and dynamic keys.

Nishanth [24] introduces SwiftEnc, a lightweight hybrid system that can be used effectively to encrypt bulk data. The main aim of this work was to produce a cipher text in a faster time than those of asymmetric algorithms.

A hybrid cryptography approach with the specification of two geometrical shapes i.e. ellipse and the rectangle has been presented in [25]. Two geographical shapes are taken as the cover to place the information and the series of geometric transformation operations are defined to encode the information, which uses the properties of ellipse, rectangle and symmetric-key algorithm, the algorithm is based on 2-d geometry using property of ellipse and rectangle. The work includes hybrid geometric shapes and alternative transformation so that more information security will be achieved. Also the work includes dynamic generation of key; it will provide more robustness against information size.

E-commerce is a prominent field in the future of business and security in e-commerce is becoming an important issue to transfer business from physical online stores [26]. Both protecting payment web application users and application systems need an integration of, technical, managerial, and physical controls. This chapter proposed a hybrid cryptographic system that combines both RSA and symmetric key algorithms.

In [27] Saeed and Al-Khalidi presented a new post-quantum unconditionally hiding commitment system which can support zero-knowledge protocols and allow to refresh the binding property over time. The design of the proposed system relies on the approximate shortest vector problem and lattice problem. Nguyen et al. [28] presented a new hybrid system whose bendiness relies on the discrete logarithm and approximate shortest vector problems.

A new hybrid crypto concept is proposed which is the combination of new symmetric and message digesting function (MD-5) has been presented in [29]. Moreover, the security and performance of the proposed technique are calculated and the presented results showing the performance of the proposed technique. In [30], the proposed architecture integrates the cryptographic algorithms, Advanced Encryption Standard algorithm (Symmetric) and the Hash function, SHA-2 to improve the data security to a greater extent.

## 3 Demonstration of a New Hybrid Scheme

Security is an important issue for communication via communication networks. With the increasing use of the Internet, the need of authentication and unconditionally secure encryption schemes is an important issue. Many governmental and commercial organizations are no longer willing to send their important information unencrypted over an the Internet network.

1. Target Objective

The main objective is to establish a secure cryptographic environment to sensitive applications such that their users can communicate securely, comfortably, and fast. Under this closed system environment, we are in need to a cryptosystem that achieve an ultimate security. We propose a cryptographic architecture that provides the following advantages:

(a) Using the tunneling technology to build the VPN (closed system) will enable using the infrastructure of the Internet. Tunneling makes extranet subscribers connectivity a viable option.
(b) An ultimate securely cryptosystem that provides the confidentiality, authentication, and data integrity.

2. Proposed Scheme Demonstration

Hybrid systems are based on the integration of the features of both the symmetric and the asymmetric cryptosystems. Furthermore, in our Proposed Hybrid scheme, the one-time pad cipher is applied. Therefore, we can explain the two main cycles of the scheme as follows:

(a) Sending Cycle

   i. *Washing Process*: In which, we wash (encrypt) the one-time pad key portion using AES algorithm and a cryptographic random session key using *ANSI X9.17* standard generator.
   ii. *Key Exchange Process*: In which, we encrypt the AES session key using the RSA public-key algorithm.
   iii. *Digital Envelope Process*: In which, we encrypt the message (file) with OTP, and complete our message formatting to be ready to transmit to the receiver supported with electronic digital signature of the sender.

(b) Receiving Cycle

   i. *Key Extraction Process*: In which, we extract the encrypted AES session key using the RSA public-key algorithm.
   ii. *Washing Process*: In which, we wash (encrypt) the one-time pad key portion using AES algorithm and the extracted AES session key.
   iii. *Digital Envelope Opening Process*: In which, we open our message to extract the different message segments. We decrypt the receiving message (file) and validate the digital signature of the sender and the integrity of the message.

3. Terminology Applied During the Demonstration

The following terminologies shown in Table 1 can be applied throughout the demonstration of our proposed scheme.

**Table 1** The terminology applied

| Terminology | Meaning |
|---|---|
| M, C | The plaintext and ciphertext messages |
| $OTP_e$ | The washed one–time pad |
| $AES^{-1}$ | AES decryption algorithm |
| Kid | The 256–bit AES session key |
| $RSA^{-1}$ | RSA decryption algorithm |
| KR, KP | The RSA private and public keys |
| KRA, KPA | The private and public keys of the sender A |
| KRB, KPB | The private and public keys of the receiver B |
| SHA | The secure algorithm SHA-1 |
| SIG | The signature (message digest encrypted by KRA) |
| CK | The Kid and the Scheme Pointers encrypted by the KPB |
| Env | The resulting digital envelope |
| $Digest_r$, $Digest_g$ | The received and generated digest respectively |
| + ⊕ | Concatenation and Vector sum modulo two operation |

## 4. Operational Description

The Proposed Hybrid scheme provides confidentiality and authentication that can be used for the Internet sensitive applications and file storage.

(a) Authentication

In our scheme we use the Secure Hash Algorithm (SHA-1) to calculate the fingerprint of the message [15]. Figure 1 depicts the digital signature which provided by the proposed system. The *SHA-1* and *RSA* are used to provide a secure digital signature system. The formulation of the authentication process in the scheme is as follows:

(i) sending cycle:

$$SIG = RSA_{KRA}(SHA\,[M])$$
$$Env = M + SIG$$

(ii) receiving cycle: to validate the sender and the integrity of the received message:

$$Env = M + SIG$$
$$Digest_g = SHA\,[M] \ldots \textit{generated by the receiver}$$
$$\begin{aligned} Digest_r &= RSA^{-1}_{KPA}[SIG] \\ &= RSA^{-1}_{KPA}(RSA_{KRA}[SHA[M]) \\ &= SHA[M] \end{aligned}$$

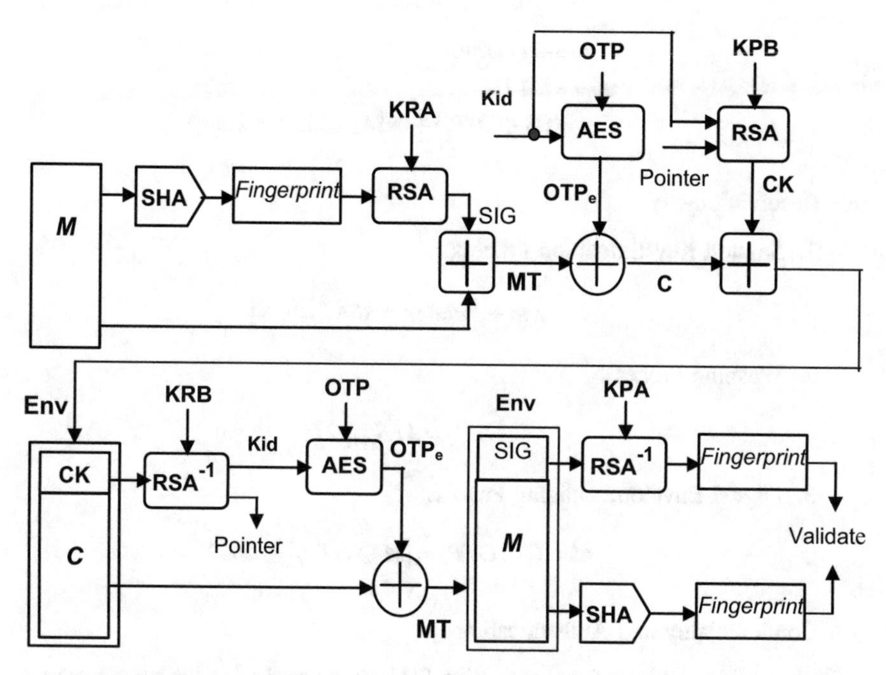


**Fig. 1** Authentication and confidentiality

(b) Confidentiality

In the proposed system in this paper, the massage has been encrypted by using the OTP string as a key with a length equal to the message length, using the vector sum modulo two operations. The receiver has the same copy of the OTP that he can use it to decrypt the received message after washing the OTP key portion. The AES session key is recovered by the RSA receiver's private key after striping it off from the message then used it in washing process. The formulation of the confidentiality is as follows:

(i) Sending cycle

1. Washing Process:

$$OTP_e = AES_{kid}[OTP]$$

2. Key Exchange Process:

$$CK = RSA_{KPB}[kid + Pointer]$$

3. Digital Envelope Process:

$$
\begin{aligned}
C &= M \oplus OTP_e \\
Env &= C + CK \\
&= (M \oplus OTP_e) + (RSA_{KPB}[kid + Pointer])
\end{aligned}
$$

(ii) Receiving cycle

1. Session Key Extraction Process:

$$Kid + Pointer = RSA_{KRB}^{-1}[CK]$$

2. Washing Process:

$$OTP_e = AES_{kid}[OTP]$$

3. Digital Envelope Opining Process:

$$M = C \oplus OTP_e = [M \oplus OTP_e] \oplus OTP_e$$

(c) Confidentiality and Authentication

Both confidentiality and authentication [31] can be applied to the same message. First, the signature of a message M is generated and appended to M. The OTP key is selected and washed using AES algorithm. Both the message and the signature together are encrypted with $OTP_e$ using the vector sum modulo two operations to generate the encrypted message C. The session key, kid, and the OTP pointer are encrypted using RSA and sent with C in a special formatting. The formulation of the confidentiality and the authentication is as follows:

(i) Sending cycle

1. Washing Process:

$$OTP_e = AES_{kid}[OTP]$$

2. Key Exchange Process:

$$CK = RSA_{KPB}[kid + Pointers]$$

3. Digital Envelope Process:

$$
\begin{aligned}
SIG &= RSA_{KRA}(SHA[M]) \\
MT &= M + SIG \\
C &= OTP_e \oplus MT = OTP_e \oplus [M + SIG] \\
&= OTP_e \oplus (M + RSA_{KRA}(SHA[M])) \\
Env &= C + CK
\end{aligned}
$$

(ii) Receiving cycle

1. Session Key Extraction Process:

$$Kid = RSA_{KRB}^{-1}[CK]$$

2. Washing Process:

$$OTP_e = AES_{kid}[OTP]$$

3. Digital Envelope Opening Process:

$$\begin{aligned} MT &= C \oplus OTP_e = (OTP_e \oplus (M + SIG)) \oplus OTP_e \\ &= M + SIG = M + RSA_{KRA}(SHA[M]) \\ Digest_g &= SHA[M] \ \ldots \textit{generated by the receiver} \\ Digest_r &= RSA_{KPA}^{-1}[SIG] = RSA_{KPA}^{-1}(RSA_{KRA}[SHA[M]]) \\ &= SHA[M]. \end{aligned}$$

Then, the two fingerprints, $Digest_g$ and the $Digest_r$, are compared for matching to validate the sender and the integration of the received message.

Before using OTP to encrypt the message, the OTP string is encrypted through the washing process using AES algorithm with a random 256–bit session key. Each OTP key and session key is used only once for each massage. To protect the used session key, it is encrypted with the RSA receiver's public key, and sent together with the encrypted massage.

(d) Transmitted Message Format

Figure 2 illustrates the transmitted message formatting, which yields the following segments:

(i) Multicasting Segment: It includes one entry for each recipient. The sender may broadcast the message to all the recipients or multicast some recipients. It includes:

- The identifier of the receiver public key, SHA (KPB) that was used by the sender to encrypt the AES session key. One entry for each recipient.
- The AES session key and the OTP pointer, encrypted by the receiver public key, KPB (Kid, Offset, Length). One entry for each recipient. The OTP pointer is composed of The starting offset of the OTP, from which an OTP string is used to encrypt the message and The length of the used OTP portion,
- The identifier of the sender public key, SHA (KPA), for which the corresponding private key was used to encrypt the message digest. Again, the public-key identifier is the SHA hash code of the key.

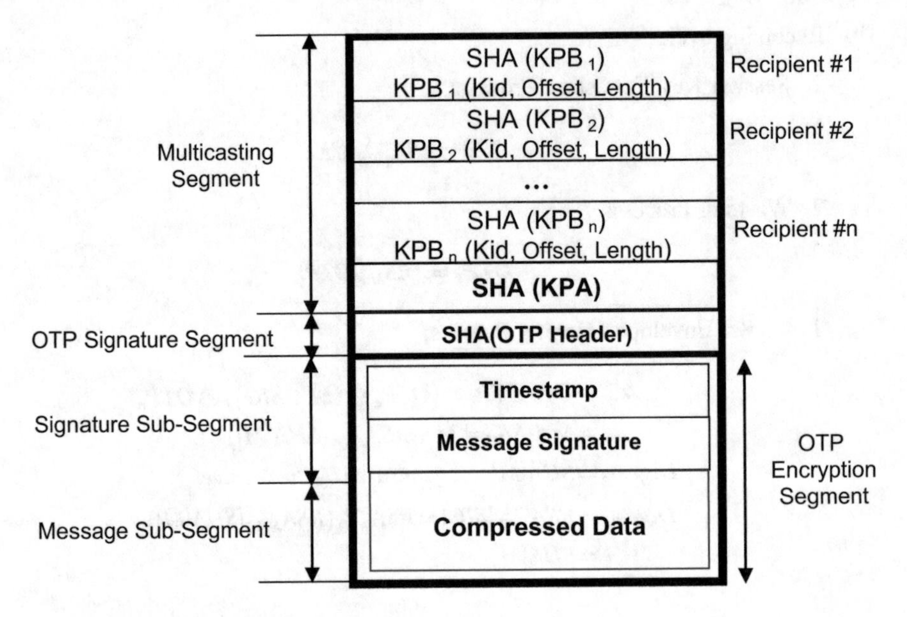


**Fig. 2** The format of the transmitted message

- OTP Signature Segment: The signature segment includes: The hash code of the OTP Header, to identify the OTP to be used by the receiver to decrypt the message.

(ii) OTP Encryption Segment: It includes two sub-segments:

- Signature Sub-segment: It includes the timestamp and message signature.
- Message Sub-segment: It includes the timestamp (the time at which the encryption was made) and the compressed data to be stored or transmitted.

5. Theory of the Proposed Scheme

Hybrid cryptosystem is currently widely used, it combines the public-key and the secret-key cryptosystems, to gain the benefits of the public-key: the strength, unforgeable digital signature, and key management, and secret-key: the security and the performance. It is good and strong but still not ultimate. We are in need to a cryptosystem that achieve an ultimate security for sensitive applications. The security of cryptographic algorithms can be measured in many different ways, stressing different metrics [32–35]. The unicity distance can be applied to approach the security strength of the proposed scheme components.

The unicity distance of the overall system should be large enough to provide perfect secrecy. The individual components of the overall system are as shown in Fig. 3. The basic components of the proposed scheme, for which we can calculate the unicity distance, are the AES, and the OTP cipher. For the RSA algorithm, calculating the unicity distance is not worthy. The security of RSA depends on a

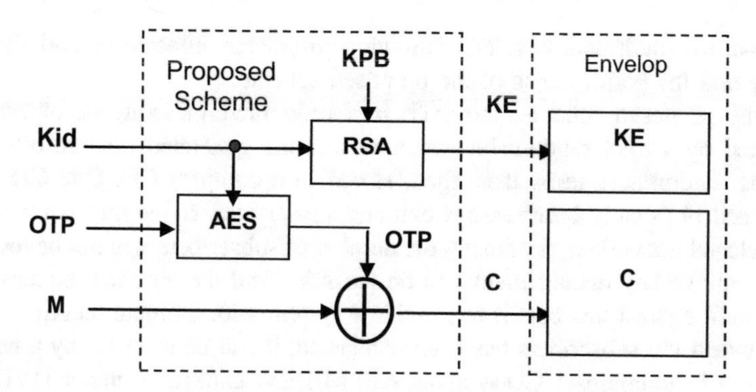


**Fig. 3** Components of the proposed scheme

hard problem that is factoring of large composite numbers which is NP-hard problem.

Unicity Distance for the One-time Pad Cipher: The One-time Pad can be considered as a cipher with a period d equal to the message length. For s possible characters (All the alphabets, digits, and special characters i.e., s = 256 characters), we can calculate the unicity distance for the OTP cipher as follows:

$$U_{OTP} = H(K)/D = Log_2 s^d/D$$

where, H (K) is the entropy of the keys, d is the message length, and D is the redundancy of the language, we have:

$$\begin{aligned} U_{OTP} &= (Log_2 s/D)d \\ &= (Log_2 256)/(log_2 256 - 2(r))d \\ U_{OTP} &= 1.4\, d\ characters, \end{aligned}$$

*where* $r = 1.2$.

This means that, we need at least 1.4 times the length of the ciphered message.

Unicity Distance for the AES Cipher: The AES algorithm enciphers a 128-bit blocks (16 characters) using a 256-bit key. We can calculate the unicity distance of the AES algorithm as follows:

$$\begin{aligned} U_{AES} &= H(K)/D = (Log_2 2^{256}/(log_2 256 - 2.4)) \\ U_{AES} &= 45.714286\ characters \end{aligned}$$

This means that, the unicity of the AES is about three blocks.

6. The Proposed Scheme Parameters

The proposed scheme employs a four level of key hierarchy. This key hierarchy makes the scheme very difficult to break. These keys are One–time pad key, cryptographic random session key, RSA Private/Public key pairs and Pass phrase–based traditional key. The following parameters affected potentially on the secrecy and the performance of the proposed scheme.

In the proposed scheme, the OTP is a truly random sequence of bits. It is generated by a truly random bit generator and the generated random bits should pass the randomness tests. It is, then, stored in a compact CD. One CD will be exchanged physically between our extranet subscribers. Since our extranet represent a closed network environment, the number of subscribers will not be too much, that the above key management will be feasible. And the cost will be acceptable. The benefit against this cost is to provide a system with ultimate security. When a CD between the subscribers has been exhausted, it can be changed by a new one. The size of the compact CD is about 650 MB and expecting, under DVD Technology, to be of several GB's in the near future. This size seems to be reasonably enough for too many exchanged messages. The compact CD with this size can be practical, and physically exchanged securely between subscribers. As the size of the OTP increases, the cycle lifetime increases. The pointer that is mentioned in our scheme is in fact a dynamic pointer and consists of the following three parts:

(1) The starting offset on the sender CD from which the OTP ring is taken to encrypt the message by the sender.
(2) The length of the used OTP portion.
(3) The OTP Header, which is the fingerprint (SHA hashed value) for one Kbytes from the beginning of the used OTP ring. This can identify the correct OTP ring to decrypt the message correctly without any mistakes.

The AES Session key will be used only once per a message. The session key is a sequence of cryptographic random numbers generated using ANSI X9.17 [2] standard cryptographic random bit generator. It is generated each time a transmission is required. It is used to wash (encrypt) the OTP using the AES. The RSA algorithm is strong enough to protect the session key.

The security of RSA depends wholly on the problem of factoring large composite numbers. As the processing power capabilities increase, this becomes apparent. The increase in size of the RSA parameters becomes the sole solution to cover this attack. In our scheme, this attack is covered by implementing the RSA algorithm with variable key length parameters up to 8192 bits, which is seemed to be sufficient for future decade. Each participant entity should maintain his private key encrypted on a protected media like smart card or optical card token.

We need to generate the RSA keys, which are p, q, n, e, d, and u. The two main keys are p and q, from which the other keys can be derived. p and q must have certain criteria in order to achieve maximum security for the generated keys, hence ensuring the security of the encryption and decryption using these keys. These criteria are as follows:

(1) p and q must be prime, to ensure the proper formulation of the mathematical parameters which depending on them.
(2) p and q must be random numbers; in order to ensure the security of the keys, so that any attacker cannot guess the generated keys or know them if they were

generated starting from a given number, or using a fixed mathematical formula.

(3) p and q must not be too close together. These are because the public parameter key, n, is the product of p and q. If p and q were close together, an attacker can start searching for p or q from the square root of n, leading to fast finding of p or q.

Calculating the difference between p and q and finding if that difference is less than a given value does checking for closeness of p and q. That value is not a constant value but instead changes when p and q gets larger. The larger the values of p and q the larger that limit must be. So, the checking is done on the "relative difference" of the two numbers. The relative difference is the ratio of the value of the difference between the two numbers, p and q, and the value of the smaller of the two numbers ($q$). In our proposed system, the relative difference must be larger than 1/128 (0.0078). That is,

$$(p-q)/q > 1/128,\ (\delta = p-q),\ \delta/q > 1/12,\ \delta > q/128,\ log_2(\delta) > log_2(q) - 7.$$

This condition must be satisfied in order to ensure that the two numbers are not close together. If we have the two *RSA* parameters $p$ and $q$, we can easily derive the other parameters $n$, $e$, $d$, and $u$ using simple mathematical formulas as follows:

$$\Phi(n) = (p-1)(q-1),\ G(n) = GCD\,(p-1,\ q-1),\ F(n) = \Phi(n)/G(n)$$

Calculating the encryption parameter $e$: $GCD(e, \Phi(n)) = 1$. Calculating the decryption parameter $d$ *and* $u$ *and* $n$ *as follows*: $e \cdot d \bmod F(n) = 1$, $p \cdot u \bmod q = 1$ and $N = p \cdot q$.

The best way to compute $e$ is to use $F(n)$ because it tends to generate the smallest possible value for $d$ that will match up with the two parameters $e$ and $n$. This means that the decryption computations will be faster.

## 4 Results and Analysis of the Proposed Scheme

The strength of AES symmetric algorithm [2] or the strength of RSA asymmetric algorithm does not bound the security of the scheme. The strength of RSA is proved in this chapter. The RSA attack, the factoring problem, seems to be effective due to the increase in processing speeds in today's computing systems. Increasing the length of the RSA parameters can cover this attack. In our scheme, we implement the RSA algorithm with variable key length parameters that can deal with length up to 8192 bit [2].

AES algorithm is a 128-bit iterative block cipher with a 256-bit key and fourteen rounds. The security of AES algorithm depends on the use of three types of arithmetical operations on 32-bit words. One important principle in the design of

AES is to facilitate analysis of its features against cryptanalysis. AES is immune from differential cryptanalysis. In addition, no linear cryptanalytic attacks on AES have been reported and there is no known algebraic weakness in AES. The strength of AES is also guaranteed through the use of a session key only once per message. A 256-bit key seems to be reasonable enough.

If we assume, in the worst case, that the session key has been compromised, then we still have the OTP, which is exchanged, securely between our extranet subscribers. On the other hand, the vector sum modulo two operation employed is fast enough that it does not add any overhead or affect the performance of the scheme. As a conclusion, the OTP provides the security [36] of the scheme if AES or RSA keys have been compromised. On the other hand, if the OTP has been compromised the RSA and AES can guarantee the security of the scheme because we apply the washing process on the used OTP portion before the message encryption process.

The scheme employs the SHA to generate the message fingerprint. It provides 160-bit message digest. It is very hard to lie under the Birthday attack.

## 5 Conclusion

This chapter presents a new cryptosystem scheme for multiple secure and attack-resistant solutions for the applications of Internet and the future of IoT. Classification of existing security algorithms has been presented in this paper depending on their key bootstrapping approach to design and implement a secure communication system. A VPN can be built using tunneling technology, and include a set of subscribers representing the members of a closed system. Such environment is necessary for the sake of protection of the sensitive applications. We have demonstrated a new hybrid cryptosystem for IoT. It combines two standard cryptosystems, the RSA, and the AES, together with the one-time pad OTP. The security strength of the scheme has been measured and demonstrated. The one-time pad presents an ultimate security. The scheme has been implemented on a commercial PC, allowing the portability, maintainability, and availability for IoT applications.

## References

1. Dorothy, E., Denning, R.: Purdue University, Cryptography and Data Security. Addison-Wesley Publishing Company (1983)
2. Schneier, B.: Applied Cryptography, Protocols, Algorithms, and Source Code in C, 2nd edn. Wiley (1996)
3. Atkins, D., Buis, P., Hare, C., Kelly, R., Nachenberg, C., Anthony Nelson, B., Phillips, P., Ritchey, T., Sheldon, T., Snyder, J.: Internet Security, Professional Reference. New Riders Publishing, Indianapolis (1997)
4. Charles, P.: Security in Computing. Prentice-Hall International Inc. (1989)

5. Brassard, G.: Lecture Notes in Computer Science, Modern Cryptology. In: Goos, G., Hartmanis, J. (eds.) (1988)
6. Goos, G., Hatmanis: Modern Cryptology, A Tutorial, Lecture Notes in Computer Science. Springer, Heidelberg (1988)
7. El Gammal, T.: A PUBLIC Key Cryptosystem and a Signature Scheme Based on Discrete Logarithms. IEEE Trans. Inf. Theory **IT-31**(4) (1985)
8. Stalling, W.: Network and Internetwork Security. Printic Hall (1995)
9. Carl Ellison, M.: The Nature of a Usable PKI. Computer Networks (1999)
10. National Institute of Standards and Technology (NIST): The digital signature standard, proposal and discussion. Commun. ACM **35**(7), 36–54 (1994)
11. ElGendy, M.M., Dakroury, Y.H., El-Hennawy, M.E., Helail, F.A., Kouta, M.M.: A proposal for a new unconditionally secure hybrid cryptosystem. In: The Proceedings of the 35th Annual Conference on Statistics, Computer Science, and Operation Research, Part (111), Cairo, Egypt, November, pp. 115–129 (2000)
12. National Institute of Standards and Technology (NIST): FIPS publication 180. Secure Hash Stand. (SHS) (1993)
13. National Institute of Standards and Technology (NIST): Announcement of Weakness in the Secure Hash Standard (1994)
14. Robshaw, M.J.B.: MD2, MD4, MD5, SHA and other hash functions. Technical report TR-101, version 4.0, RSA Laboratories (1995)
15. Chabaud, A., Joux, A.: Differential Collisions in SHA-0, Centre d'Electronique de l'Armement CASSI/SCY/EC, F-35998 Rennes Armee, France, Advances in Cryptology—CRYPTO '98, Lecture Notes in Computer Science, vol. 1462, pp. 56–71 (1998)
16. National Institute of Standards and Technology (NIST): The digital signature standard, proposal and discussions. Commun. ACM **35**(7), 36–54 (1992)
17. Roman, Rodrigo, Zhou, Jianying, Lopez, Javier: On the features and challenges of security and privacy in distributed internet of things. Comput. Netw. **57**, 2266–2279 (2013)
18. Chatterjee, T.: Symmetric key Cryptosystem using combined Cryptographic algorithms—Generalized modified Vernam Cipher method, MSA method and NJJSAA method: TTJSA algorithm, 978-1-4673-0125-1 IEEE, p. 1179 (2011)
19. Baek, J.: Compact identity-based encryption without strong symmetric cipher. In: ASIACCS'11, March 22–24, 2011, Hong Kong, China, pp. 61–70. ACM 978-1-4503-0564-8/11/03
20. Acharya, B.: image encryption using index based chaotic sequence, M sequence and gold sequence. In: ICCCS11, Rourkela, Odisha, India, pp. 541–544, ACM 978-1-4503-0464-1/11/02, 12–14 Feb 2011
21. Kusters, R.: Computational soundness for key exchange protocols with symmetric encryption. In: CCS'09, Chicago, Illinois, USA, pp. 91–100, ACM 978-1-60558-352-5/09/11, 9–13 Nov 2009
22. Gentry, C.: Fully homomorphic encryption using ideal lattices. In: STOC'09, Bethesda, Maryland, USA, pp. 169–178, ACM 978-1-60558-506-2/09/05, May 31–June 2 2009
23. Ateniese, G.: Provably-secure time-bound hierarchical key assignment schemes. In: CCS'06, Alexandria, Virginia, USA, pp. 288–297, ACM 1-59593-518-5/06/0010, October 30–November 3, 2006
24. Nishanth, R.B., Ramakrishnan, B., Selvi, M.: Improved Signcryption Algorithm for Information Security in Networks. Int. J. Comput. Netw. Appl. (IJCNA) **2**(3) (2015)
25. Alagl, Y.S., El-Alfy, E.S.M.: SwiftEnc, hybrid cryptosystem with hash-based dynamic key encryption. In: The 7th International Conference on Information Technology, ICIT (2015)
26. Malhotra, R.: A hybrid geometric cryptography approach to enhance information security. J. Netw. Commun. Emerg. Technol. (JNCET) **3**(1) (2015)
27. Saeed, Q., Al-Khalidi, Y.: E-Commerce Business Using Hybrid Combination Based on New Symmetric Key and RSA Algorithm, vol. 20, no. 1, pp. 59–71. National Chengchi University and Airiti Press Inc. Securing (2014)
28. Nguyen, K.T., Laurent, M., Oualha, N.: Survey on secure communication protocols for the internet of things. Ad Hoc Netw. **32**, 17–31 (2015)

29. Cabarcas, D., Demirel, D., Göpfert, F., Lancrenon, J., Wunderer, T.: An unconditionally hiding and long-term binding post-quantum commitment scheme. Cryptol. ePrint Arch. Rep. **628** (2015)
30. Sinha, S.K., Shrivastava, M., Pandey, K.K.: A new way of design and implementation of hybrid encryption to protect confidential information from malicious attack in network. Int. J. Comput. Appl. **80**(3), 0975–8887 (2013)
31. Kalra, S., Sood, S.K.: Secure authentication scheme for IoT and cloud servers. Pervasive Mobile Comput. **24**, 210–223 (2015)
32. Shannon, C.: Communication theory of secrecy systems. Bell Syst. Tech. J. **28**, 656–715 (1949)
33. Beker, H.: Cipher Systems: The Protection of Communications. Wiley (1983). ISBN 0-471-89192-4
34. Menezes, A., et al.: Handook of Applied Cryptography. CRC Press, New York (1997)
35. Schneier, B.: Crypto-gram (1998)
36. Moreira, N., Molina, E., Lázaro, J., Jacob, E., Astarloa, A.: Cyber-security in substation automation systems. Renew. Sustain. Energy Rev. **54**, 1552–1562 (2016)

# A Practical Procedure for Collecting More Volatile Information in Live Investigation of Botnet Attack

**Yashar Javadianasl, Azizah Abd Manaf and Mazdak Zamani**

**Abstract** Nowadays because of the growth of internet usage in all over the world, users of this global service are faced with many different threats. Attackers are trying to improve their methods in order to penetrate the users' machines to misuse their systems and their information. Most of the cyber-crimes are the result of one attack to a user or a network of many users. One of the important attacks in this area is Botnet which is controlling some compromised computers by an attacker remotely in terms of specific victim. This study tries to propose and implement a procedure in order to extract information and footprints of infected system with Botnet in order to reconstruct the Botnet attack and prepare a digital evidence package which shows the malicious activities and malicious files of this attack to present in a court.

**Keywords** Botnet · Investigation · Detection · Attack · Digital forensics · Digital evidence · Live investigation · Volatile information

## 1 Introduction

All the time security researchers (white hats) and creators of malwares and viruses (black hats) have competition on trying to discover and detect the malwares and avoiding the detection of malwares. One of the most famous and latest threats in the security area which called Botnet, is a collection of compromised and infected computers which are controlled by a bot master (herder) remotely. The original type

Y. Javadianasl (✉) · A.A. Manaf
AIS, UTM, Jalan Semarak, Kuala Lumpur, Malaysia
e-mail: jyashar2@live.utm.my

A.A. Manaf
e-mail: azizah07@ic.utm.my

M. Zamani
Kean University, NJ, USA
e-mail: mzamani@kean.edu


A.E. Hassanien et al. (eds.), *Multimedia Forensics and Security*,
Intelligent Systems Reference Library 115, DOI 10.1007/978-3-319-44270-9_17

of Botnets which used IRC (Internet Relay Chat) as their command and control channel (C&C) are used in DDOS (Distributed Denial of Service) attack [1]. White hats started to try to detect Botnet attack at network layer using network traffic aside from doing the virus scanning with famous anti viruses or signature based techniques at host based level. In this method they try to discover malicious IRC communications from transformed traffic over the network [2].

Because of using new techniques by Botnets and making its structure more complicated, the methods and paradigms of forensic investigation on Botnet should be changed. The most of Botnet creators try to use anti-forensic techniques such as rootkit and also there are many difficulties in investigating Botnet in a network layer [3], so the Botnet investigators faced with difficult problems in detecting the Botnets and reconstruct their activities on infected machines. On the other hand some information related to Botnet activities is not accessible in network layer and needs to consider on infected hosts [4]. A procedure for investigating the Botnet activities on an infected machine to extract volatile information in a live investigation will be proposed in this research [5].

## 2 Background of Problem

Botnets are one of the growing threats which are used by more attackers every day. Researches show that nowadays more than 40 % of computers which are connected to the internet are one member of Botnets. The researchers classified the research of Botnet into three main category which includes of understanding Botnets structure and behavior, detecting and tracking Botnets especially over the network, and defending against Botnets [6].

The study in analyzing and understanding the Botnets structure and behavior includes the methods which they use to communicate with their servers, the methods which they used to run and execute on infected systems and how they can control the infected system to attack to the specific victim and so on. The study on detecting the Botnet focused on the methods, which can be used to detect Botnet over the networks and also track their malicious activities. Finally the research studies in defending against Botnets covered the prerequisite measures and countermeasures to avoid the Botnet entrance and also to defend against Botnets to decrease or neutralize their malicious activities [6].

Nevertheless, those research studies do not fulfill needing a forensic investigation on Botnet attack. The objective of forensic research is to restore and reconstruct the criminal activity and using as a proof to the judge [7]. Generally, a forensic detective should adhere to the electronic forensic techniques to protect the reliability (accuracy) of evidence. Moreover, it is needed that data acquisition and analysis be repeated at any time [8]. Collected (Gathered) information should be analyze by relational procedures to describe the hierarchy of activities that happened at the criminal scene (at the scene of crime) [7].

Bot-herders are regularly shifting and adjusting the framework of their Botnets at the network level to make ever more solid control systems, and to prevent present recognition methods. However, the interaction actions and features of Botnets at the regional device level are constant, comparable and informative [4], even so, the need for forensic research on live system has become more because live 'forensics' provides useful information that cannot be acquired when the system is not on. Anyway this process cannot be repeated anytime in any situation and also it may be affected by some anti-forensic methods such as rootkit [9].

## 3 Problem Statement

One of the most serious problems related to the Botnet investigation is use of anti-forensic techniques which makes the Botnet investigation more difficult for investigators or example rootkit which hamper the forensic investigator's analysis [3]. On the other hand because of the lack of information which is gained in examining Botnet in network level for preparing a digital evidence, this neglects the importance and potential advantages of examining an infected host at the local level [4]. Moreover, the investigation on an infected machine needs a procedure to collect some more volatile information which is not repeatable so the need of live investigation before turning off the system is apparent [9], but the most important point of the live investigation is to increase the repeatability of investigation on information so it needs to extract and store enough volatile information and then analyze the running processes on physical memory of the infected system [4].

Also investigation procedure should be usable on popular types of Botnet samples which use new C&C channels such as HTTP and the investigation should be independent of previous knowledge of Botnet and network investigation.

## 4 Compare Host-Based and Network-Based Botnet Investigation

In this part two main approaches in investigating the Botnet will be compared and its results will be presented in a comparison table. It was mentioned in previous parts of this study the other problems of network-based approach were reviewed.

The Table 1 shows the comparison between these two main approaches of investigating on Botnet based on the investigation on different methods of both approaches and their results and outcomes [10].

## 5 Related Works

### *5.1 Internet Forensics on the Basis of Evidence Gathering with Peep Attacks*

Kao and Wang offered an agenda forensic action in order to research an abomination, which is committed with the awful cipher that we call it Peep. [11]. Whilst a

**Table 1** Comparison between host-based and network-based investigation

| Type | Comparison | | | | | | |
|---|---|---|---|---|---|---|---|
| | Unknown Botnet detection | C&C channel protocol dependent | Encrypted Bot detection | Network link speed | User friendly | Needs network traffic | Rootkit detection |
| Network-based | Yes | Limited approaches proposed | No | Mostly not tested | No | Yes | No |
| Host-based | Yes | Yes | Yes | Irrespective | Yes | No | Yes |

dual of Peep would be kind of Trojan horse, that would be another sort of Botnet. Peep browser and peep applicant is a sort of infected computers with P2P network protocol. Peep delivers the action, which attackers are able to admission the infected computer's file system. Meanwhile, it will let the attackers to take down and rob info and data via a command relayed medium server [12].

Investigator represents a happening of a Peep attack in order to clarify the certain appearance of agenda evidence. It would be quite similar with the structure of the Botnet, which is described in current chapter. The Botnet is linked together and complete in hierarchical. Wang and Kao took this method; therefore, they can explain the connection among the responsibility of a malicious network and the intention of an intruder. To describe this connection between them, the authors categorical three tiers of the Peep Botnet. The first one would be regularly a compromised client computer, the second one would be also different kinds of servers, and then the third one would be an attacker's computer. In order to rely on their attack command, keep stolen info and hide their actions, attackers will use the second tier. All in all, rebuilding the event could be useful for representing an incentive of the crime and the crime performance as well.

Wang and Kao have described how they continue with the Peep attack research, also explaining two other steps of test. These 2 steps contain on-line test of sniffing network traffic and off-line test. The first step of off-line test contains consecutive commands in order to research a victim's system. Researcher would act a simple forensic test like gathering network settings, checking the system clock, recognizing strange files and testing running process. Therefore, once an infected system boots up, researcher should check all the settings of the program. At the end of the examination, it has to recognize suspicious malwares through using computer forensic method like Hash analysis or scanning with antivirus devices [13].

The Authors test the network packets in order to specify the network traffic throughout the on-line test, the second step of the test. It would be sort of advised for researcher to the network traffic related to the criminal activities and to the area where the attackers receive network connectivity. Packet sniffing would be one of the best successful ways for testing the info, which has been transferred among observe the intrusion activities and infected hosts [11]. The benefit of this method was variable based on the goal of malware. In the research, the on-line research was using for specifying whether or not precious info was stolen as the key purpose of this issue malware.

This investigation on Peep attack is an acceptable sample for administering Botnet forensic investigation. It will support researchers with a whole anatomy of Botnet analysis from analyzing an infected host till disclosing the criminal activities. In a meanwhile, by using this technique, researchers can find out the intentions of the crime and criminal attitude. Authors describe the multiple tiers of the Peeps Botnet and stages of researching Peep attack, though; also they have not supported practical instructions for implementing the research in their article. Meanwhile, since this research has done manually, it will be tough for others to replicate [14].

## 5.2 Botnet Analysis

A research that was carried out by Ard [15], goes through two steps or stages in its investigation. According to Ard, two different steps are required and necessary in every Botnet investigation for identifying the author of Botnet and detecting the digital fingerprints. The first step is to analyze and inspect the malware itself, which entails an investigation of binary file. A run time evaluation for the discovery of particular network information might also be included in this examination. The other step or stage is concerned with the tracking of resources, which includes the identification of controllers, the IRC servers, and the DNS name register. However, proper processes to obtain the digital evidences in order to preserve the evidence's integrity for the investigator, was not provided by this research [15, 16].

## 5.3 Zombie Networks: An Investigation into the Use of Anti-forensic Techniques Employed by Botnets

This study looks into finding out what anti forensic techniques are being employed by Botnets during the life cycle of Botnet. Through a group of controlled experiments, some Botnets were inspected within a "safe" environment by a dynamic employment of the malware as well as a statistic code analysis [17].

In every experiment, various kinds of anti-forensic techniques that are currently being employed were recorded, and an attempt for discovering the time in the life cycle of Botnet when it was employed, was made.

These experiments indicated that Botnets employ various anti forensic techniques which are a grave challenge and obstacle for the forensic investigator. A catalogue about these techniques was created containing the challenge that might be posed to the analyst by every technique [17].

They focus on using anti-forensic techniques which are used by Botnet and analyze the behavior of each group of Botnets based on their anti-forensics techniques but they didn't try to extract information of digital evidences and also they didn't focus on host-based investigation to acquire volatile information and also they didn't use any popular Botnet sample in their investigation and didn't try to detect rootkit as an anti-forensic technique.

## 5.4 SLINGbot: A System for Live Investigation of Next Generation Botnets

Another proposed framework is SLINGbot through which the structure of harmless Botnets are facilitated by various C&C constructs in an attempt to allow the researchers to create imitated ground truth in a repeatable, safe, and controlled fashion for present and potential future Botnet threats. It is possible to employ this imitated ground truth for the characterization of different Botnet C&C constructs and after that create efficient defensive techniques [18].

SLINGbot holds some advantageous points in Comparison with present approaches to this issue. SLINGbot allows repeatable, controllable experiments through the employment of potential future and present Botnet C&C techniques; and it also is extensible motivating the employment of Botnet modules' shared libraries [18].

SLINGbot is presently being modified and adjusted to be strategically positioned in the Department of Homeland Security/National Science Foundation financing and funding the Tested laboratory of cyber-Defense Technology Experimental Research (DETER) [18].

Right now, SLINGbot is being utilized and employed for the characterization of different C&C architectures.

This framework just can help the researchers in investigating on Botnet area, it will not provide instruction to detect Botnet or extract digital evidences in Botnet attack.

## *5.5 A Host-Based Approach to Botnet Investigation*

A host oriented method in Botnet investigation is the detection and monitoring of the Botnet as well as its harmful activities on an internal host rather than doing on a network [8].

One good investigation approach is proposed in [5]. Most Botnet studies focused on the detection of Botnet existence, realizing the behaviors of them and finding ways for breaking them down. Most researchers used advanced technology for performing network level investigation [19].

Nevertheless, this study is different from most others. The researchers stressed on the value of digital footprint which can be regained from one infected host. This approach is based on the host. This is quite simple and easy. Since the network protocol collated from the host that is infected by the bot, harmful data is given for helping the network based investigation. As such, the researchers proposed the Botnet investigation that is host based and this could complement the investigations that are network based and gives clear data regarding the Botnet's complete structural design [20].

They went in detail on their ideas via assuming that the bot herder normally uses a hierarchical method for controlling their bots. Investigators who were part of the analysis of Botnet can come across bots at the last level of the hierarchy. Harmful binaries indicated at this phase can have crucial data for identifying the next phase. Next assumption is that host based method is aimed more when compared with the network level investigation. In the latter, the investigator has to keep in mind that the data connected from numerous network traffic volumes for spotting the position of the C&C server. Even though the host based investigation has minimal data generation by one or many bot clients, where an infected host is needed. They did research and gathered evidentiary data from the host that has been infected. The assumptions were tested in a forensic manner. As such, the researchers at first exposed the C&C server data for estimating the Botnet size and expanded the system of dysfunction. Later researchers analyzed the formation of Botnet as well as functions with the C&C data. They did investigate the binary code of the malware for understanding the possible threats as well as propagation approaches. Finally the investigation strategy was put forward for tracing the bot header [5].

For pursuing the bot herder, analysis of an adulterated system was used in 2 segments. The first one was a reside analysis regarding the machine that is adulterated. For the lab ambience, they had a LAN environ which was made of an adulterated ambition systems and a machine of the researcher. The system hub acts as an aperture to affix the web and shifting the entire network traffic to the program. The machine of the researcher is established with the software for catching the information regarding network that is made by the ambition machine. Apart from this software, the system acclimated by the researcher involves a strange hard drive for ensuring collection of data.

This topology is widely used for the transportation of live data to the machine of the researchers from the target machine. Researchers would gain a memory snapshot from the running target machine for reaching the entirety of the potential evidentiary information; it has to be done upon checking the traffic in the network. This analysis offline focuses on exploring the location of bot malware on the aim machine and also for recognition of the doubtful phenomena [5].

As an additional perspective, researchers propose the rebooting of the ambition machine. That information has been made would be different from the one that was calm on the antecedent study. As such the aim of this is that almost every bot are associated to the C&C server in an automatic way. The below phase of this study is similar to the general forensics procedure which contains grabbing of the disk image of the target machine with the objective of later court proceedings [16].

From this study, the researcher proposes that their host based approach, openly investigating a host adulterated by Botnet, is one sort of analytical action for the analysis of Botnet and gives an accepted practice towards this accomplishment [5]. In adverse to the research of system level, its demonstrations increase the analysis' ability as well as better after effects of recognition of the Botnet ascendancy server. Also, the evidence which has been collected from an adulterated host is clearer than if it is made in a class environment. The ban with this approach is identified, even though specifically there would be attainment of information regarding the adulterated host which may be used during extra expression. In this study, as based on the host based approach, the ambition system got to be switched off and for grabbing forensically by established forensic methods.

## 5.6 *Acquiring Digital Evidence from Botnet Attacks: Procedures and Methods*

Increasing and enhancing the repeatability of live forensic investigation as well as the accuracy and correctness of the digital evidence is the main aim of the presented approach. Moreover, the presented approach employs different and varied information types in order to elevate the efficiency of the Botnet investigation [21, 22].

The presented approach is assessed with an experiment that has two stages which are forensic investigation and malware collection. In the stage of collecting malware, Botnet samples are collected by the researchers and he or she tries to gain a better understanding of the gathered samples [13]. In the next sage, a forensic investigation is performed by the researcher on a host that has been infected with a Botnet in an attempt to try and figure out the answers to the key research questions [23, 24]. Although, it seems that, the outcomes of the host-based inspection was not enough for the reconstruction of the entire Botnet incident [24]. The researcher focuses on the weak points by the integration of two information types; the first one being created in the malware collection stage and the second derived from an infected host [21].

According to this study, the most efficient approach to a Botnet incident forensic investigation is to integrate the external information with the internal ones. The internal information such as the outcomes of malicious activities and the existence of Botnet can be provided by an infected host during a forensic investigation. And the external information created by the Sandbox services and honeypot system can offer an explanation for the external malicious activities which are done outside the infected host system [21].

This research worked on IRC sample Botnet and didn't propose any special way to detect rootkit feature using by new Botnets and also based on the mentions in this study, the investigation on a real situation didn't considered in this study.

## 6 Initial Findings

Based on the investigations on the different methods and procedures in investigation on Botnet attack, Table 2 shows the six chosen related works in this area briefly and mentions the weaknesses and advantages of each method and also will provide an overview on important points of this study.

## 7 Methodology Framework

The procedure that will be provided in this study is collecting all the information related to the topic, find out the previous methods and researches which is related to the topic, analyze them and adopt some useful features and adapt them with the objectives of the project and finally improve an efficient general procedure in live investigation using physical memory image on an infected machine with Botnets.

Because of the different problems in investigating the Botnets in network layer such as huge needed network traffics and logs and spending much time, the need for other procedures is explained. As mentioned in previous methods, it is clear that investigating host-based on infected system can be more cost effective and efficient so one part of proposed procedure in this study includes the host-based

**Table 2** Initial findings

| Year | Title | Achievement | Advantages | Weaknesses |
|---|---|---|---|---|
| 2007 | Internet forensics on the basis of evidence gathering with peep attacks | Covers Botnet forensic investigation | • With examining on an infected host system, this method will present the structure of botnet investigation completely | • A practical instruction is not provided<br>• Didn't cover rootkit detection |
| 2007 | Botnet analysis | Explains two needed important steps in any Botnet investigation | • Intends to identify and detect Botnet authors | • Needed procedures for obtaining the digital evidence and keep the integrity of it, is not provided |
| 2009 | Zombie networks: an investigation into the use of anti-forensic techniques employed by Botnets | Examines the anti-forensic techniques in Botnet investigation | • Complete overview on different types of anti-forensic techniques<br>• Use different types of C&C server types in their case studies | • Didn't cover rootkit<br>• Didn't use popular types of Botnet samples<br>• A practical instruction on host-based investigation is not provided |
| 2009 | SLINGbot: a system for live investigation of next generation Botnets | Presented SLINGbot framework which facilitates the construction of benign Botnets | • Enable researchers to generate simulated ground truth in a controlled, safe, and repeatable manner for current and potential future Botnet threats<br>• Make the investigation on Botnet more easier | • It will not provide instruction to detect Botnet or extract digital evidences in Botnet attack |
| 2010 | A host-based approach to Botnet investigation | Host-based investigation approach on Botnet attack | • Recover information from infected host<br>• Perform memory capturing<br>• Perform live investigation<br>• Increase efficiency of investigation and clear result for identifying Botnet | • Extraction of volatile information and avoiding the alteration of extracted information is not focused<br>• Popular Botnet samples are not covered<br>• Anti-forensic techniques are not covered |
| 2011 | Acquiring digital evidence from Botnet attacks: procedures and methods | Conduct an investigation on Botnet attack based on host-based approach | • Increasing the repeatability and accuracy of digital forensics live investigation<br>• Combine internal and external information on Botnet investigation | • Just worked on IRC Botnet<br>• Didn't cover rootkit detection<br>• The investigation is based on previous knowledge |

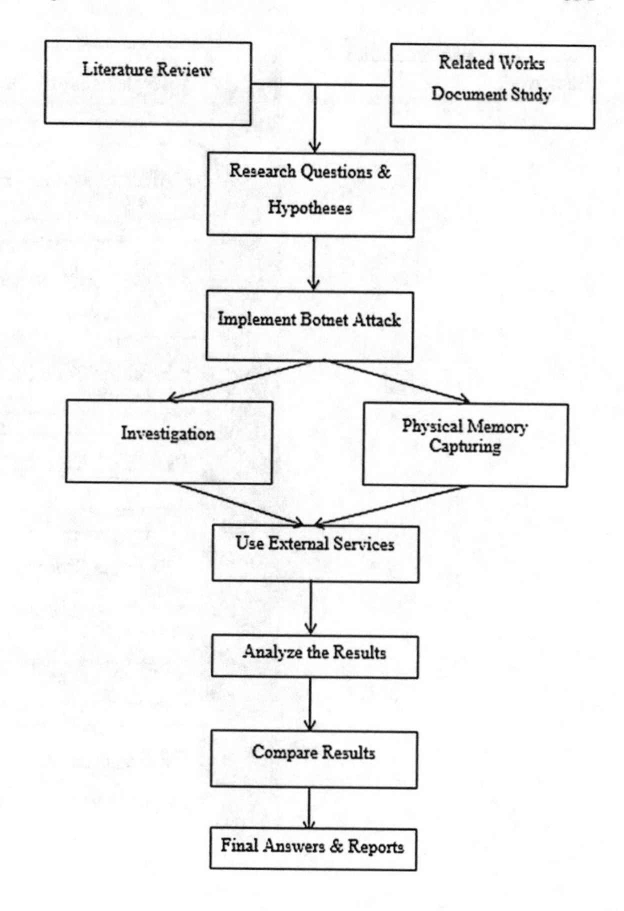


**Fig. 1** Methodology framework

investigation. Moreover, new anti-forensic techniques such as rootkit, opens a new research field in Botnet investigation so the proposed method should cover the rootkit detection, on the other hand the proposed procedure should be executable on different versions of windows operating system and most important windows XP so the main attack on this procedure will occurred on windows XP and also the attack should be based on the popular and common Botnet so for this purpose 'Zeus' which is a most popular in this area and uses rootkit technique is the case study of this project. All the mentioned points will be implemented in a lab situation close to the real situation in order to consider unexpected events (Fig. 1).

## 8 Proposed Procedure Framework

See Fig. 2.

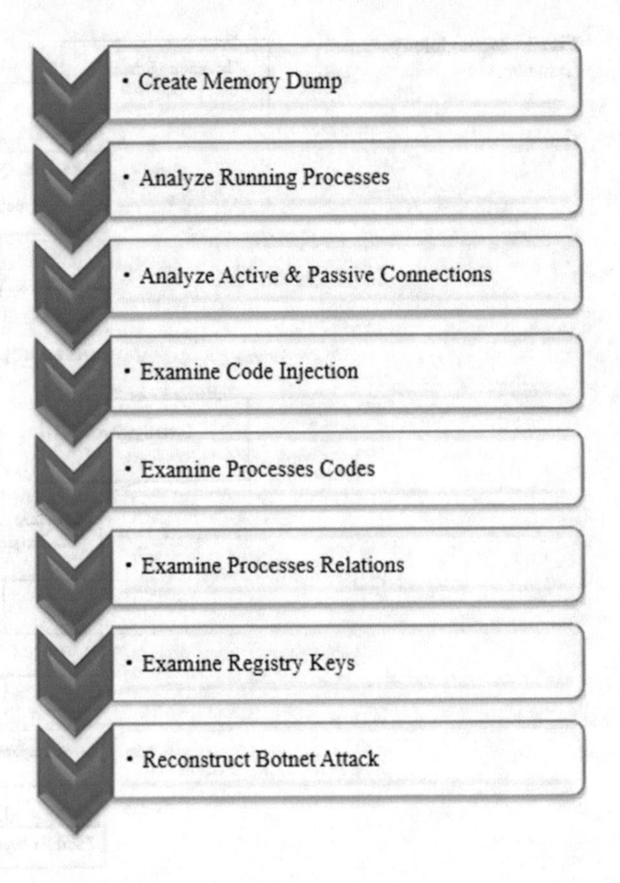


**Fig. 2** Proposed procedure framework

## 9 Attack Implementation

The first part is related to the implementation of the attack on victim system which based on the objectives of this study includes the implementation the Botnet attack using chosen sample Bot.

### *9.1 Choosing Botnet Sample*

According to the main goal of this research which is an improvement in investigating Botnet attack, it's needed to choose a new and popular Botnet sample which uses new techniques and methods in its structures and functionalities so based on the studies about different types of Botnets, Zeus Botnet has been chosen as a sample Botnet for this study in order to implement an attack [25].

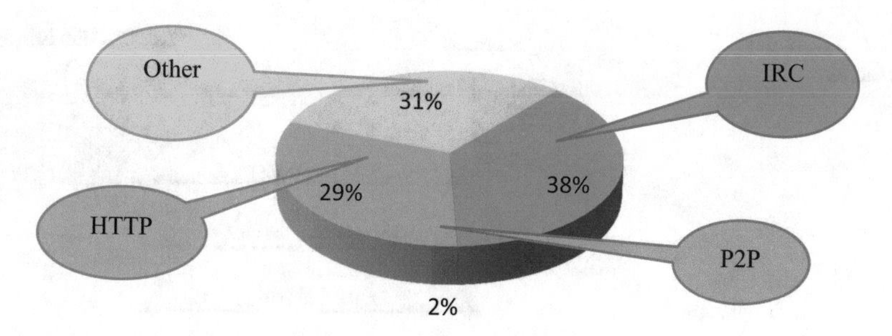


**Fig. 3** Popularity of HTTP

Based on the investigation on related works of this study most of the researches have worked on the IRC Botnets in this area so the scope of this study is based on one of the new methods which is used in communication protocol between Bots and command and control servers and based on the studies on popularity of the different types of new methods which are used by Botnets, HTTP protocol is more popular than P2P and other methods. The diagram below shows the popularity of HTTP in comparison with P2P and other methods [26].

According to the literature review of this study on different samples of Botnet and their features and also the objective of this study to cover the weaknesses of related works in this area, ZEUS Botnet can be a good case study for this research because of the using HTTP as a communication protocol and also using rootkit feature in its structure. Moreover because of ability of disabling the anti-virus and firewall of the system this kind of Botnet is really important case in terms of investigation [27] (Fig. 3).

## 9.2 *Implementation of Botnet Attack*

In this part after choosing the Botnet sample, the implementation of attack will be done. In order to implement this attack, the study on its structure and functional behavior is needed. So first of all an overview on its functional environment will be presented in order to provide different steps requirements.

Based on the Botnet structure, one system as victim and one system as a command and control server, is needed. Also Zeus botnet as same as other types of Botnets needs theses infrastructures in order to accomplish. The diagrams below show the functional environment and data transactions of Zeus Botnet [28] (Figs. 4 and 5).

Zebot, also recognized as Zeus, is a malware package that is being sold or is being traded by underground forums. This package consists of Web server files (SQL templates, images, PHP) that can be employed as the C&C server and a builder that has the ability to create an executable bot. Although Zbot is a generic back door that gives and offers an unauthorized user full control, the initial purpose of Zbot monetary gain through the online credentials like online banking, email,

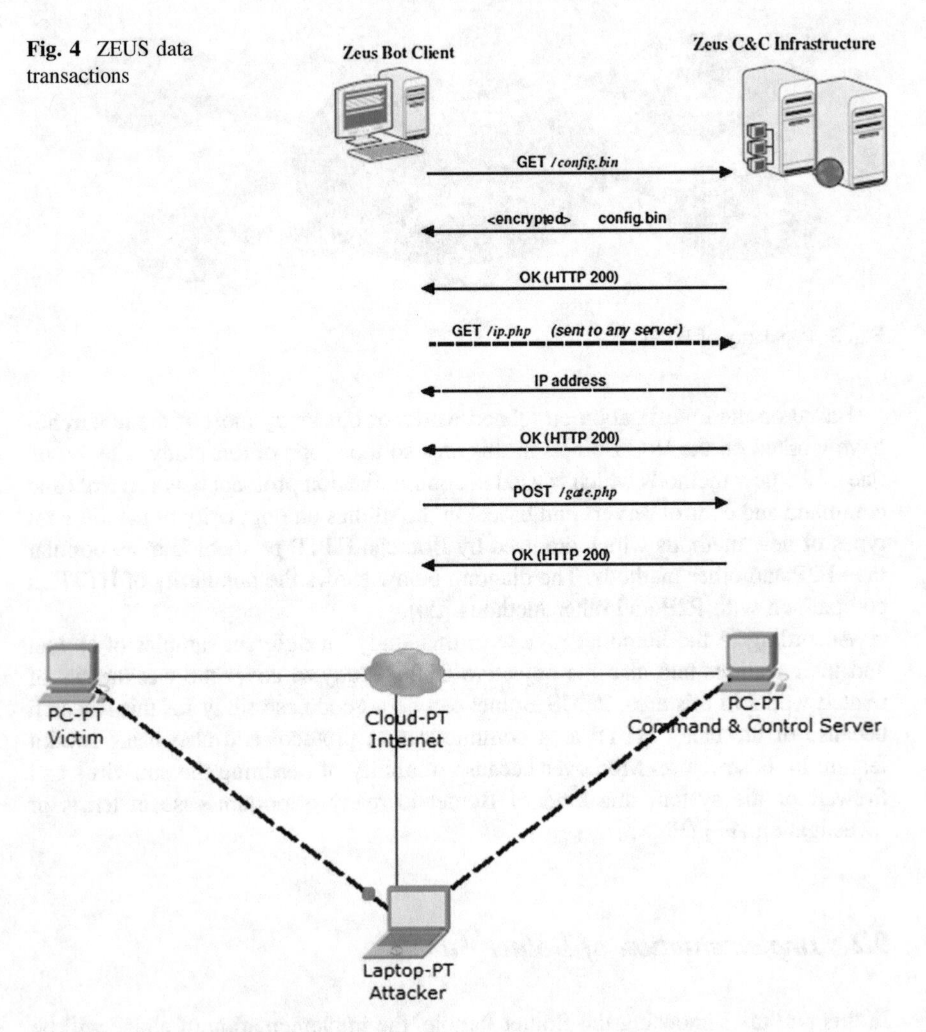


**Fig. 4** ZEUS data transactions

**Fig. 5** Botnet functional environment

FTP, along with other online passwords so first of all, the source codes of the Zeus Botnet was downloaded [28].

Based on the source codes of Zeus and its functional environment, a command and control server is needed so a virtual machine based on the Windows 7 operating system was chosen in order to run apache server. For covering both apache server to run PHP and MYSQL database to execute database file, the XAMP Server was installed on Windows operating system.

The control panel of command and control server installed on apache server and the configuration of Botnet file was applied then added to the source code in order to compile it and make an executable file in order to send to the victim system [29].

The text below shows one part of configuration of Botnet executable files which includes the IP address of command and control server.

```
entry "StaticConfig"
 :botnet "btn4"
 timer_config 60 1
 timer_logs 1 1
 timer_stats 20 1
 url_config "http://192.168.0.100/xampp/1/cfg.bin"
 url_compip "http://192.168.0.100/xampp/1/ip.php" 4096
 encryption_key "2343242566564"
 :blacklist_languages 1049
end
entry "DynamicConfig"
 url_loader "http://192.168.0.100/xampp/1/bt.exe"
```

After finishing the configuration, it was added to source code with using an executable builder, then after compiling the file named "bt.exe" is ready which can be sent to the victim system with different methods.

One of the most popular ways in order to run the executable files is using in crack files and key generators for cracked softwares. In this manner, the executable file will be added to the crack files or key generators then whenever the user wants to execute those files, the Botnet file also will be executed. Anyway the manner which is chosen to send the malware to the victim system is not the scope of this study.

For the purpose of victim system, another virtual machine which is based on Windows XP operating system was chosen. The firewall of the operating system is enabled by default.

Both of the virtual machines are connected to the internet with using same access point so the IP addresses of both systems are in same range. Victim system is using DHCP but the command and control was assigned by static IP address.

After making sure that the Botnet is active and the bot is connected to the command and control server, it is time to try to gain some information from the victim computer which can shows the main ability of Zeus Botnet in harvesting information from the victim computer and sending them to the attacker. For this purpose, imagine the situation which the user of the victim computer wants to open one website for example Facebook to log in.

Whenever the user enters the information in username and password field of this website and clicks on the log in button, the information will be transferred to the command and control server so it's the ability of Zeus Botnet in harvesting information from victim computer as a key logger.

It shows the infected system is totally connected to the command and control server and is transferring the information to the attacker so the attacker can put any command on the command and control server to remotely control the victim system to establish any kind of attack. So in this the implementation of Botnet attack is already done.

## 10 Preparing Memory Dump

According to acknowledgments and examining on previous works as mentioned during the definition of methodology of this research, footprints of the malicious code on the physical memory can be an useful evidence, so in this step is preparing an image which is captured from physical memory of the infected system before turning off the system because the methodology of this research is based on the live investigation to collect more volatile information so turning off the system can effects on the footprints of the attack.

There are many ways in order to taking image capture from physical memory. In this research one of the latest and most efficient ways, named “DumpIt” is chosen.

### *10.1 DumpIt*

Memory forensic is turning out to become a necessary part of incident response and digital forensics. A convenient way of taking a snapshot from the host is required by the researchers in cases where it is though that a system has been infected or compromised. This has been made quite easy by MoonSols’ new toolkit called DumpIt, even in cases when the individual dealing with the infected or compromised computer is not technical.

DumpIt is toolkit combined of two trusted tools which are win64dd and win32dd that are integrated into one that are executable. DumpIt has been created to be used by a non-technical user through a USB drive. The DumpIt executable has only to be double clicked by the users; this simple act enables the running of the tool. After that, DumptIt will take a snapshot from the memory of the host and save the information into the folder that the DumpIt executable was placed.

DumpIt furnishes the investigator with an easy way of attaining a Windows system’s memory image, even when the person who is investigating is not physically present in front of the target computer. The process of employing this toolkit is so easy that even a user with least bit of experience can perform it. It does not fit to every scenario; however, the acquisition of memory will undoubtedly become much easier in many situations.

The mentioned tool was executed on victim operating system to capture physical memory as shown in picture below and the captured image was saved on the specific folder.

This process can be an automated process and for example in specific times based on the planned schedule, a captured image can be taken from different systems in a network and can be shared with the investigator to check the processes of the physical memory.

After preparing the image from physical memory, next step is investigation on the physical memory image and try to extract footprints of Botnet attack from physical memory (Fig. 6).

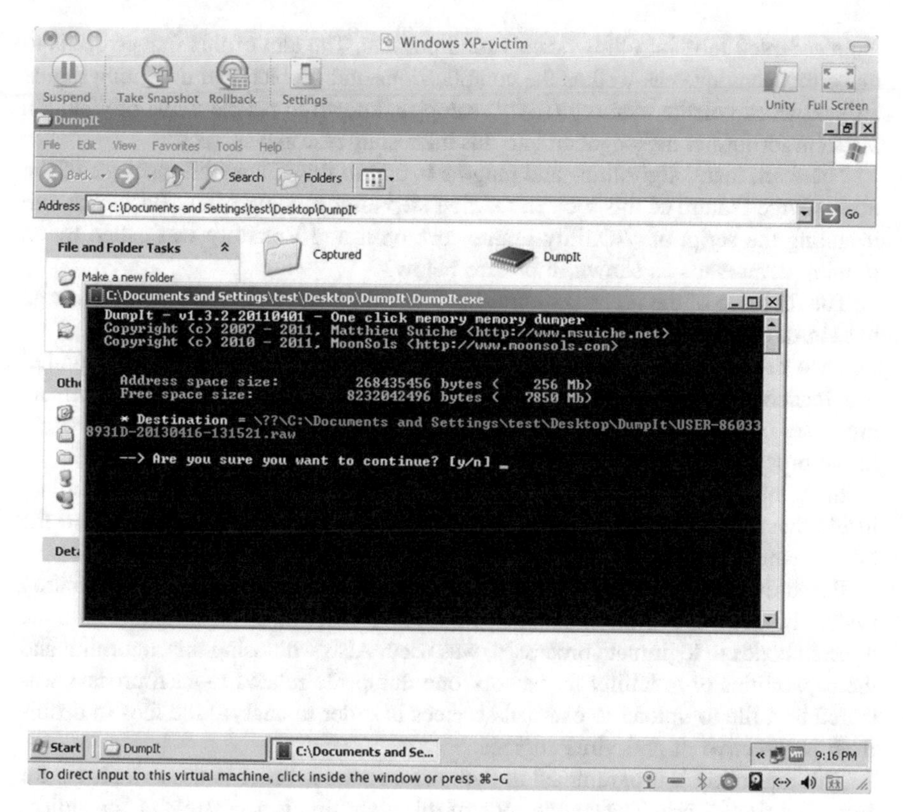


**Fig. 6** DumpIt UI

# 11 Investigation on Physical Memory Image

Now it's time to investigate the physical memory image to extract the evidences and footprints related to the Botnet attack from the physical memory. For this purpose another virtual machine based on Linux Backtrack is chosen because of better environment to run the investigation tool.

The physical memory captured file was copied to the Linux Backtrack to import to the investigation tool which is named Volatility Framework as mentioned in methodology of this research.

## *11.1 Volatility Framework*

Volatility Framework is an entire open group of selected tools that have been implemented in Python under the GNU General Public License, for the purpose of extracting digital artifacts from volatile memory (RAM) samples. The techniques of extracting are carried out independently of the system investigation; however, they preset an unprecedented look into the system's runtime state. The aim of this framework is to make the techniques as well as the complications and problems of extracting digital artifacts from volatile memory (RAM) samples, known to people while providing a platform for further investigation into this interesting research subject.

There are many algorithms and plugins to add to this framework because of the open source feature of this tool. In the first step after running Linux Backtrack and installing the script of Volatility Framework on it, the Volatility was called by its running commands as shown in picture below.

The first step of the investigation is loading the captured physical memory image to gain the information of the captured image.

Then the running processes of the physical memory when it had been captured was loaded to the environment of the volatility framework to check the legitimate processes of the operating system and also find out if there is any abnormal process in the process list.

In the next step the processes checked for find out if there is any hidden activity inside them then the active connection when the image was captured and also the other connections from the starting time of the infected system was checked.

Based on the possibility of the hidden codes in legitimate processes of operating system because of rootkit feature so "MalFind", which is an algorithm to find the injected codes to legitimate processes, was used. Also with using this algorithm and the capabilities of volatility framework one dump file related to each process was stored as a file to upload to external services in order to analyze the files in details from perspective of anti-virus engines.

This algorithm is programmed in Python language in order to use as a plugin for Volatility framework. The source code of this algorithm is accessible in appendices of this research. It was added to the Volatility framework.

Then with using one external service the dump file of each process was checked in order to find the status of the files according to the information stored in databases of anti-virus engines. For this purpose VirusTotal which is a popular analyzer service is chosen to check the memory dump of each process.

Then relationships between the processes of the operating system, was considered in order to find the main activity and its sub-activities in lower layers to find out the sequential activities for reconstructing the Botnet attack.

In final step of investigation, important registry keys which are related to the abnormal processes and also the status of operating system firewall were checked in order to complete the extracted evidences to reconstruct the Botnet attack.

All of the outputs of the implementation and the results of the implementation will be presented in next part in order to analyze and discuss about the results (Fig. 7).

## 12 The Results of Investigation on Infected Host

Before starting the investigation on physical memory image, the running processes on operating system was checked using task manager of Windows XP, but based on rootkit feature which is used by Zeus Botnet, the activities of the executable file is

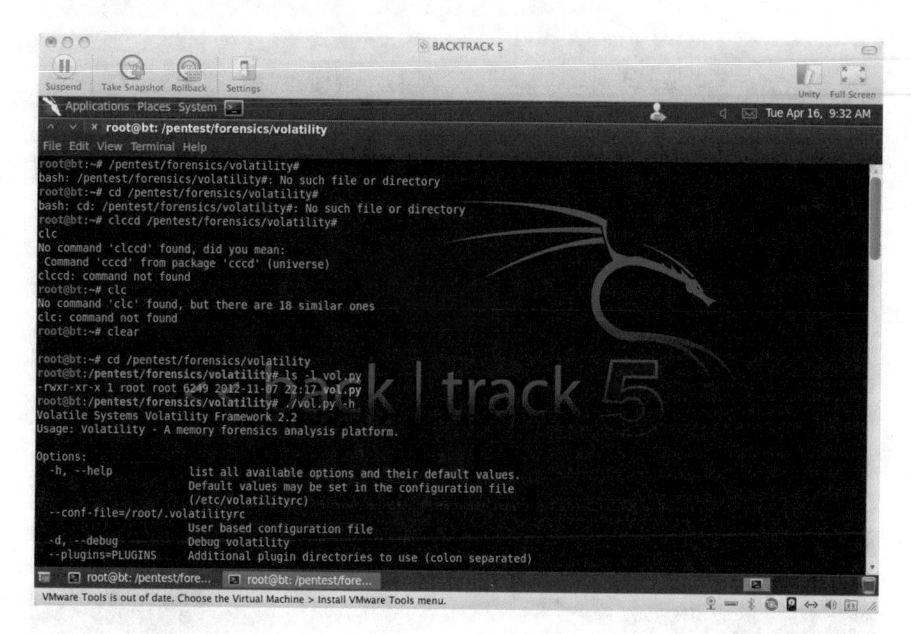


**Fig. 7** Volatility framework UI

hidden so it is not possible to find the malicious executable file using running processes of the O.S. As shown in picture below, all the running processes are Windows legitimate processes and there isn't anything abnormal (Fig. 8).

Next part was based on the importing the captured file to the investigation tool and try to load the information related to the captured memory dump. The picture below shows the information of memory dump of infected system in Volatility Framework environment (Fig. 9).

Then the list of running processes when the physical memory image had been taken, was loaded. The name of the executable file of each process and also the process Id is apparent in this list as shown in the picture below but again as is clear, there isn't any non-legitimate activity or any abnormal executable files in the list below so the progress of the investigation should be continued (Fig. 10).

It seems all the processes are legitimate processes but they may hide any malicious codes inside themselves so it was tried to go deeper for this purpose so in next step the possibility of the hidden activity inside each process was checked as shown in picture below (Fig. 11).

If any process has false for first three values so it can have potential to hide any malicious codes or activity but anyway there is nothing interesting here so it should be continued. The next step in investigating is checking the connections between the victim system and outside. As shown in picture below we found one connection which is established by process ID 860 which according to the list of processes is related to svchost.exe so because this is legitimate process of operating system, it cannot be a footprint of any attack but it has potential. So the investigation should be processed in deeper layers (Fig. 12).

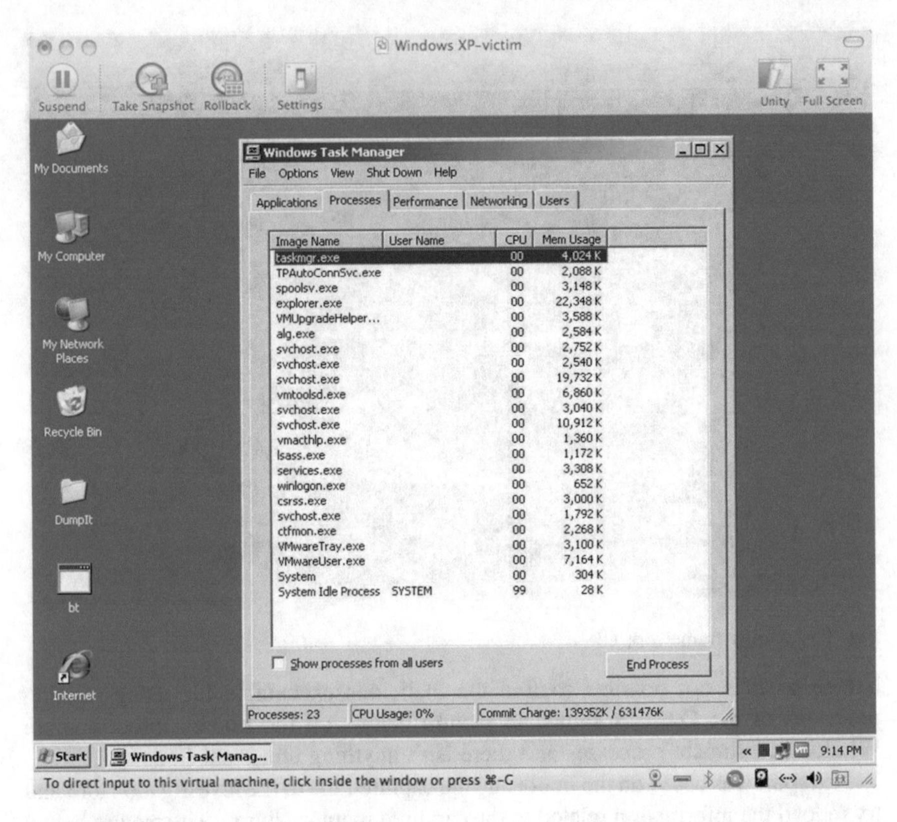


Fig. 8 Running processes of windows

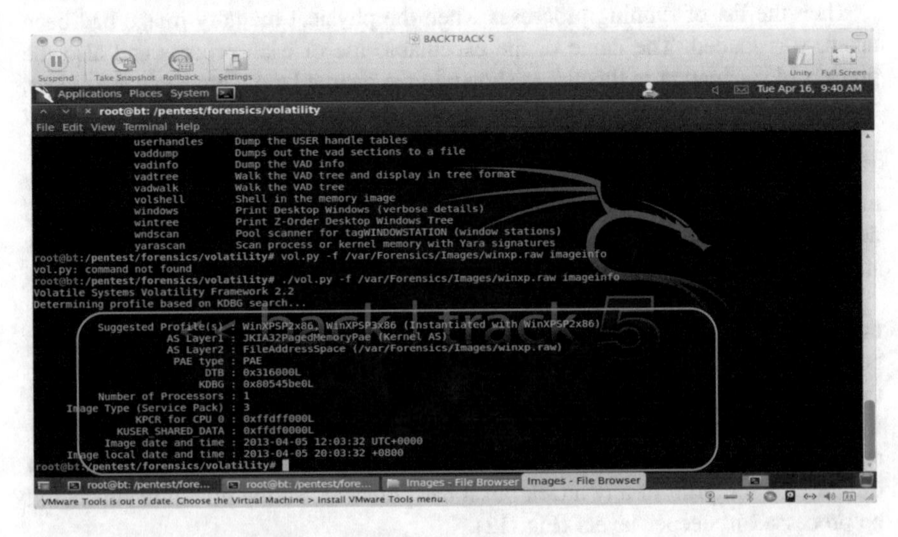


Fig. 9 Information of physical memory of infected system

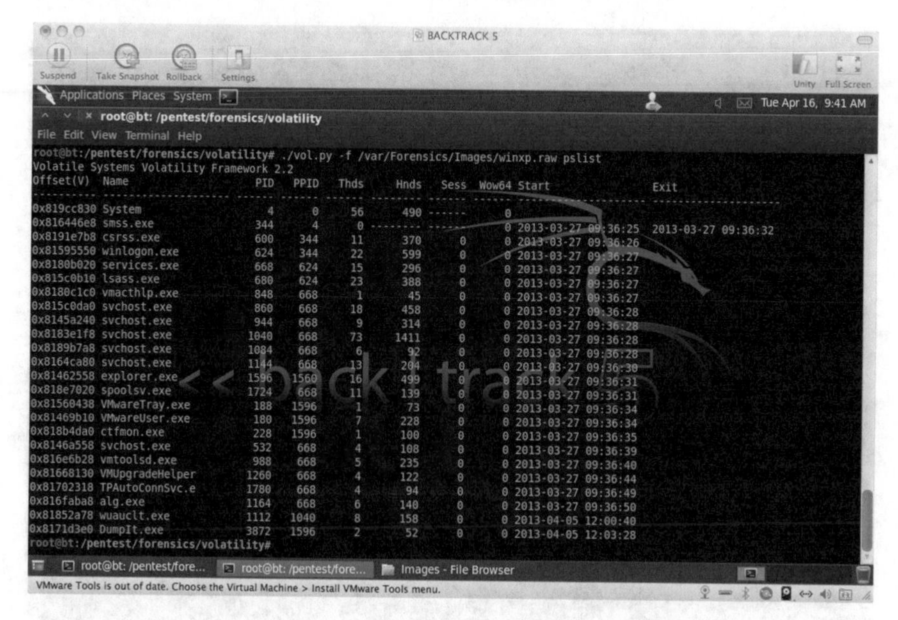


**Fig. 10** The list of running processes

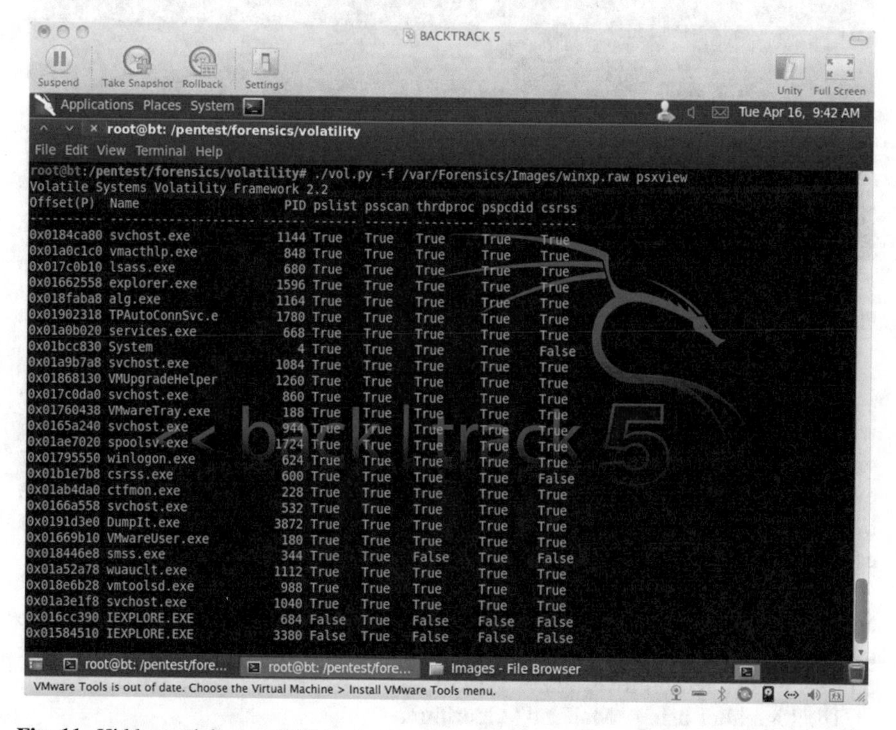


**Fig. 11** Hidden activity possibility

```
root@bt:/pentest/forensics/volatility# ./vol.py -f /var/Forensics/Images/winxp.raw connections
Volatile Systems Volatility Framework 2.2
Offset(V)  Local Address              Remote Address             Pid

0x818626c8 10.10.166.160:3975         10.10.167.3:80             860
root@bt:/pentest/forensics/volatility#
```

Fig. 12 Active connection

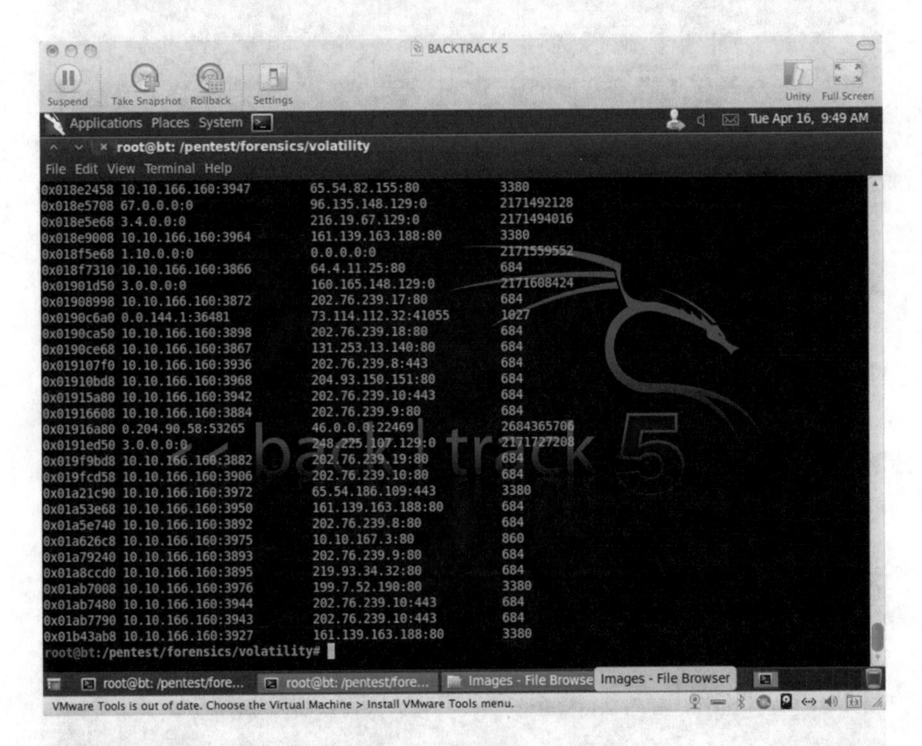


Fig. 13 List of connections

Previous step shows the active connection while the image of physical memory was captured but in this step all the connections which was established after turning on the system was checked in order to find any illegal connection but again everything seems normal and there isn't any suspicious point as shown in picture below. This list will help the investigator to find the connections which is established by each process in operating system (Fig. 13).

Based on the description on implementation part, after using "MalFind" algorithm as a plugin for Volatility Framework to detect the injected codes into the normal processes of the system this log was gained and also one dump file for each process was stored on investigator system.

The logs after using "MalFind" algorithm:

```
Process: winlogon.exe Pid: 624 Address: 0x396c0000
        Vad Tag: VadS Protection: PAGE_EXECUTE_READWRITE
        Flags: CommitCharge: 4, MemCommit: 1, PrivateMemory: 1, Protection: 6

        0x396c0000  ff b5 f0 fd ff ff e8 16 62 fa ff e9 48 3f ff ff   ........b...H?..
        0x396c0010  ff 75 08 8b 06 8b 08 68 70 5d 9d 7c 50 ff 51 24   .u.....hp].|P.QS
        0x396c0020  e9 21 0c fc ff ff 75 0c 56 ff 15 58 1a 9c 7c e9   .!....u.V..X..|.
        0x396c0030  20 ac fa ff 33 c0 40 e9 3f af fa ff ff 75 08 e8   ....3.@.?....u..

        Process: winlogon.exe Pid: 624 Address: 0x2cf30000
        Vad Tag: VadS Protection: PAGE_EXECUTE_READWRITE
        Flags: CommitCharge: 4, MemCommit: 1, PrivateMemory: 1, Protection: 6

        0x2cf30000  00 00 00 00 00 00 00 00 00 00 00 00 00 00 00 00   ................
        0x2cf30010  00 00 00 00 00 00 00 00 00 00 00 00 00 00 00 00   ................
        0x2cf30020  00 00 00 00 00 00 00 00 00 00 00 00 00 00 00 00   ................
        0x2cf30030  00 00 00 00 00 00 00 00 00 00 00 00 00 00 00 00   ................

        Process: winlogon.exe Pid: 624 Address: 0x52e30000
        Vad Tag: VadS Protection: PAGE_EXECUTE_READWRITE
        Flags: CommitCharge: 4, MemCommit: 1, PrivateMemory: 1, Protection: 6

        0x52e30000  5c f5 06 00 c8 f5 06 00 8f 04 44 7e 30 88 41 7e   \.........D~0.A~
        0x52e30010  ff ff ff ff 2a 88 41 7e a0 8e 42 7e b0 94 08 00   ....*.A~..B~....
        0x52e30020  fc 65 3f 77 4c 00 01 00 18 00 00 00 01 00 00 00   .e?wL...........
        0x52e30030  00 00 00 00 24 54 58 00 ab 8e 42 7e 00 00 ff ff   ....STX...B~....

        Process: winlogon.exe Pid: 624 Address: 0x60a30000
        Vad Tag: VadS Protection: PAGE_EXECUTE_READWRITE
        Flags: CommitCharge: 4, MemCommit: 1, PrivateMemory: 1, Protection: 6

        0x60a30000  00 00 00 00 00 00 00 00 00 00 00 00 00 00 00 00   ................
        0x60a30010  00 00 00 00 00 00 00 00 00 00 00 00 00 00 00 00   ................
        0x60a30020  00 00 00 00 00 00 00 00 00 00 00 00 00 00 00 00   ................
        root@bt:/pentest/forensics/volatility# ./vol.py -f /var/Forensics/Images/winxp.raw malfind
        Volatile Systems Volatility Framework 2.2
        Process: csrss.exe Pid: 600 Address: 0x7f6f0000
        Vad Tag: Vad  Protection: PAGE_EXECUTE_READWRITE
        Flags: Protection: 6

        0x7f6f0000  c8 00 00 00 bd 01 00 00 ff ee ff ee 08 70 00 00   ............p..
        0x7f6f0010  08 00 00 00 00 fe 00 00 00 00 10 00 00 20 00 00   ................
        0x7f6f0020  00 02 00 00 00 20 00 00 8d 01 00 00 ff ef fd 7f   ................
        0x7f6f0030  03 00 08 06 00 00 00 00 00 00 00 00 00 00 00 00   ................

        Process: winlogon.exe Pid: 624 Address: 0x45460000
        Vad Tag: VadS Protection: PAGE_EXECUTE_READWRITE
        Flags: CommitCharge: 4, MemCommit: 1, PrivateMemory: 1, Protection: 6

        0x45460000  00 00 00 00 00 00 00 00 00 00 00 00 00 00 00 00   ................
        0x45460010  00 00 00 00 00 00 00 00 00 00 00 00 00 00 00 00   ................
        0x45460020  00 00 00 00 00 00 00 00 00 00 00 00 00 00 00 00   ................
        0x45460030  00 00 00 00 00 00 00 00 00 00 00 00 00 00 00 00   ................

        Process: winlogon.exe Pid: 624 Address: 0x13890000
        Vad Tag: VadS Protection: PAGE_EXECUTE_READWRITE
        Flags: CommitCharge: 4, MemCommit: 1, PrivateMemory: 1, Protection: 6

        0x13890000  00 00 00 00 00 00 00 00 00 00 00 00 00 00 00 00   ................
        0x13890010  00 00 00 00 00 00 00 00 00 00 00 00 00 00 00 00   ................
        0x13890020  00 00 00 00 00 00 00 00 00 00 00 00 00 00 00 00   ................
        0x13890030  00 00 00 00 00 00 00 00 00 00 00 00 00 00 00 00   ................

        Process: winlogon.exe Pid: 624 Address: 0x23ec0000
        Vad Tag: VadS Protection: PAGE_EXECUTE_READWRITE
        Flags: CommitCharge: 4, MemCommit: 1, PrivateMemory: 1, Protection: 6

        0x23ec0000  00 3f 73 6e 65 78 74 63 40 3f 24 62 61 73 69 63   .?snextc@?Sbasic
        0x23ec0010  5f 73 74 72 65 61 6d 62 75 66 40 44 55 3f 24 63   _streambuf@DU?Sc
        0x23ec0020  68 61 72 5f 74 72 61 69 74 73 40 44 40 73 74 64   har_traits@D@std
        0x23ec0030  40 40 40 73 74 64 40 40 51 41 45 48 58 5a 00 3f   @@@std@@QAEHXZ.?
0x60a30030  00 00 00 00 00 00 00 00 00 00 00 00 00 00 00 00   ................

        Process: winlogon.exe Pid: 624 Address: 0x69d40000
        Vad Tag: VadS Protection: PAGE_EXECUTE_READWRITE
        Flags: CommitCharge: 4, MemCommit: 1, PrivateMemory: 1, Protection: 6

        0x69d40000  00 00 00 00 00 00 00 00 00 00 00 00 00 00 00 00   ................
        Dx69d40010  00 00 00 00 00 00 00 00 00 00 00 00 00 00 00 00   ................
        0x69d40020  00 00 00 00 00 00 0aba r 0 00 00 00 00 00 00 00 00 00   ................
        0x69d40030  00 00 00 00 00 00 00 00 00 00 00 00 00 00 00 00   ................
        Process: winlogon.exe Pid: 624 Address: 0x7e980000
        Vad Tag: VadS Protection: PAGE_EXECUTE_READWRITE
        Flags: CommitCharge: 4, MemCommit: 1, PrivateMemory: 1, Protection: 6

        0x7e980000  8b 85 2c f3 ff ff 48 48 74 11 83 e8 07 68 00 01   ......HHt....h..
        0x7e980010  00 00 75 4c 68 a8 06 00 00 eb 4a 6a 02 e8 1d f4   ..uLh.....Jj....
        0x7e980020  fe ff 3b c3 74 29 68 04 01 00 00 8d 8d f4 f4 ff   ..;.t)h.........
        0x7e980030  ff 51 68 8c 00 00 00 50 ff 15 28 16 40 76 85 c0   .Qh....P..(.@v..

        Process: explorer.exe Pid: 1596 Address: 0x2330000
        Vad Tag: VadS Protection: PAGE_EXECUTE_READWRITE
        Flags: CommitCharge: 1, MemCommit: 1, PrivateMemory: 1, Protection: 6

        0x02330000  00 00 00 00 00 00 00 00 00 00 00 00 00 00 00 00   ................
        0x02330010  00 00 33 02 00 00 00 00 00 00 00 00 00 00 00 00   ..3.............
        0x02330020  10 00 33 02 00 00 00 00 00 00 00 00 00 00 00 00   ..3.............
        0x02330030  20 00 33 02 00 00 00 00 00 00 00 00 00 00 00 00   ..3.............
```

As it seems in mentioned parts, there are some things wrong related to winlogon. exe file. MalFind algorithm detected some injected codes in legitimate processes of operating system. So the rootkit feature is detected here, but it is not enough to prepare complete evidence, so more information is needed.

A dump file is saved for each process after running MalFind so it can be used for uploading to the external services in order to find which kind of malwares it is possible to inject its code into the process. The dump files related to four mentioned parts in log file were uploaded to the VirusTotal and the results showed that the first process is clean but next three mentioned processes were detected by "McAfee-GW-Edition" as "Heuristic.Behaves Like.Exploit.CodeExec.I". The pictures below show the results of VirusTotal external service (Fig. 14).

Based on McAfee report, it shows the probability of existence a malicious code in web application related to HTTP protocol but as the report of VirusTotal shows, the other services couldn't detect anything wrong in the uploaded dump files so it shows that it is not enough evidence here so the progress should be continued more.

Next step is analyzing the relationships between running processes to find which processes are controlled by winlogon.exe which has potential to include injected malicious codes based on previous steps. The picture below shows the relationship tree between running processes (Fig. 15).

The winlogon.exe is controlling services.exe which includes many legitimate running processes so winlogon.exe may effect on the other processes also. With reviewing previous steps, a connection was active using Pid 860 which is related to svchost.exe so it can be under control with winlogon.exe so it shows the possibility

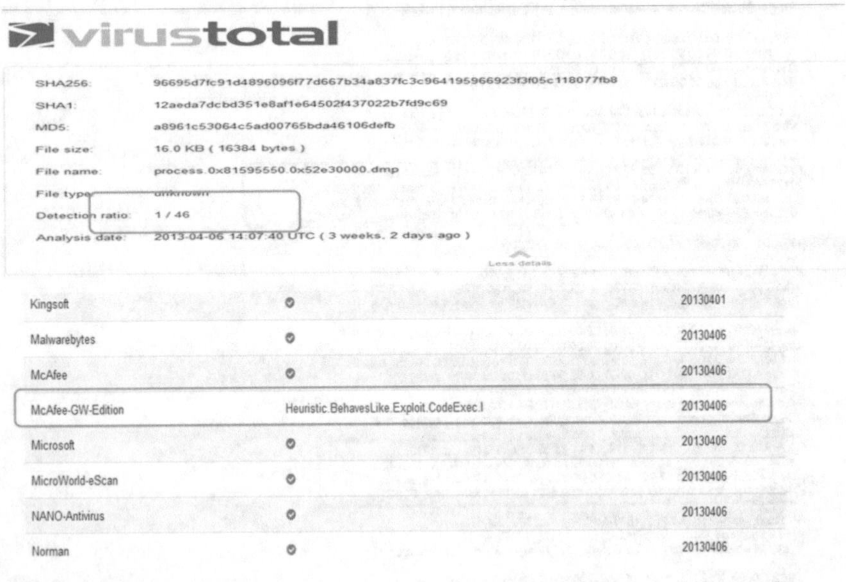

virustotal

SHA256: 96695d7fc91d4896096f77d667b34a837fc3c964195966923f3f05c118077fb8
SHA1: 12aeda7dcbd351e8af1e64502f437022b7fd9c69
MD5: a8961c53064c5ad00765bda46106defb
File size: 16.0 KB ( 16384 bytes )
File name: process.0x81595550.0x52e30000.dmp
File type: unknown
Detection ratio: 1 / 46
Analysis date: 2013-04-06 14:07:40 UTC ( 3 weeks, 2 days ago )

Less details

| | | |
|---|---|---|
| Kingsoft | | 20130401 |
| Malwarebytes | | 20130406 |
| McAfee | | 20130406 |
| McAfee-GW-Edition | Heuristic.BehavesLike.Exploit.CodeExec.I | 20130406 |
| Microsoft | | 20130406 |
| MicroWorld-eScan | | 20130406 |
| NANO-Antivirus | | 20130406 |
| Norman | | 20130406 |

**Fig. 14** VirusTotal results

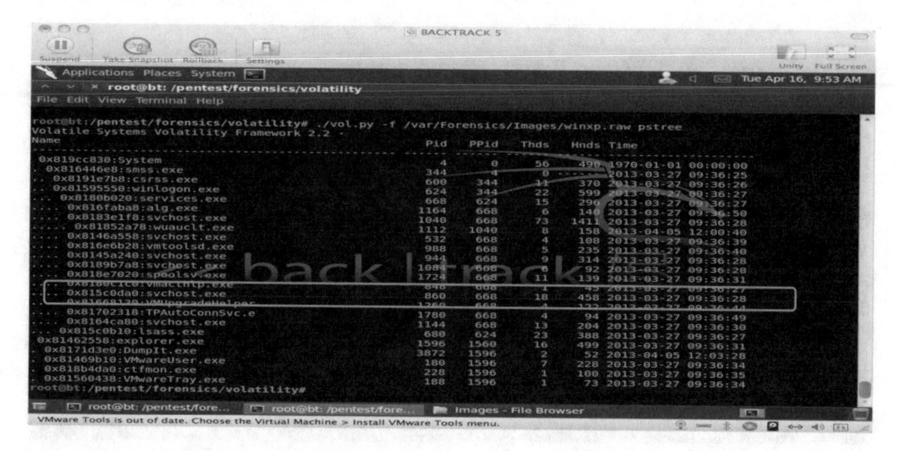


**Fig. 15** Relations between processes

of illegal active connection with svchost.exe but it is not enough so the progress should be continued in order to exactly find the malicious binaries.

Another place for detecting the footprints of main system files and running processes are registry files or in other words registry keys so in order to detect the executable files which effects on winlogon.exe, it's better to analyze the registry key which is related to the winlogon.exe to detect which executable files are effecting. The pictures below show the process of this analysis (Fig. 16).

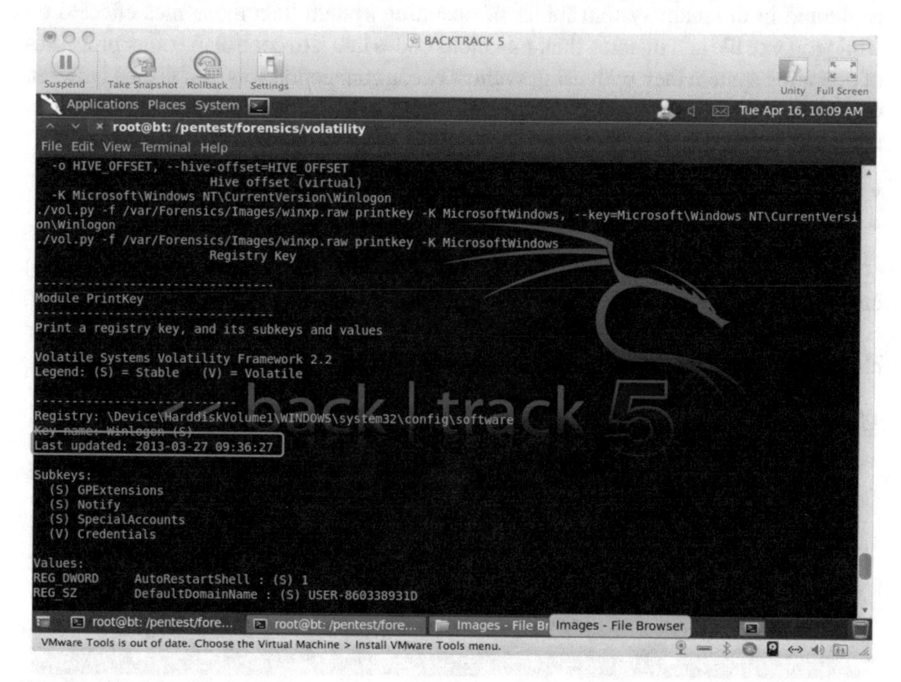


**Fig. 16** WinLogon registry key

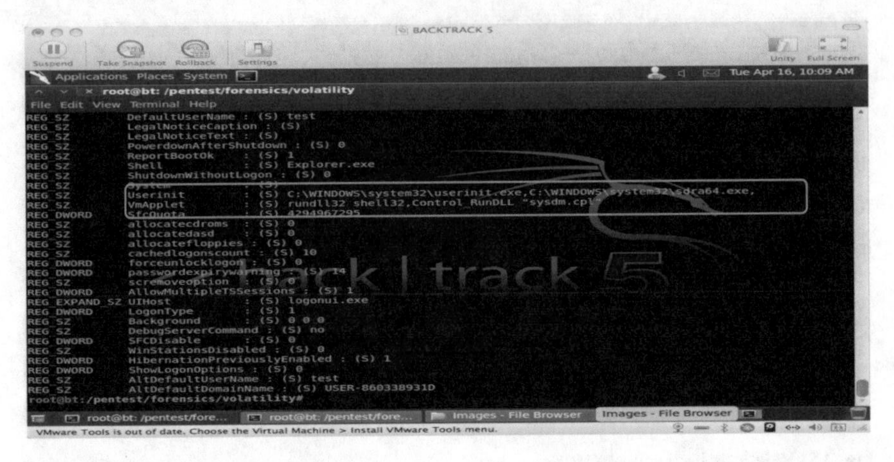


**Fig. 17** Registry key investigation result

As same as winlogon.exe this process was done for svchost.exe also to find which executable files are effecting, and finally two executable files were found on system folders of operating system which are not legitimate files so based on the acknowledgments of the Botnet sample it verified the correctness of analysis.

The picture below shows the result of analyzing the registry keys of winlogon. exe (Fig. 17).

So it indicates that as it's clear in mentioned part of the picture, two executable files are stored in the main system folder of operating system then those files effected on winlogon.exe file and injected their malicious codes into it to control the other processes of operating system then with using svchost.exe, a connection was established between the victim system and attacker's command and control server to transform the information. The executable files which were stored after executing the Botnet file, are:

- C:\WINDOWS\system32\userinit.exe
- C:\WINDOWS\system32\sdra64.exe

These are not legitimate executable files so it showed that the Botnet file (bt.exe) created these executable files which injected their malicious codes to winlogon.exe.

In final step the registry key of status of firewall of operating system was checked and the result is shown in picture below (Fig. 18).

```
root@bt:/pentest/forensics/volatility# ./vol.py -f /var/Forensics/Images/winxp.raw printkey -K "ControlSet001\Services\SharedAccess\Parameters\Firewallpolicy\StandardProfile"
Volatile Systems Volatility Framework 2.2
Legend: (S) = Stable   (V) = Volatile

----------------------------
Registry: \Device\HarddiskVolume1\WINDOWS\system32\config\system
Key name: StandardProfile (S)
Last updated: 2013-03-27 07:57:56

Subkeys:
  (S) AuthorizedApplications

Values:
REG_DWORD    EnableFirewall   : (S) 0
root@bt:/pentest/forensics/volatility#
```

**Fig. 18** Registry key of firewall status

Based on the result of this process is clear that the firewall of the system has been disabled after running the executable file of Botnet because last update time of the registry key shows the change of the status of firewall. This is the last evidence against Botnet attack in this sample.

## 13 Findings of Investigation

After finishing the investigation on infected system physical memory, some important information related to the Botnet attack was gained so some important digital evidences were prepared. In this part a general overview of processes which were done during the implementation of investigation and also the results of each part will be considered.

In the next part the proposed procedure of this study will be compared with other procedures and other approaches based on the findings and results of this procedure. Some parameters were chosen to use in comparison part and finally one numerical value will be assigned to the performance of each procedure in two aspects and the results will be presented on bar charts.

The diagram below shows the sequential stages in investigation procedure and also the output results of each step is shown in this diagram (Fig. 19).

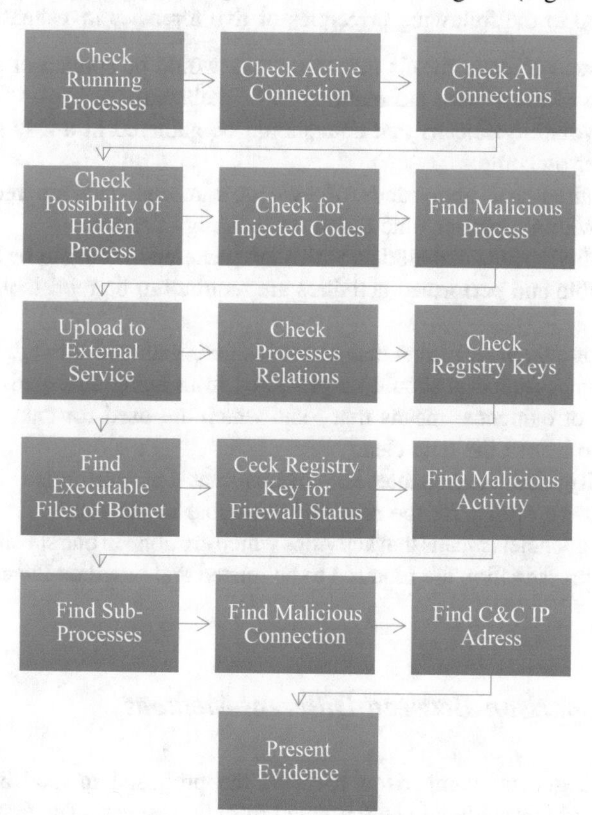


**Fig. 19** Overview of investigation processes

## 14 Investigation Evaluation

In DFRWS 2006, digital forensics investigator of FBI Live forensics emphasized that investigation turn into one of the most significant area in investigation of digital forensics because of increase in the size of data storage, central internet cases and critical requirement of investigation. Toolboxes are required to be developed based on essential investigation process so that the requirement of live forensics investigation is supported. A lot of freeware IT security together with incident response tools have been improved before the improvement in live forensics process. The majority of them were utilized for obtaining specific piece of information. Some forensics investigator began developing their investigation process in order to execute these series of tool [30].

Nevertheless, every toolkit has got its own process of executing. Majority of them were developed on the basis of experience of developer. Therefore, it is not easy to declare whether the gathered evidence of live forensics is sound forensically. Live forensics investigation data which are presented to the court are required to be testified case by case when there is no attestation and certification process [30].

Live forensics investigation, as emphasized in a study by Ieong and Chau [30], can be sum up to the following principles of live forensics investigation [30]:

- Completeness of data: means that those data would be spoiled or affected after shut down of the system are required to be gathered.
- Order of volatility: means that data should be gathered in a way that does not affect other outcome.
- Time required and importance of evidence: means data are required to be gathered within a proper time based on their significance.
- Repeatability: means that all data which are gathered in order to be tested should be accessible and performed activities are required to be repeatable as much as possible.
- Integrity of evidence: means that data which are gathered through investigation of live digital forensics should be protected from being damaged.
- Accuracy of evidence: means that tools which are used for data gathering are required to record the data clearly.
- Verifiability and Reasonableness: means that the performed activities are required to be proper for the case and verifiable in the court.
- Case dependencies: means that activities which are done in one specific live digital forensics investigation are required to be related and based on the case (Fig. 20).

### *14.1 Comparison Between Different Methods*

In this part a general comparison between the proposed method and the other related works of this study will be presented from two aspects. The first one is based

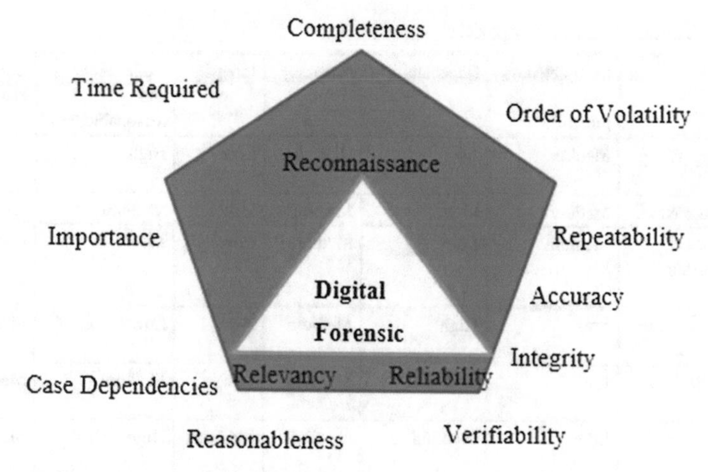


**Fig. 20** Digital forensics investigation principles

on general parameters which are important in any investigation in digital forensics area.

Based on the literature review on evaluation method on digital investigation, some important principles were chosen to use in comparison different methods and procedures which were discussed during this study with proposed procedure of this study and also in second part of the comparison, the results of this study will be compared with the other related works based on the outcomes and contributions of different procedures.

In the progress of this part of the study, the comparison tables will be assigned with a numerical value to draw a chart for making the comparison results more clear (Tables 3, 4 and 5).

The numerical values are based on the range of the assigned values of each parameters and then with using a formula the average of each methods based on chosen principles will be calculated which shows the performance of each method in comparison with proposed procedure.

Finally, calculated values will be shown in bar charts to give a better overview of the effectiveness of the proposed method in comparison with other related works in this area.

All assigned values are based on the extracted information in reviewing the related works documents. The weaknesses and recommendations for future works of each study and also the contributions and mentioned results of each one is the basis of this comparison and the differences between low, medium and high values is based on the quantity of each principle in each method.

$$\mathbf{P = 100 \sum_{i=1}^{n} \frac{x_i}{3n}} \tag{1}$$

**N = Number of Principles**
**$x_i$ = Assigned Value**
**P = Performance**

**Table 3** Digital forensics comparison

| | Completeness and reliability | Repeatability | Accuracy | Integrity | Verifiability and reasonableness | Ease of use and review |
|---|---|---|---|---|---|---|
| Shiuh-Jeng Wang method | Medium | Low | High | Low | High | Low |
| Chris Ard method | Medium | Low | Medium | Low | Medium | Medium |
| Jeremy David Brazier Annis method | Medium | Low | Medium | Low | Medium | Low |
| Alden W. Jackson method | Low | High | Medium | High | Low | Medium |
| Frank Y.W. Law method | Low | Low | Medium | Low | Medium | Medium |
| Junewon Park method | Low | Medium | Medium | Medium | High | Medium |
| Proposed procedure | Medium | High | High | High | Medium | Medium |

Equation 1: performance calculation.

In each part of comparison, some of principles were chosen and also one numerical value was assigned to each principle in order to calculate the performance of these methods and procedures.

Based on the value for each principle in each table with using the mentioned formula the performance of each method in each table was calculated and all the results are shown in charts below to differentiate between different methods and show the improvement of this study.

The charts show that the proposed method of this study has an improvement in the type and number of extracted volatile information and also the quality of the information based on the principles of digital forensic investigation was improved in comparison with other methods (Figs. 21 and 22).

## 15 Contributions of This Study

The contributions of this study can be classified into some points from different aspects:

- Proposing a procedure for collection more volatile information of Botnet attack
- Collecting information is independent of network-based approach
- Collected information can reconstruct Botnet activity hierarchy completely
- Reconstructing the Botnet activity is independent of previous acknowledgments of Botnet malware

**Table 4** Botnet investigation comparison

| | Executable files of Botnet | Registry keys changes | Malicious activities | Malicious processes | Malicious connections | Irrespective of network-based | Rootkit detection |
|---|---|---|---|---|---|---|---|
| Shiuh-Jeng Wang method | Low | Low | Low | Low | High | Low | No |
| Chris Ard method | Low | Low | Low | Low | High | Low | No |
| Jeremy David Brazier Annis method | No | High | Medium | Low | High | Low | No |
| Alden W. Jackson method | No | Medium | Medium | Medium | High | Medium | No |
| Frank Y.W. Law method | Medium | Medium | Low | Low | Medium | High | No |
| Junewon Park method | High | High | High | Medium | High | High | No |
| Proposed procedure | High | High | High | High | High | High | Medium |

**Table 5** Assigned values

| | No | Low | Medium | High |
|---|---|---|---|---|
| Value ($x_i$) | 0 | 1 | 2 | 3 |

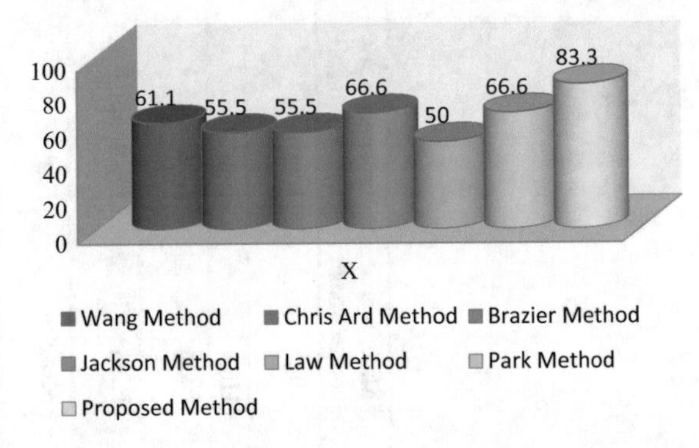


**Fig. 21** Performance in digital forensics comparison

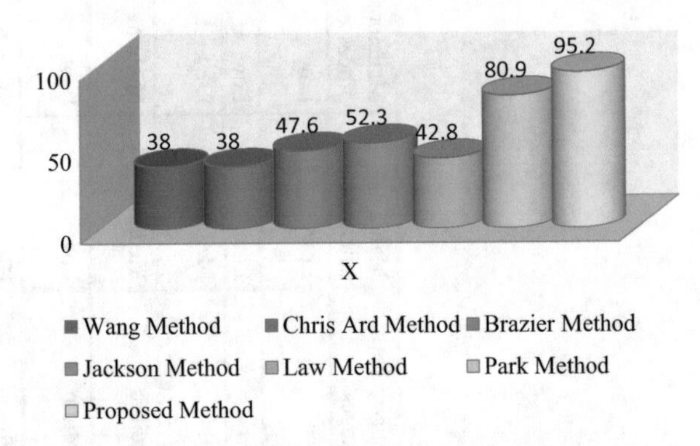


**Fig. 22** Performance in Botnet investigation comparison

- Finding the executable files of Botnet malware and ability to remove the malware from infected system
- Detecting rootkit technique in the malicious activities of Botnet
- Detecting the IP address of command and control server of Botnet.

## 16 Conclusion

Because of the growth of using internet services over the world and the growth of the attacks on users' machines, detecting and defensing against the attacks is one of the most important issues in security area.

Botnet attack is a popular attack nowadays. The detection of this kind of attack is really difficult because of using special features in its structure. Because of many difficulties in investigating the Botnet over the network and lack of enough information in this kind of approach another approach called host-based was chosen in this study.

This study focused on one of the most popular samples of Botnet named ZEUS which uses HTTP protocol as a C&C channel and also uses rootkit feature which is an anti-forensic technique.

Based on presented procedure in this study with extracting some volatile information in live investigation using host-based approach and analyze the information and find out the relations of activities, researcher could reconstruct the Botnet attack in order to prepare the digital evidence package and also with finding the executable files of Botnet and removing them researcher could present a procedure to clean infected system from Botnet malware.

## References


1. Zeidanloo, H.R., et al.: Botnet detection based on traffic monitoring. In: 2010 International Conference on Networking and Information Technology (ICNIT). IEEE (2010)
2. Boe, B.: Super Awesome Project Name Here (2009)
3. Brand, M., Valli, C., Woodward, A.: Malware Forensics: Discovery of the Intent of Deception (2010)
4. Law, F.Y.W., et al.: A host-based approach to botnet investigation? Digital Forensics Cyber Crime 161–170 (2010)
5. Silva, S.S., et al.: Botnets: a survey. Comput. Netw. **57**(2), 378–403 (2013)
6. Zhu, Z., et al.: Botnet research survey. In: Computer software and applications, 2008. In: COMPSAC'08, 32nd Annual IEEE International. IEEE (2008)
7. Casey, E.: Handbook of Computer Crime Investigation: Forensic Tools and Technology. Academic Press (2004)
8. Hay, B., Nance, K., Bishop, M.: Live analysis: progress and challenges. IEEE Secur. Priv. **7** (2), 30–37 (2009)
9. Adelstein, F.: Live forensics: diagnosing your system without killing it first. Commun. ACM **49**(2), 63–66 (2006)
10. Ilavarasan, E., Muthumanickam, K.: A Survey on host-based Botnet identification. In: 2012 International Conference on Radar, Communication and Computing (ICRCC). IEEE (2012)
11. Wang, S.J., Kao, D.Y.: Internet forensics on the basis of evidence gathering with peep attacks. Comput. Stand. Interfaces **29**(4), 423–429 (2007)
12. Cavalca, D., Goldoni, E.: An open architecture for distributed malware collection and analysis. In: Open Source Software for Digital Forensics, pp. 101–116. Springer (2010)
13. Britz, M.T.: Computer Forensics and Cyber Crime: An Introduction, 2/E. Pearson Education India (2009)
14. Ligh, M., et al.: Malware Analyst's Cookbook and DVD: Tools and Techniques for Fighting Malicious Code. Wiley (2010)
15. Ard, C.: Botnet analysis. Int. J. Forensic Comput. Sci. **2**(1), 65–74 (2007)
16. Feily, M., Shahrestani, A., Ramadass, S.: A survey of botnet and botnet detection. In: SECURWARE'09. Third International Conference on Emerging Security Information, Systems and Technologies, 2009. IEEE (2009)

17. Annis, J., et al.: Zombie networks: an investigation into the use of anti-forensic techniques employed by botnets (2008)
18. Jackson, A.W., et al.: SLINGbot: a system for live investigation of next generation botnets. in Conference For Homeland Security, 2009. In: CATCH'09. Cybersecurity Applications and Technology. IEEE (2009)
19. Brand, M., Valli, C., Woodward, A.: Malware forensics: discovery of the intent of deception. J. Digital Forensics Secur. Law **5**(4), 31–42 (2010)
20. Hay, B., Bishop, M., Nance, K.: Live analysis: progress and challenges. IEEE Secur. Priv. **7** (2), 30–37 (2009)
21. Junewon, P.: Acquiring Digital Evidence from Botnet Attacks: Procedures and Methods. AUT University (2011)
22. Tabish, S.M., Shafiq, M.Z., Farooq, M.: Malware detection using statistical analysis of byte-level file content. In: Proceedings of the ACM SIGKDD Workshop on Cybersecurity and Intelligence Informatics. ACM (2009)
23. Wang, P., Sparks, S., Zou, C.C.: An advanced hybrid peer-to-peer botnet. IEEE Trans. Dependable Secure Comput. **7**(2), 113–127 (2010)
24. Wang, L., Zhang, R., Zhang, S.: A model of computer live forensics based on physical memory analysis. In: 2009 1st International Conference on Information Science and Engineering (ICISE). IEEE (2009)
25. Zeidanloo, H.R., Manaf, A.A.: Botnet command and control mechanisms. In: Second International Conference on Computer and Electrical Engineering, 2009. ICCEE'09. IEEE (2009)
26. Zeidanloo, H.R., Manaf, A.B.A.: Botnet detection by monitoring similar communication patterns (2010). arXiv:1004.1232
27. Zeidanloo, H.R., et al.: A proposed framework for P2P Botnet detection. IACSIT Int. J. Eng. Technol. **2**, 161–168 (2010)
28. Zeidanloo, H.R., et al.: A taxonomy of Botnet detection techniques. In: 2010 3rd IEEE International Conference on Computer Science and Information Technology (ICCSIT). IEEE (2010)
29. Ibrahim, L.M., Thanoon, K.H.: Detection of Zeus Botnet in Computers Networks and Internet (2012)
30. Ieong, R.: Freeware Live Forensics Tools Evaluation and Operation Tips (2006)

Printed in the United States
By Bookmasters